Textile Reinforced Concrete

T0187991

Modern concrete technology series

A series of books presenting the state-of-the-art in concrete technology

Series Editors

Arnon Bentur
National Building Research Institute
Faculty of Civil and Environmental Engineering
Technion-Israel Institute of Technology
Technion City, Haifa, Israel

Sidney Mindess
Department of Civil Engineering
University of British Columbia
Vancouver, British Columbia, Canada

1. **Fibre Reinforced Cementitious Composites**
   A. Bentur and S. Mindess

2. **Concrete in the Marine Environment**
   P.K. Mehta

3. **Concrete in Hot Environments**
   I. Soroka

4. **Durability of Concrete in Cold Climates**
   M. Pigeon and R. Pleau

5. **High Performance Concrete**
   P. C. Aïtcin

6. **Steel Corrosion in Concrete**
   A. Bentur, S. Diamond and N. Berke

7. **Optimization Methods for Material Design of Cement-based Composites**
   Edited by A. Brandt

Textile Reinforced Concrete

Alva Peled, Barzin Mobasher, and Arnon Bentur

CRC Press
Taylor & Francis Group
Boca Raton London New York

CRC Press is an imprint of the
Taylor & Francis Group, an **informa** business

A SPON PRESS BOOK

CRC Press
Taylor & Francis Group,
6000 Broken Sound Parkway NW, Suite 300,
Boca Raton, FL 33487-2742

First issued in paperback 2019

© 2017 by Taylor & Francis Group, LLC
CRC Press is an imprint of Taylor & Francis Group, an Informa business

No claim to original U.S. Government works

ISBN-13: 978-1-4665-5255-5 (hbk)
ISBN-13: 978-0-367-86691-4 (pbk)

Visit the Taylor & Francis Web site at
http://www.taylorandfrancis.com

and the CRC Press Web site at
http://www.crcpress.com

Contents

Authors

Alva Peled is associate professor at the structural engineering department, Ben Gurion University of the Negev, Israel and member of RILEM TC-TRC-Materials Characterization, Analysis, and Structural design of Textile Reinforced Concrete.

Barzin Mobasher is professor of Structural Materials Engineering at Arizona Structural Materials Engineering at Arizona State University, USA, former chair of ACI's committee 544 on Fiber Reinforced Concrete and chair of RILEM TC-TRC-Materials Characterization, Analysis, and Structural design of Textile Reinforced Concrete.

Arnon Bentur is a professor emeritus of Civil and Environmental Engineering and former vice president and director general of the Technion, Israel Institute of Technology, and is former president of RILEM.

Chapter 1

Introduction

1.1 Structure, properties, and application

Textile-reinforced concrete (TRC) has emerged in recent years as a new and valuable construction material (Brameshuber, 2006, 2010; Curbach and Heeger, 1998; Hegger et al., 2006a). It is made with a continuous textile fabric that is incorporated into a cementitious matrix consisting of a Portland cement binder and small-size aggregates. This material can be categorized as a strain-hardening or strain-hardening/deflection-hardening cement composite, using the classification outlined in Figure 1.1.

The strain-hardening cement composites can be obtained by a variety of modes, ranging from special formulations using short and dispersed yarns to continuous textile fabrics. The advancement in the developments of such composites is based to a large extent on a more in-depth understanding of the micromechanics of such systems, enabling the achievement of strain hardening with the use of a modest content of fiber reinforcement, less than 10% by volume, and in many instances even going below 5% (Bentur and Mindess, 2007; Reinhardt and Naaman, 2007; Toledo Filho et al., 2011). This, of course, provides a much more cost-effective composite, making its application much more feasible.

Strain-hardening composites consisting of discrete short reinforcement (yarns, fibers) require special tailoring of the reinforcement and the matrix to achieve the strain-hardening effect through the control of micromechanical interactions. This requires special control of the particle grading of the matrix and the use of high-performance chemical admixtures. In TRC, the sophistication is in the textile fabric, with regard to its geometry and the type of yarns used, or hybridization of yarns of different properties to optimize reinforcing efficiency (Gries et al., 2006) (Figure 1.2).

The cementitious matrix for TRC, although of high quality, is not the kind that requires very stringent formulation or use of ingredients that are drastically different than conventional ones. Thus, with a largely conventional cementitious matrix, it is possible to obtain an impressive range of

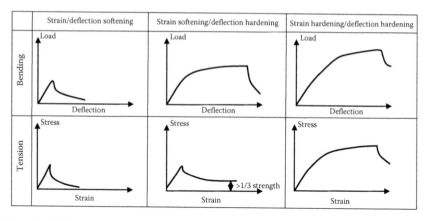

	Strain/deflection softening	Strain softening/deflection hardening	Strain hardening/deflection hardening
Bending	Load / Deflection	Load / Deflection	Load / Deflection
Tension	Stress / Strain	Stress / Strain >1/3 strength	Stress / Strain

Figure 1.1 Schematic description of strain softening, strain hardening (in tension), and deflection hardening (in bending) on FRC composites. (From Bentur, A. and Mindess, S., *Fibre Reinforced Cementitious Composites*, Taylor & Francis, London, U.K. 2007; Adapted from Naaman, A.E., Strain hardening and deflection hardening fibre reinforced cement composites, in A.E. Naaman and H.W. Reinhardt (eds.), *Fourth International Workshop on High Performance Fibre Reinforced Cement Composites, HPFRCC 4*, RILEM Publications, Paris, France, 2003, pp. 95–113.)

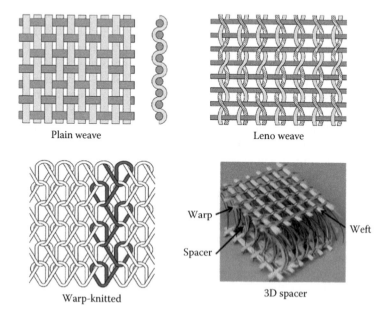

Plain weave

Leno weave

Warp-knitted

Warp
Weft
Spacer

3D spacer

Figure 1.2 Fabrics of different types.

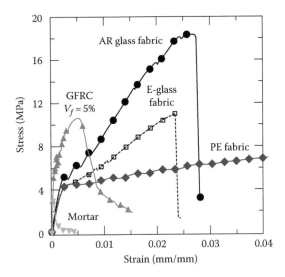

Figure 1.3 Stress–strain curves of TRC of different fabrics compared to unreinforced matrix and sprayed fiber–cement composite, GFRC. (After Peled, A. and Mobasher, B., *ACI Mater. J.*, 102(1), 15, 2005.)

mechanical properties, having elastic–plastic behavior at one end and a very effective strain hardening at the other (Figure 1.3).

Strain-hardening cement composites made with short yarns have the advantage that their production is simple and site friendly as compared to the production of components with TRC. Yet, a major advantage of TRC is the continuity of the yarns providing inherently high efficiency and reliability, which is so essential for structural and semistructural applications. This advantage is further enhanced when considering the flexibility in producing textile fabrics that can be tailored for structural performance, static and impact, by optimization of the use of high-performance yarns, such as carbon, and control of their orientation.

The production of prefabricated TRC components of structural and semistructural nature can take place by hand lay-up (Figure 1.4a) or mechanized processes (Figure 1.4b and c, Brameshuber et al., 2006). The more simple hand lay-up techniques are much more attractive for on-site repair and retrofit (Curbach et al., 2006).

The matrix in TRC is usually in the range of 0.40–0.45 w/c ratio, which is the low end of conventional reinforced concrete, making the TRC somewhat more durable from a matrix performance point of view. A major advantage with regard to durability is the crack control in TRC, characterized by multiple cracking upon loading, with cracks being less than 100 µm wide, usually about 50 µm (Ahmed and Mihashi, 2007; van Zijl et al., 2012). This provides enhanced durability performance compared to conventional concrete

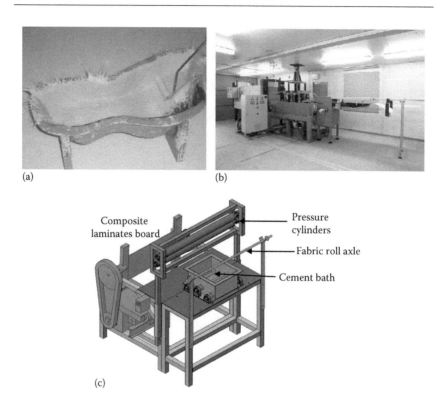

Figure 1.4 Production processes of TRC: (a) hand lay-up—wetting textile with roller (Vrije Universiteit, Brussels, Belgium), (b) production machine for the manufacture of U-shaped products, and (c) production by the pultrusion process. (a: From Brameshuber, W. et al., Production technologies, Chapter 5, in W. Brameshuber (ed.), *Textile Reinforced Concrete: State of the Art Report, RILEM TC 201-TRC*, RILEM Publications, Paris, France, 2006, pp. 57–81; b: From Hegger, J. et al., *Beton- und Stahlbetonbau*, 99(6), 482, 2004; c: From Peled, A. and Mobasher, B., *Mater. Struct.*, 39, 787, 2006.)

where cracks are limited in width to the range of 0.1–0.3 mm. The penetrability of a matrix containing cracks of ~50 μm is not much different from that of an uncracked matrix, as at this width scale cracks tend to self-heal.

This class of cement composites, and in particular TRC, which lends itself more to structural and semistructural applications, paves the way for innovation in construction, which is facilitated by the high tensile strength and ductility of these composites. The innovation that can be fostered is by making it feasible to design components of a thin cross section to replace the heavier conventional reinforced concrete, such as profiles for bridges (Figure 1.5a) and curtain walls for the facades of buildings, which may also combine mechanical and physical functionality with flexibility of shaping, to meet architectural design requirements (Figure 1.5b).

(a)

(b)

Figure 1.5 Structural and architectural components from TRC: (a) profile for a bridge (Technical University, Dresden, Germany) and (b) curtain walls for facades, office building, Arnhem, NL (Fydro B.V., NL). (a: From Hegger, J. et al. (eds.), in *First International Conference in Textile Reinforced Concrete (ICTRC1)*, Aachen, Germany, 2006a; b: From Hegger, J. et al., Applications of textile reinforced concrete, Chapter 8, in W. Brameshuber (ed.), *Textile Reinforced Concrete: State of the Art Report, RILEM TC 201-TRC*, RILEM Publications, Paris, France, 2006b, pp. 237–270.)

TRC can also be incorporated into a structure as an external layer to provide strength and enhanced durability. This is of particular interest for applications in new structures as well as repair and retrofit of existing structures (Figure 1.6).

1.2 Sustainability aspects of construction with TRC

The unique combination of high-level mechanical performance and long-term resistance to degradation effects is of particular interest in the modern era, when new technologies to provide sustainability are becoming a major

Figure 1.6 Repair and retrofitting with TRC by applying an external layer of TRC on concrete components. (From Lieboldt, M. and Mechtcherine, V., *Cement Concr. Res.*, 52, 53, 2013.)

driving force for innovation in the construction industry. The sustainability attributes offered by TRC spans over a wide range:

1. Potential for making components with considerably smaller amounts of material (conventional concrete construction is "material-rich" technology in the negative sense—cross sections of 100–300 mm compared to less than 50 mm for TRC)
2. Components that can provide service life that is expected to be much longer than conventional concrete
3. Provide the basis for technologies to extend the life-span of existing structures that are undergoing deterioration or need upgrading of their mechanical performance to withstand higher loads, static and dynamic

Within the context of sustainability, TRC shows greater advantage relative to a range of construction materials, not just concrete but also high-performance materials such as steel and wood, which have favorable mechanical performance, as quantified in terms of tensile strength, ductility, and strength/weight ratio. The sustainability of materials is often quantified in terms of their energy and CO_2 emission (energy and carbon footprints), in units of embodied energy per unit weight (MJ/kg) and embodied carbon per unit weight (kg-CO_2/kg). Typical values for common construction materials are provided in Figure 1.7, including calculated values for TRC assuming 5% of glass fabric reinforcement, as well as reinforced concrete assuming 2% by volume of steel reinforcement. The values presented are normalized per unit volume of the material. It can be seen that the values for the reinforced concrete and the TRC are similar, and much smaller than any other construction material. However, when considering a component made of these materials, one should take into account the cross section (volume) that needs to be designed for, and the resulting volume of material.

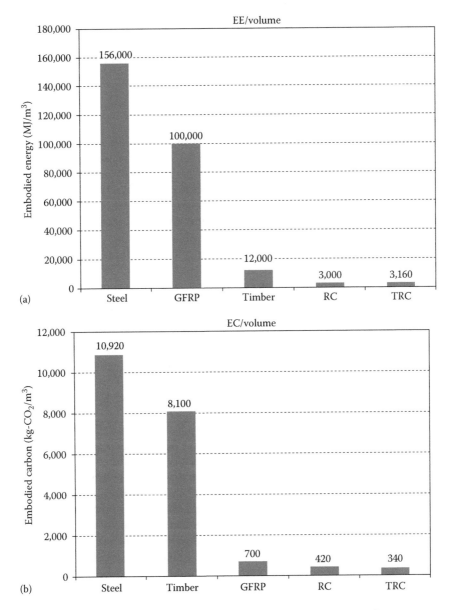

Figure 1.7 Embodied energy (a) and embodied carbon (b) per unit volume of construction materials.

The unique mechanical performance of TRC lends itself to innovation in construction, to develop thin section components, which can be considerably lighter and consume less materials than conventional reinforced concretes. The present book is intended to foster such innovation by presenting an overview of TRC from the material and component aspects, providing the necessary background, insight, and quantitative information and guidelines to facilitate its efficient application and design.

1.3 Innovation with TRC

This book addresses a combination of scientific, engineering, and economic aspects of using TRC materials, including the innovative use of composites in construction materials, integrated structural analysis and design tools, and lightweight and energy-efficient materials. The objective is to address new specifications, analysis, and design guidelines so that material models can be directly integrated into structural analysis software. Emphasis is placed on developing alternative solutions that allow for sustainable development of infrastructure systems using mechanics of materials, statistical process control, and design for durability.

The use of composites such as geotextiles, geomembranes, and textile for repair and retrofit is gaining popularity for its direct role in reducing construction time and labor costs. Many structural elements can be reinforced with TRC as a substitute for conventional reinforcement (Soranakom et al., 2008). In addition to cost savings in terms of construction efficiency, labor, and material, the quality aspects of the final product are quite important as serviceability issues begin to dominate the performance. For example, crack control and distributed cracking with a much smaller crack width will ultimately lead to durability enhancements.

Materials, methodologies, and test methods to enhance the durability of infrastructure systems are addressed to arrive at fundamental sustainability solutions. Current directives on the reduction of maximum cement requirements and increases in the use of supplementary products can help in meeting the sustainability criteria and reduce the carbon footprint considerably. From a durability perspective, both early-age cracking and drying shrinkage reduce the load-carrying capacity and accelerate deterioration, resulting in increased maintenance costs and reduced service life. The use of TRC materials will have a direct impact in this area, since additional strength and ductility reduce the section sizes, while strain-hardening behavior reduces the potential for early-age cracking and drying shrinkage.

TRC is an economical material in architectural cladding panels, sandwich membranes subjected to flexure, permanent formwork, 3D-shaped architectural elements, and shotcrete and groundwork for mining, tunneling, and excavation support applications. There are numerous fiber types available for commercial and experimental use. Steel, glass, and synthetic have been used in lieu of traditional welded wire fabric reinforcement (ACI Committee 544, 1996). Hybrid reinforcement systems utilizing two or more fibers in a single textile, or multiple layers of different fiber textiles, have also been used to optimize composite performance for strength and ductility (Mobasher, 2011). Many structural elements can now be reinforced with textiles as a partial or total substitute for conventional reinforcement (Soranakom et al., 2008).

In addition to a general overview of the sustainability aspects of TRC use in civil infrastructure systems, the mechanical properties of different textile types and their reinforcing methodology tested under different loading conditions are important for designing with these new materials. Specifically, the response under tensile and flexural tests is used in static, creep, fatigue, and high-speed conditions to measure material properties that are needed in the design of various sections. Due to the experimental and analytical limitations of available test methods (Gopalaratnam et al., 1991; Gopalaratnam and Gettu, 1995; Soranakom and Mobasher, 2007), procedures for design of flexural TRC members based on calculated residual strength are presented based on the strain-hardening behavior of the material.

These proposed developments are supported by experimental and theoretical derivations and show considerable cost savings due to composite use, which in turn increases the load-carrying capacity and ductility. Examples include pipes, canal lining, shotcrete, slope stability applications, thin and thick section panels, sandwich construction, precast retaining wall sections, and many other applications where heavy reinforcement and high flexural and shear loading cases are applied.

The main advantage of micro- over macroreinforcement is in the integration of microstructural toughening for the enhancement of mechanical behavior, especially when low-modulus, low-cost yarns are used. This is also combined with shape-forming TRC material fabrics as reinforcements. Modern textile technology offers a wide variety of fabrics with great flexibility in fabric design and control of yarn geometry and orientation (Figure 1.2). This flexibility allows engineering of composite materials for various product categories. The practical use of TRC composites requires an industrial cost-effective production process, and pultrusion and extrusion processes as industrial manufacturing options are natural candidates for this purpose (Figure 1.4). They can be geared toward a relatively simple setup using low-cost equipment while assuring uniform production and good control of laminates alignment. Development of a dense microstructure to achieve intimate interaction between the matrix and the fabric, as seen for example in Figure 1.8, using novel fabrication techniques, can serve as the basis for production of high-performance fabric-cement composites.

Figure 1.8 Microstructure of TRC. (From Peled, A. and Mobasher, B., *ACI Mater. J.*, 102(1), 15, 2005.)

References

ACI Committee 544, State-of-the-art report on fibre reinforced concrete, Publication ACI 544.1R-96, American Concrete Institute Committee 544, American Concrete Institute, Farmington Hills, MI, 1996.

Ahmed, S. F. U. and Mihashi, H., A review on durability properties of strain hardening fibre reinforced cementitious composites, *Cement and Concrete Composites*, 29, 265–376, 2007.

Bentur, A. and Mindess, S., *Fibre Reinforced Cementitious Composites*, Taylor & Francis, London, U.K., 2007.

Brameshuber, W. (ed.), *Textile Reinforced Concrete: State of the Art Report, RILEM TC 201-TRC*, RILEM Publications, Paris, France, 2006.

Brameshuber, W. (ed.), *International RILEM Conference on Material Science— Second ICTRC—Textile Reinforced Concrete—Theme 1*, Paris, France, 2010.

Brameshuber, W., Nrameshuber, W., Brockman, T., Mobasher, B., Pachow, U., Peled, A., Reinhadt, H. W., Kruger, M., and Wastiels, J., Production technologies, Chapter 5, in W. Brameshuber (ed.), *Textile Reinforced Concrete: State of the Art Report, RILEM TC 201-TRC*, RILEM Publications, Paris, France, 2006, pp. 57–81.

Curbach, M. and Heeger, J. (eds), Sachstandbericht zum Einsatz von Textilien im Massivbau (German State of the Art Report for the use of technical textiles in massive structure, DAfStb, Heft 488, Beuth, Berlin, 1998.

Curbach, M., Ortlepp, R., and Triantafillou, T. C., TRC for rehabilitation, Chapter 7, in W. Brameshuber (ed.), *Textile Reinforced Concrete: State of the Art Report, RILEM TC 201-TRC*, RILEM Publications, Paris, France, 2006, pp. 221–236.

Franzén, T., Shotcrete for underground support—A state of the art report with focus on steel fibre reinforcement, in P. K. Kaiser and D. R. McCreath (eds.), *Rock Support in Mining and Underground Construction, Proceedings of the International Symposium on Rock Support*, Sudbury, Ontario, Canada, Balkema, Rotterdam, the Netherlands, 1992, pp. 91–104.

Gopalaratnam, V. S. and Gettu, R., On the characterization of flexural toughness in fibre reinforced concretes, *Cement and Concrete Composites*, 17(3), 239–254, 1995.

Gopalaratnam, V. S., Shah, S. P., Batson, G. B., Criswell, M. E., Ramakrishnan, V., and Wecharatanan, M., Fracture toughness of fibre reinforced concrete, *ACI Materials Journal*, 88(4), 339–353, 1991.

Gries, T., Roye, A., Offermann, P., Engler, T., and Peled, A., Textiles, Chapter 3, in W. Brameshuber (ed.), *Textile Reinforced Concrete: State of the Art Report, RILEM TC 201-TRC*, RILEM Publications, Paris, France, 2006, pp. 11–27.

Hegger, J., Brameshuber, W., and Will, N. (eds.), in *First International Conference in Textile Reinforced Concrete (ICTRC1)*, Aachen, Germany, 2006a.

Hegger, J., Will, N., Aldea, C., Brameshuber, W., Brockmann, T., Curbach, M., and Jesse, J., Applications of textile reinforced concrete, Chapter 8, in W. Brameshuber (ed.), *Textile Reinforced Concrete: State of the Art Report, RILEM TC 201-TRC*, RILEM Publications, Paris, France, 2006b, pp. 237–270.

Hegger, J., Will, N., Schneider, H. N., and Kolzer, P., New structural elements made of textile reinforced concrete—Applications and examples, *Beton- und Stahlbetonbau*, 99(6), 482–487, 2004.

Lieboldt, M. and Mechtcherine, V., Capillary transport of water through textile-reinforced concrete applied in repairing and/or strengthening cracked RC structures, *Cement and Concrete Research*, 52, 53–62, 2013.

Mobasher, B., *Mechanics of Fibre and Textile Reinforced Cement Composites*, CRC Press, Boca Raton, FL, 2011, p. 480.

Naaman, A.E., Strain hardening and deflection hardening fibre reinforced cement composites, in A. E. Naaman and H. W. Reinhardt (eds.), *Fourth International Workshop on High Performance Fibre Reinforced Cement Composites, HPFRCC 4*, RILEM Publications, Paris, France, 2003, pp. 95–113.

Peled, A. and Mobasher, B., Pultruded fabric-cement composites, *ACI Materials Journal*, 102(1), 15–23, 2005.

Peled, A. and Mobasher, B., Properties of fabric-cement composites made by pultrusion, *Materials and Structure (RILEM)*, 39, 787–797, 2006.

Reinhardt, H. W. and Naaman, A. E. (eds.), in *Fifth International RILEM Workshop on High Performance Fiber Reinforced Cement Composites (HPFRCC5)*, 2007.

Soranakom, C. and Mobasher, B., Closed-form moment-curvature expressions for homogenized fibre reinforced concrete, *ACI Materials Journal*, 104(4), 351–359, 2007.

Soranakom, C., Mobasher, B., and Destreé, X., Numerical simulation of FRC round panel tests and full scale elevated slabs, in P. Bischoff and F. Malhas (eds.), *Deflection and Stiffness Issues in FRC and Thin Structural Elements*, ACI SP-248-3, American Concrete Institute, Farmington Hills, MI, 2008, pp. 31–40.

Toledo Filho, R. D., Silva, F. A., Koenders, E. A. B., and Fairbairn, E. M. R. (eds.), in *Second International RILEM Conference on Strain Hardening Cementitious Composites (SHCC2-Rio)*, 2011.

van Zijl, G. P. A., Wittman, F. H., Oh, B. H., Kabele, P., Toledo Filho, R. D., Fairbarin, E. M. R., Slowik, V. et al., Durability of strain hardening cement based composites (SHCC), *Materials and Structures*, 45, 1447–1463, 2012.

Vandewalle, M., *Dramix: Tunnelling the World*, 3rd edn., N.V. Bekaert S.A., Zwevegem, Belgium, 1993.

Chapter 2

Textiles

2.1 Introduction

Textiles constitute an essential part of everyday life, a position they have held since ancient times. As remarkable today as across history, the breadth of their applicability and ubiquitous utilization in everyday products includes clothing, ropes and sail cloths, lightweight materials, furnishings and decorative ornaments, symbolic items such as flags, printing, medicine, and construction. There is evidence of the production of textiles for clothing, packaging, and storage dating as far back as 8,000 and 7,000 BC, respectively, in Mesopotamia and Turkey. Among the earliest records of fabric-making techniques are looped needle netting or knotless netting and sprang, which generally comprise rows of interlocked loops that control fabric density. These techniques survived through centuries as they enabled the fabric to be shaped to make head and feet coverings without any cut or waste material. While the first known textiles used in the West originated from Turkish Anatolia and dated to 6,000 BC, netting and allied structures have been dated earlier (Wild, 2003), making textiles one of the first engineering materials and products made of natural materials.

At first glance, today's textile industry has little in common with regard to the wide range of raw materials utilized from wool and silk to vegetable and modern artificial fibers. However, the fiber, the textile industry's basic unit, is the link between those materials, as they are all produced from fibers that are combined into yarn or thread that, in turn, is then assembled together into a finished fabric. Basically, a fiber is a long, fine filament whose diameter is generally of the order of 10 μm and whose length-to-diameter aspect ratio is usually at least 1,000, but in the case of continuous fibers, that value increases virtually to infinity (Bunsell, 1988). Owing to its fineness, the fiber has high flexibility that has been exploited by humans and even animals from their beginnings to make cloths, ropes, shelter, and nest. In addition to its flexibility, the fiber's strength is extraordinary, often possessing a Young's modulus far greater than the same material in its bulk form. Finally, the beneficial, fundamental qualities of the fiber—strength, stiffness in tension, and flexibility in bending—are often combined with low weight.

In the history of composite materials production, the first fiber-like reinforcing unit was straw, which was used to reinforce clay bricks to prevent cracking. Likewise, bamboo shoots have also been used to reinforce mud walls since prehistoric times. In modern society, composite materials were first used in the 1930s when glass fibers were exploited as the reinforcement for resin-based matrices. The development of new fibers such as carbon, boron, and aramid since the 1970s has significantly increased the applications of composites.

This chapter provides an overview of basic concepts of textiles, including fibers, yarns, and fabrics, with a focus on the main parameters that influence cement-based matrices for textile (fabric)-reinforced concrete (TRC) applications. Some basic definitions of textile nomenclature are introduced as follows.

Textile nomenclature

Fiber—Class of materials that comprises continuous filaments or discrete, elongated pieces (staple fibers) used for yarns, fabrics, etc.

Filament—A single fiber of indefinite length; a continuous fiber.

Yarn—A continuous monofilament or multifilament made of natural or man-made fibers intended for use in a textile construction such as knitting, weaving, or another fabric. The fibers in the yarn may or may not be twisted together.

Tex—A unit of fiber fineness: the weight in grams of 1,000 m long yarn; the lower the number, the finer the yarn (weight per unit length, i.e., 1 g/1,000 m = 1 tex).

Decitex (dtex)—A unit of fiber fineness: the weight in grams of 10,000 m long yarn (1 g/10,000 m = 1 dtex).

Denier—A unit of fiber fineness: the weight in grams of 9,000 m long yarn (1 g/9,000 m = 1 denier).

Strand—An ordered assemblage of several hundreds to thousands of single filaments that are normally used as a unit, bundled together to prevent splitting.

Roving—A glass strand in which untwisted parallel glass filaments are assembled into larger diameter threads that are then used for glass fabrics and mats.

Fabric—A manufactured textile structure made of fibers, yarns, or both, which is assembled by any of a variety of means—weaving, knitting, tufting, felting, braiding, or bonding of webs—to give the structure sufficient strength and other properties required for its intended use. The yarns can be arranged in two, three, or more dimensions.

Mesh—A fabric (two- or three-dimensional structure) with open structure. In an open structure, the yarns or strands are sufficiently separated to leave interstices in the fabric.

Textile—Any type of material made from fibers or other extended linear materials such as thread or yarn. A textile can be a fabric or any other manufactured product constructed from fibers and yarns, as well as garments and other articles fabricated from fibers, yarns, or fabrics when the product retain the characteristic flexibility and drape of the original fabrics.

2.2 Fiber materials

Made from a mass of raw materials, fibers are commonly divided into two main groups or types: natural and man-made (Figure 2.1). Natural fibers include those produced by plants, animals, and geological processes. A wide variety of plants are used for their fibers, including cotton, pulp (cellulose), agave (sisal), coconut, hemp, flax, jute, kenaf, piassava, and others. Animal fibers can be of silk or of wool from sheep or goats, for example. Geological mineral fibers include basalt and asbestos. Among the man-made fibers are those, such as glass, produced from inorganic materials and those made from synthetic materials like petrochemicals used to produce polymers such as carbon, aramid, polyethylene (PE), polypropylene, polyvinyl alcohol (PVA), and others. Fibers can also be made of ceramics, for example, boron, silicon carbide, or aluminum oxide, and metal, typically steel, for concrete applications.

Until the end of the nineteenth century, only natural fibers, the most important being cotton, wool, flax, and silk, were available and used. The desire in the late 1800s to produce cheap fibers as substitutes for silk was the driving force behind the initial attempts to develop synthetic fibers.

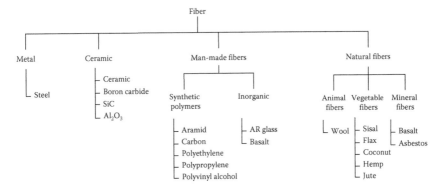

Figure 2.1 Fiber types.

Despite efforts since the Renaissance to produce artificial silk fibers, however, it was not until the end of the nineteenth century that real progress was made in synthetic fiber development. Indeed, the beginning of the synthetic fiber revolution is marked, in 1885, by the successful production of the first rayon fibers in France (Moncrieff, 1979). In 1899, an artificial silk was successfully manufactured when it was discovered that natural cellulose can be dissolved in cuprammonium hydroxide, and the resulting mixture could then be extruded through holes to produce filaments (Coleman, 2003).

The advent of commercial glass and polyamide in the United States in the 1930s sparked the development of a whole new range of synthetic fibers. Fifty years after the arrival of these first synthetic fibers, many other organic fibers have been introduced with Young's modulus values 35 times greater than those of the first synthetic fibers and strength-to-weight ratios five times larger than that of steel. Synthetic fibers such as boron and carbon have up to two to four times improved Young's moduli compared to steel and at much lower densities (Moncrieff, 1979).

The characteristics of composite materials are defined both by the properties and amounts of the fibers used and their arrangement within the composite. The requirements of each composite, therefore, determine the fiber material and the type of textile used in its fabrication. Thus, to ensure that cement-based composites retain their reinforcing effects for their entire lifetime, the fiber material must be able to withstand the alkaline medium of the Portland cement matrix virtually indefinitely without suffering a decline in its properties or performance. Further requirements of the fiber are that it possesses small relaxation under permanent load, a well-defined and uniform adhesion between reinforcement and matrix, low cost, and the potential for easy processing using existing textile machinery.

2.2.1 Man-made fibers

2.2.1.1 Basic production principles

Many man-made fibers (i.e., glass, polymers, and metallic) are formed by drawing them in a spinning process that provides a fiber structure of fine diameter. The chains, molecules, or grains, depending on fiber material, are aligned, which confers on them high structural properties. In general, spin drawing starts with a material in a molten state or in solution that is extruded through a die (spinneret) containing tiny holes of about 1–2 mm in diameter (Figure 2.2a). The jets, or liquid streams, of the molten material or solution emerging from the spinneret are then solidified. Depending on fiber material, hardening of the liquid stream can then be achieved by cooling, hot air, or immersion in a coagulation bath. Three main spinning processes are common (Figure 2.2b): (1) *dry spinning* (the molten/solution streams emerging from the spinneret are solidified in a dry environment

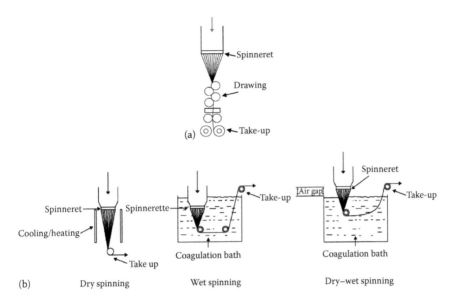

Figure 2.2 Spinning and drawing process: (a) general process and (b) three main spinning processes.

under either cooling conditions for molten materials or hot conditions for solution materials to induce solvent evaporation), (2) *wet spinning* (the spinning nozzle is located in a coagulation bath, and the streams of extruded spinning solution pass directly from the spinneret into the coagulation bath for material solidification and solvent removal), and (3) *dry–wet spinning* (an air gap is used between the spinneret and the coagulation bath). The solidified fibers in many cases are stretched to produce the fine filaments that constitute the end product. The stretching process is done through a set of rollers whose speeds can be varied to obtain filaments of different diameters (Figure 2.2a). The filaments can then be cut into shorter lengths or assembled together or separately wound on a bobbin for the production of monofilament yarns or multifilament bundles that contain hundreds to thousands of filaments.

The filament drawing and stretching processes essentially produce fibers whose tensile properties are significantly improved relative to those of their bulk materials. For glass fibers, their brittleness dictates that filament strength is determined by the size of the defects it contains, most of which are on the filament surface as it is easily damaged. The drawing process deforms the molten material and reduces its diameter to one-hundredth of its original value before drawing, which increases the surface area. The movement of material from the filament interior to its surface facilitates the elimination of surface defects and greatly increases filament strength.

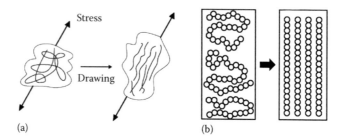

Figure 2.3 Stretching of polymer chains during the spinning process and drawing: (a) contribution of general drawing and (b) chain arrangements before and after drawing.

Although during the drawing process, glass fibers are generally considered to be isotropic (equal properties in all directions), for polymer materials, the stretching process orients their macromolecular structures more or less in the drawing direction (Figure 2.3), parallel to the fiber axis, to produce long, continuous filaments. The higher the degree of molecular orientation in the drawing direction, the higher the Young's modulus and fiber strength and the lower the strain to failure. During the drawing process, the structures of drawn polymer fibers become anisotropic (unequal properties in the different directions), with lower strengths in their transverse direction in which the weak atomic bonds linking the polymer chains are more prone to failure. For metallic materials, the drawing process elongates the grains along the fiber axis (i.e., drawing axis), leading to improved tensile properties in that direction.

Based on Griffith's theory, fiber tensile strength depends on—in addition to chain orientation and the number of surface defects—the length, c, of the defect within the material. Defects like microcracks effectively concentrate and magnify applied stresses (Figure 2.4) such that the longer the defect, the greater the actual stress concentration at the crack tip. In terms of fiber structure, the diameter of the fiber is the upper bound length for formation of a microcrack, that is, the maximum crack length in an individual fiber is limited by the fiber's diameter. This means that the smaller the fiber spacing,

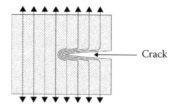

Figure 2.4 Concentration of applied stresses at the crack tip.

the smaller the size of the defect. Thus, in a fiber structure, only very short length defects are present, ensuring that the stress concentrations at the crack tip are relatively low.

Tensile strength calculations of fibers assume that (1) the surface defects have been eliminated in the drawing process, (2) the polymer fibers are in the correct chain orientation, and (3) the lengths of any microcracks are the shortest possible, that is, within the upper limit, which is the fiber diameter. The fiber is then considered structurally perfect, and therefore, its tensile strength—controlled mainly by the strengths of the chemical bonds between the atoms within the material—is expected to be close to its theoretical value, that is, for a fiber with hardly any defects.

The theoretical tensile strength of a material free of defects depends on the chemical bonding between its atoms (Equation 2.1):

$$G = \frac{(1/2)pda}{A} = \frac{1}{2}\sigma\varepsilon a = \frac{\sigma^2 a}{2E}, \quad \sigma = \sqrt{\frac{2GE}{a}} \tag{2.1}$$

where
  $G = 2\gamma$ (surface energy per unit area)
  p is the force
  A is the cross section
  σ is the stress
  ε is the strain
  E is the Young's modulus
  a is the distance between atoms

Calculations, for example, of the theoretical tensile strength of a steel material—for which $E = 200$ GPa, $G = 1$ J/m², and $a = 2$ Å—show that the tensile strength obtained in the drawing process is $\sigma = 40$ GPa, which is about one-fifth that of the Young's modulus for steel, that is, $\sigma \cong E/5$. For a fiber, the smaller its diameter, the closer it is to its theoretical tensile strength—a relation that is clearly shown in Figure 2.5 for glass fibers.

Extrapolation based on Figure 2.5 to a fiber thickness value of zero shows that the theoretical tensile strength of a glass material without any defects is expected to be 11 GPa, which is relatively close to one-fifth its Young's modulus (~70 GPa), that is, $\sigma \cong 70/5$ GPa, as already described. These calculations demonstrate the upper bound tensile strength of a material, and as such, they show the clear benefit of exploiting fibers as reinforcement materials. The maximization of material tensile strength by the fiber structure is the result of the drawing and stretching processes entailed in fiber production insofar as these produce, in general, a structurally perfect fiber with hardly any defects, a thin diameter, and, for some fibers, alignments of their polymer chains. An analysis of the influence of the drawing process on

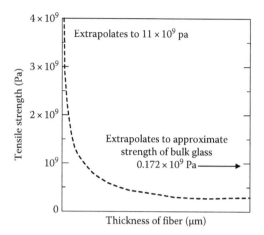

Figure 2.5 Tensile strength of a glass fiber relative to its diameter. (Adapted from the results of Griffith, A.A., *The Phenomena of Rupture and Flow in Solids*, The Royal Society, London, U.K., 1920.)

tensile behavior showed that higher drawing rates resulted in greater tensile strengths with enhanced brittle behavior (Bunsell, 1988).

2.2.1.2 Fiber types

Detailed information for several fiber materials that were examined for their applicability in TRC applications is presented. The main properties of some selected fibers are given in Table 2.1.

2.2.1.2.1 GLASS

The commercial production of glass fibers for composite materials began around 1940 (Gupta, 1988). Since then, glass fibers have become a common reinforcement component in polymer-based composite materials, for which demand is increasing. In cement-based matrices, glass fibers constitute a widely used reinforcing fiber. The advantages of glass fibers include their high specific strength (ratio of tensile strength to density) and high specific modulus per unit cost. The basic ingredient in all glass fibers is silica (SiO_2), and other oxides are added to modify the network structure. Because the atoms of silicon, oxygen, and other elements in glass fibers are arranged randomly in long, three-dimensional (3D) network structures (Figure 2.6a), glass fibers are amorphous (noncrystalline) and isotropic (equal properties in all directions). To facilitate more efficient fiber production, low-valence atoms such as Ca, K, and Na can be introduced to break some of the covalent

Table 2.1 Properties of selected fibers

Fiber type	Tensile strength (MPa)	Modulus of elasticity (GPa)	Ultimate strain (%)	Density (g/cm³)
Glass (AR)	2,500	70	3.6	2.78
Carbon	3,500–6,000	230–600	1.5–2.0	1.60–1.95
Aramid	3,000	60–130	2.1–4.0	1.4
Polyethylene	250	1.4–2.2	10–15	0.95
Polypropylene	140–690	3–5	25	0.9–0.95
Dyneema (HDPE)	2,000–3,500	50–125	3–6	0.97
PVA	880–1,900	25–41	6–10	1.3
Basalt	3,000–4,840	79.3–93.1	3.1	2.7
Asbestos	620	160	—	2.55
Sisal	600–700	38	2–3	1.33
Steel	1,200	200	3–4	7.85
Ceramic (boron/silicon)	800–3,600	360–480	0.79	2.4–2.6

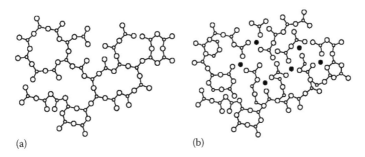

(a)   (b)

Figure 2.6 (a) 3D network structure of glass fibers and (b) modified glass fiber network structure showing low-valence atoms such as Ca, K, and Na in black.

chemical bonds between silicon and oxygen (Figure 2.6b). Although under these conditions the resultant network structure, characterized by lower stiffness, is favorable for the fiber production process, there is a notable reduction in the fiber's modulus of elasticity.

The manufacture of glass fibers includes the melting of a dry mix of silica sand and other ingredients in a furnace at temperatures of up to about 1,350°C (Gupta, 1988). The molten glass is extruded through a set of nozzles (spinneret) as a vertical jet in the dry-spinning process (Figure 2.2b). The jet is simultaneously cooled, and its tension is attenuated to the required level by rapidly drawing the jet through wheels that rotate at speeds between 25 and 150 m/s into filaments ranging from 9 to 27 μm in diameter. The physical properties of the glass fibers (density, chemical durability, and Young's modulus) depend on the temperature and the stress history during fiber processing. The individual filaments are then covered in a protective coating (sizing) before they are collected into a strand to prevent abrasion and to limit static friction between them. For subsequent processes, sizing is critical to ensure yarn properties and adhesion of the yarn with the matrix, which together confer more efficient stress propagation between the filaments and the matrix. Each strand comprises an assembly of 400–2,000 filaments that are combined without twisting to form a yarn. The resulting strands can be used in their continuous form to produce fabric for reinforcements, or they can be cut into shorter lengths. The former is of interest in TRC composites.

Several types of glass fibers, differing by their compositions have been developed and are available. For polymer-based composites, E-glass (high electrical resistivity), S-glass (high strength), and C-glass (chemical resistant) fibers are available, among which E-glass is the most economical and widely used (for polymer-based composites). However, E-glass fibers are sensitive to the alkaline environment typical of cement-based matrices (pH ~ 12.5–13.5), which is a result of the hydration products such as calcium hydroxide [$Ca(OH)_2$] and alkalis that are present in the pore solution of the hardened cement paste matrix. This instability led to the development of alkali-resistant (AR) glass filament yarns specifically designed for use in composite material products incorporating Portland cement. The alkali resistance of AR filaments is achieved by adding 15% zirconia (ZrO_2) by mass to the glass fiber. Some typical glass fiber compositions are presented in Table 2.2. The durability and sustainability of glass-based TRC are discussed further in Chapter 9.

AR glass filament yarns with high proportions of zirconia were fabricated industrially for the first time by Nippon Electric Glass (NEG) Co., Ltd. in 1975. The basic mechanical properties of AR glass and several other glass fibers are provided in Table 2.3. The density of AR glass filament, about 2.8 g/cm^3, is relatively high compared to that of carbon (1.75 g/cm^3) or aramid (1.4 g/cm^3) fibers.

Because of their good adhesion with cement-based matrices and their relatively low cost, which provides a good cost–performance ratio, AR glass filaments are currently the most common glass reinforcing material used in TRC applications.

2.2.1.2.2 CARBON

Carbon fibers were first developed in the 1970s for the aerospace industry, for which quality was more critical than the cost. A significant reduction in the cost of carbon fibers in the 1980s eventually increased the scope of their use to nonaerospace applications (Fitzer and Heine, 1988).

Table 2.2 Compositions of various glass fibers (% weight)

	E	S	C	AR (CEMFILL)	AR (ARG)
SiO_2	52–56	64.3–65	60–65	71	60.7
Al_2O_3	12–16	24.8–25	2–6	1	
B_2O_3	18–13		2–7		
CaO	16–25	0.01	13–16		
MgO	0–6	10–10.3	3–4		
Li_2O				1	1.3
K_2O			0–2		2
Na_2O	0–1	0–0.27	7.5–12	11	14.5
TiO_2	0–0.4			2.2	
ZrO_2				16	21.5
Fe_2O_3	0.05–0.4	0–0.2			
F_2	0–0.5				

Sources: Adapted from Glass industry, January 1987, p. 27; Gupta, P.K., Glass fibres for composite materials, Chapter 2, in A.R. Bunsell (ed.), *Fibre Reinforcements for Composite Materials*, Composite Materials Series, Elsevier, Amsterdam, the Netherlands, Vol. 2, 1988, pp. 20–71; Lowenstein, K.L., *The Manufacturing Technology of Continuous Glass Fibers*, Elsevier, Amsterdam, the Netherlands, 1983, p. 38; Matovich, M.A. and Pearson, J.R.A., *Industrial & Engineering Chemistry Fundamentals*, ACS Publications, Washington, DC, Vol. 8, Chapter 4, 1969.

Table 2.3 Mechanical properties of several glass fibers

	E	S	C	AR
Density (g/cm³)	2.54–2.55	2.48–2.49	2.49	2.78
Tensile strength (MPa)	3,500	4,600	2,758–3,103	2,500
Modulus of elasticity (GPa)	72.4–76	84–88	70	70
Elongation at break (%)	4.8			3.6

Sources: Adapted from Gupta, P.K., Glass fibres for composite materials, Chapter 2, in A.R. Bunsell (ed.), *Fibre Reinforcements for Composite Materials*, Composite Materials Series, Elsevier, Amsterdam, the Netherlands, Vol. 2, 1988, pp. 20–71; Majumdar, A.J. and Nurse, R.W., Glass fiber reinforced cement, Building Research Establishment Current Paper, CP79/74, Building Research Establishment, Hertfordshire, England, 1974.

Most carbon filaments are manufactured from two basic but fundamentally different materials: polyacrylonitrile (PAN) fibers and mesophase pitch (Edie, 1998). Pitch is a liquid crystalline material based on a by-product of petroleum refining and is a lower-cost precursor than PAN, which is a polymer-based material. The final properties of the carbon fiber are determined by the relative contribution of these two materials, either of which can also be used exclusively in fiber production. PAN-derived carbon fibers are of enhanced mechanical properties compared to PITCH produced carbon fibers due to improved control of the orientation of the molecular structure.

Processing the two raw materials for carbon into fibers involves heat treatments, stretching, oxidation, carbonization, and graphitization (Figure 2.7) (Fitzer and Heine, 1988). The first step in the fiber production process is the conversion of the carbonaceous precursor into fiber form through a spinneret. PAN-based carbon filaments are produced using a wet-spun process, in which the PAN solution is extruded through a spinneret and then stretched to align the polymer chains in parallel with the filament direction. Pitch-based carbon filaments are produced by the melt spinning of mesophase pitch through a spinneret, which arranges the carbon atoms in aromatic ring patterns. The aromatic molecules of the mesophase pitch are polymerized at 300°C into long, two-dimensional sheet-like structures. The filaments are then cooled to freeze the molecular orientation. The next step in carbon fiber processing, oxidative stabilization at 200°C–300°C, entails cross-linking the CN groups to form a stable, rigid structure in which some of the CH_2 groups are oxidized. For PAN-based carbon fibers, tension must be applied during this step to limit relaxation of the polymer structure and improve its molecular orientation. The already high molecular orientation of pitch-based carbon fibers at this stage precludes the use of tension during their stabilization. The next step is carbonization, in which the cross-linked precursor filaments are heated at 1,000°C–2,000°C in an inert atmosphere (i.e., no oxygen) that volatilizes nearly all of the noncarbon elements, resulting in a fiber comprising mainly carbon atoms. Although during this stage, tension is maintained on the filaments to prevent shrinking and to improve molecular orientation, the carbon atoms are not yet perfectly ordered and their tensile modulus is relatively low. A true graphite structure is obtained when the filaments are stretched at temperatures of 2,000°C and above to form graphite planes that are aligned in parallel with the filament direction (Figure 2.8), a process that markedly increases filament tensile strength.

The graphite structure is characterized by a hexagonal plane array of carbon atoms that are bound together in each plane with strong covalent bonds, while weaker van der Waals forces hold the planes together (Fitzer and Heine, 1988; Johnson, 1971). The distance between the planes (3.4 Å) is larger than that between the adjacent atoms in each plane (1.42 Å), which provides carbon fibers with highly anisotropic physical and mechanical properties.

The properties of carbon filaments can vary over a wide range depending on the degrees of crystallization and perfection, which are functions of the production process. Compared to pitch-derived fibers, PAN carbon fibers are of higher quality and are therefore more expensive (Hull, 1981; Nishioka et al., 1986). Through the process of graphitization incandescence, PAN-based carbon is used to produce filaments of high strength (HT fibers)

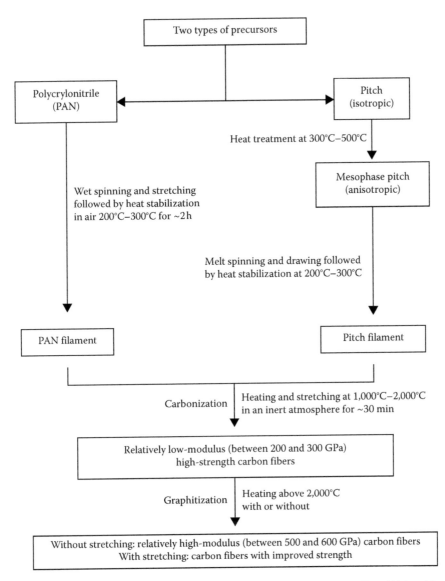

Figure 2.7 Schematic description of carbon fiber production. (From Fitzer, E. and Heine, M., Carbon fibers and surface treatment in fiber reinforcements for composite materials, Chapter 3, in A.R. Bunsell (ed.), *Fibre Reinforcements for Composite Materials*, Composite Materials Series, Elsevier, Amsterdam, the Netherlands, Vol. 2, 1988, pp. 74–148.)

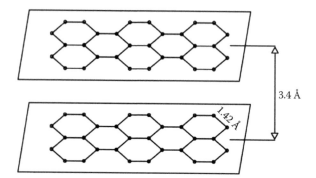

Figure 2.8 Arrangement of carbon atoms in a graphite crystal (two planes are presented).

by heating at 1,500°C–1,700°C and to produce fibers of high modulus of elasticity (HM fibers) by heating at 2,200°C–3,000°C. HT carbon fibers have strengths between 3,000 and 5,000 MPa and moduli of elasticity between 200 and 250 GPa, and for HM fibers those values are 2,000–4,500 MPa in tensile strength and 350–450 GPa in modulus.

Carbon fibers are commercially available as continuous tow and chopped (6–50 mm long) fibers. Continuous carbon yarns comprise numerous (~10,000) filaments with diameters of 7–15 μm. Tests of commercial carbon yarns have found yarn strengths of more than 2,000 MPa depending on yarn fineness. Some typical properties of PAN- and pitch-derived carbon fibers are presented in Table 2.4.

Table 2.4 Properties of carbon fibers

	PAN		Pitch
	Type 1	*Type 2*	
Diameter (μm)	7.0–9.7	7.6–8.6	18
Density (kg/m³)	1,950	1,750	1,600
Modulus of elasticity (GPa)	390	250	30–32
Tensile strength (MPa)	2,200	2,700	600–750
Elongation at break (%)	0.5	1.0	2.0–2.4
Coefficient of thermal expansion × 10^{-6}°C^{-1}	−0.5 to −1.2 (parallel) 7–12 (radial)	−0.1 to −0.5 (parallel) 7–12 (radial)	

Sources: Adapted from Hull, D., *An Introduction to Composite Materials*, Cambridge Solid State Science Series, Cambridge University Press, Cambridge, U.K., 1981; Nishioka, K. et al., Properties and applications of carbon fiber reinforced cement composites, in R.N. Swamy, R.L. Wagstaffe, and D.R. Oakley (eds.), *Proceedings of the Third RILEM Symposium on Developments in Fibre Reinforced Cement and Concrete*, Sheffield, U.K., RILEM Technical Committee 49-FTR, Lancaster, England, Paper 2.2, 1986.

Although carbon filament bundles (tows) can be used to prepare fabrics, compared with other filamentous materials, carbon filaments are much more difficult to process without deterioration during yarn and fabric production. Their sensitivity to lateral pressure makes the filaments fragile. However, modern textile fabrication technology enables the carbon fabrics to be economically produced with reproducible and predictable properties. Carbon filaments exhibit high resistance to acid, alkaline, and organic solvents, all of which are important for their use in cement-based composites, although their adhesion to cement-based material is not as good as that of AR glass.

2.2.1.2.3 ARAMID

Aramid fibers are highly crystalline aromatic polyamide fibers; their basic chemical formula is presented as follows:

The synthesis of aromatic polyamides first began in the 1950s following the rapid commercial expansion of nylon. Since the 1980s, improvements in synthesis techniques led to aramid fibers of higher strength and modulus. Synthesized from the monomers 1,4-phenylenediamine (paraphenylenediamine) and terephthaloyl chloride, aramid fiber is a polymeric aromatic amide (aramid) with alternating aromatic rings, between which are amide (–NH–) and carbonyl (–CO–) end groups (also found in nylons), as shown in Figure 2.9. The aromatic ring provides high chain stiffness (modulus) and a rigid filament. In their final state, aramid fibers are characterized by high melt temperature, excellent thermal stability and flame resistance, and low solubility in many inorganic solutions.

In 1972, the DuPont Company obtained a trademark for the brand name Kevlar, which it gave to the aramid fiber invented by its scientists (Yang, 1988). Kevlar is produced by a dry-jet wet-spinning process (Figure 2.2b) in which the polymer is dissolved in concentrated sulfuric acid for spinning and the solution is extruded through spinneret holes via an air gap into a coagulation bath. The dry-jet wet-spinning method enables the use of a highly concentrated solution to obtain a high degree of molecular orientation that confers on the aramid fibers the desired strength and modulus properties. The coagulated filaments are then washed, neutralized, and dried to form a highly crystalline, fibrous structure with high strength and high modulus. During the spinning and drawing processes, the stiff aramid molecules completely orient in the direction of the filament axis. Different types

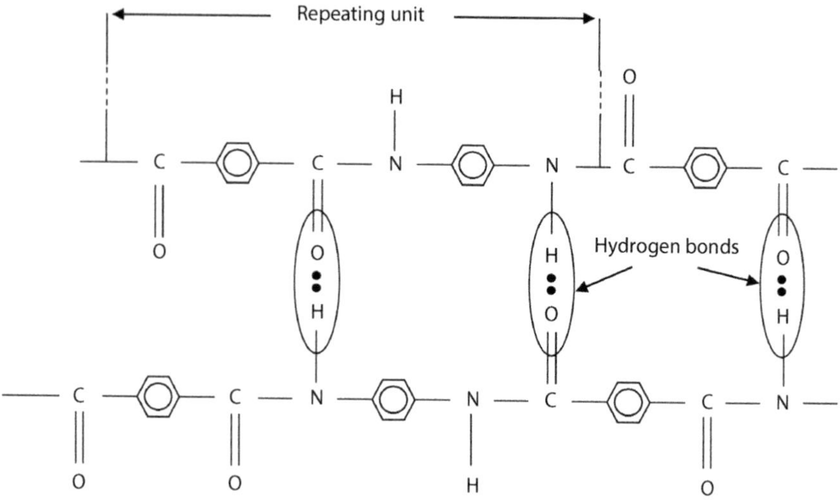

Figure 2.9 Molecular structure of aramid fiber.

of aramid fibers can be produced by altering the temperature and degree of tension during spinning such that at high temperature and under high tension, fiber crystallinity and degree of crystalline orientation increase. The structure of the aramid fiber crystal can be described as pleated hydrogen-bonded sheets, whose fibrillar morphology is highly ordered along the fiber axis, and it possesses a different skin–core structure—with uniform fibril orientation at the skin and imperfectly packed fibrils at the core—all of which have a strong influence on its mechanical properties.

Highly anisotropic, aramid is characterized by weak hydrogen bonds between hydrogen and oxygen atoms in adjacent molecules that hold the chains together in the transverse direction (Figure 2.9). Because the resulting aramid filament has much better physical and mechanical properties in the longitudinal than in the radial direction, it exhibits a low longitudinal shear modulus and poor transverse properties. However, in contrast to glass and carbon, aramid filaments fracture in a ductile manner with considerable necking and fibrillation. This property is considered beneficial for impact or dynamic loading applications (Mallick, 2007).

The structural and morphological features of the aramid fiber determine its tensile properties. Owing to its stable chain structure, aramid shows reasonable resistance to temperature compared with many other synthetic fibers. It is virtually unaffected by temperatures up to 160°C, but above 300°C, the filament will likely lose most of its strength. Aramid fibers are hygroscopic and can absorb moisture, but moisture appears to have very little effect on fiber tensile properties. At high moisture content, however, they tend to crack internally and exhibit longitudinal splitting (Morgan and Allred, 1989), an attribute that must be taken into consideration for cement-based composite applications. In addition, aramid fibers are sensitive to ultraviolet light, but this is less of a problem in composites where the fibers are immersed in a matrix.

The tensile strength of an aramid filament is about 3,000 MPa, its modulus of elasticity is from 60 to 130 GPa, and it has a low density of 1400 kg/m^3 compared with carbon and glass filaments. Kevlar, for example, is approximately five times stronger than steel on an equal weight basis. The aramid yarn produced for commercial sale comprises a bundle of several thousand filaments, each 10–15 μm in diameter.

2.2.1.2.4 POLYETHYLENE

Polyethylene (PE) fibers are made by the polymerization of ethane, which has the following polymer form:

$$-\left[CH_2 - CH_2\right]_n-$$

Two main types of PE fibers, conventional low-modulus PE and high modulus extended chain PE, are commercially available, the latter of which is known by the trade names of Spectra and Dyneema. High-modulus PE fibers, discussed in detail later, are made of a semicrystal flexible random-coil polymer by a gel-spun process.

The potential to obtain mechanical properties characterized by a modulus >220 GPa from a flexible polymer such as PE was first recognized in 1936, when Herman Mark was credited with straightening the polymer chains into extended form in the solid state (Calundann et al., 1988). Since then and until the 1980s, vast research efforts were invested in elucidating the morphological characterization of PE, with the goal of understanding the structure–property relationships of highly oriented molecular chains to enable the production of extended chain crystals with the desired mechanical properties (Geil, 1963; Peterlin, 1979; Ward, 1975; Wunderlich, 1973–1980).

The commercial production of strong, stiff fibers from flexible random-coil polymer is difficult because it requires that the high–molecular weight polymer be drawn to its full molecular extension to obtain the ultrahigh molecular orientation needed to achieve such properties. These challenges were overcome with the introduction of the gel-spinning approach in the 1980s, which enabled the commercial production of high-modulus, high-strength PE fibers from extremely high–molecular weight polymer (MW $\geq$ 10^6) (Smith and Lemstra, 1980). Fundamental to this process is the formation of a polymeric gel with minimum chain tangles from a low-concentration solution of high–molecular weight polymer. The heated polymer solution is extruded through spinnerets via an air gap into a cooling coagulating bath where it forms a gel filament. The wet filament is then transferred into an air oven where it is heated again and stretched though a super drawing process to form the high-strength, high-modulus fiber. The draw rate is necessarily slow to correctly orient and carefully extract the chains from the crystal lamellae and avoid fracturing them, an intricate process that contributes to the high cost of these fibers.

To optimize the mechanical properties of the PE fibers, their production must use a maximum polymer gel concentration of about 1%–2%. Although producing the fibers with increased gel concentrations lowers their final cost, the resulting quality of the fiber's mechanical properties is lower. Fiber modulus and tensile strength are functions of the backbone chemistry stiffness, molecular orientation, chain packing, and, to a lesser extent, the interchain interactions, as tensile strength is also strongly influenced by molecular weight. High-modulus PE fibers have extremely high molecular chain orientation along the fiber axis with fibrous orientation rated above 95% and high crystallinity of above 85% relative to conventional PE fibers, which exhibit very low orientation and fibrous crystallinity lower than 60%, conferring on the latter highly anisotropic mechanical properties. Tensile properties of gel-spun PE fibers were reported by Calundann et al. (1988). For commercial Spectra 900 and Spectra 1000, moduli of elasticity of 119 and 175 GPa, respectively, were reported, with respective tensile strengths of 2,600 and 3,000 MPa. For Dyneema, modulus elasticity in the range of 50–125 GPa and tensile strength in the range of 2,000–3,500 MPa were reported. All are with a density of 970 kg/m^3. The tensile strength of conventional PE is about 250 MPa with a modulus of elasticity of 2 GPa.

In general, gel-spun high-modulus PE fibers exhibit high ductility and, of most interest, high toughness and energy absorption compared to other fibers when incorporated in composites, making them attractive fibers for antiballistic and other impact and critical dynamic loading applications. Both conventional and high-modulus, high-strength PE fibers exhibit superior resistance to most chemicals and to the highly alkaline environment in the cement matrix, low moisture regain, and UV resistance. PE fibers have a relatively low melting point (~150°C), and they are also characterized by extensive creep behavior.

The two types of PE fibers have dramatically different costs. Conventional fibers are typically very cheap relative to other fibers used for similar applications. In contrast, the cost of high-modulus, high-strength PE is generally extremely high, above those of aramid and carbon, rendering the PE fiber a less attractive candidate for cement-based applications. Moreover, as a hydrophobic fiber, PE develops low bonding with the hydrophilic cement-based matrix, an undesirable property in composites.

2.2.1.2.5 POLYPROPYLENE

Polypropylene fibers are made by the polymerization of a propylene resin having the following chemical formula:

$$\left[CH_2 - \underset{\underset{CH_3}{|}}{CH} \right]_n$$

Polypropylene fibers can be produced in a variety of forms and shapes, including monofilament, multifilament, and fibrillated network mesh structures. The multifilament bundle consists of hundreds or thousands of filaments. The fabric structure for TRC applications can be produced with both the mono- and multifilament forms. Bundle and monofilament production is by the extrusion process, in which hot resin is passed through a spinneret with small holes and is then drawn and stretched to achieve a high degree of orientation that ensures good fiber properties (Figure 2.2). The diameters of the filaments constituting the multifilament structure can range from about 50 to 500 μm. The network fibrillated mat structure is also produced by the extrusion process, but in this case, the hot resin is extruded through a rectangular rather than a circular die to produce a film that is then drawn and split to form a mat structure (Bentur et al., 1989; Hannant and Zonsveld, 1980). The properties of the polypropylene fibers can be varied in the production process to obtain moduli of elasticity in the range of 3–5 GPa and tensile strengths of about 140–690 MPa (ACI Committee 544, 2005). Their melting point of about 160°C is relatively low, as is their density, which ranges from 900 to 950 kg/m³.

Polypropylene fibers are highly resistant to most chemicals and to the high alkaline environment of the cement matrix, which makes them attractive for use in cement-based composites. However, similar to PE, these fibers are also hydrophobic, and therefore, they bond poorly with cement-based matrices. Although they also have poor fire resistance and are sensitive to UV and to oxygen, these limitations are not critical once the fibers are embedded in the cement matrix, which effectively protects them from exposure.

2.2.1.2.6 POLYVINYL ALCOHOL

Polyvinyl alcohol (PVA) is produced by the polymerization of vinyl acetate to polyvinyl acetate (PVAc), followed by the hydrolysis of PVAc to PVA. It has the following chemical structure:

$$\left[CH_2 - \underset{\underset{OH}{|}}{CH} \right]_n$$

PVA fibers are produced by wet or dry spinning, and high strength and stiffness are obtained by adding boron, which promotes the formation of intermolecular bonds. Insoluble PVA fibers were developed in Japan for cement-based composites—in which they exhibit strong bonding with the cement matrix as both materials are hydrophilic—and they were first introduced to the market around the middle of the twentieth century. They exhibit relatively high mechanical properties with tensile strengths in the range of 880–1,900 MPa, Young's moduli of 25–41 GPa (the same order of magnitude as that of the cementitious matrix), elongation between 6% and 10%, and density of 1300 kg/m³. PVA fibers are produced in a variety of diameters, from 10 up to 670 μm and they are used in both thin element applications and concrete structures (Li et al., 2002; Peled et al., 2008b).

Used in the concrete field for over 20 years, PVA fibers provide good performance over the long term, as they do not degrade in the highly alkaline cement environment or under hot and humid conditions, which make them durable fibers for cement-based applications. Because they do not cause chromosomal aberrations, PVA fibers are safe to use in cement reinforcement applications (Hayashi and Fumiaki, 2002; Li et al., 2004). In a multifilament form, PVA yarns can be crafted into a fabric structure for TRC applications.

2.2.1.2.7 STEEL

Attempts to use metal fibers for textile purposes were not successful due to the lack of flexibility and high weight of the fibers, precluding their use in the production of textile fabrics or as reinforcements for general composite materials. However, in the form of short fibers, steel is extensively used in concrete applications, and in a mesh form comprising thin wires, known as "ferrocement," it is used for thin section composites as the main reinforcement with a mortar matrix (Mobasher, 2011; Naaman and Shah, 1971). Although ferrocement has been widely used during the last 50 years in the construction industries in developing countries, its use in the United States, Canada, and Europe has been limited, possibly due to the high cost of labor.

Compared with conventional reinforced concrete, ferrocement has a very high tensile strength-to-weight ratio and superior cracking behavior. Ferrocement structures, therefore, can be thin and relatively light, rendering ferrocement an attractive material for the construction of boats, barges, prefabricated housing units, and other portable structures (Naaman, 2000). Although for these applications ferrocement is more efficient on a weight basis, it is usually more economical to use conventional reinforced concrete instead. Despite its higher cost, ferrocement is superior to fiberglass laminates or steel when used in unconventional structures. However, due to their higher potential to undergo corrosion compared to polymer or glass types of meshes, steel meshes constitute an overall less attractive reinforcing material compared with synthetic open fabric structures. In addition, because steel meshes are not considered textiles, they are beyond the scope of this chapter and will not be discussed further here.

2.2.1.2.8 CERAMIC

First introduced in the 1960s, ceramic fibers combine high strength and stiffness with low weight and very high temperature resistance for use in composites designed to perform at extreme temperatures. Reductions in their prices during the 1970s and 1980s made these fibers more attractive as composite ingredients (Bunsell and Simon, 1988). The types of ceramic fibers available for use in composites include boron, boron carbide, silicon carbide (SiC), and aluminum oxide (Al_2O_3). However, their highly brittle nature precludes their production via the conventional drawing process described earlier. Instead, silicon carbide or boron fiber production, for example, is generally based on a chemical vapor deposition (CVD) process, in which a very fine tungsten or carbon filament of about 17 μm is used as the substrate. The filament is introduced into a hot chamber heated to at least 1,000°C, and the vaporized boron or silicon carbide components are then added to the chamber, where they decompose and then precipitate on the tungsten or carbon filament (Franklin and Warner, 1988; Mileiko and Tikhonovich, 1997). The resultant boron or silicon carbide monofilament fiber with a carbon or tungsten core has a total diameter of 100–200 μm. The desired diameter of the fiber can be controlled during the CVD process by varying pulling speed and deposition temperature. Ceramic fiber mechanical properties of interest include densities in the range of 2200–3300 kg/m^3, tensile strengths of 2,800–3,600 MPa, and Young's moduli of 360–480 GPa. In addition, they can withstand high temperatures above 2,000°C (Mileiko and Tikhonovich, 1997). Ceramic fibers based on boron or silicon carbide are available as short fibers or as multifilament bundles.

Aluminum oxide fibers where first reported in 1978 (Bunsell and Simon, 1988). Several crystal alumina structures, each possessing different properties, have been identified. Alumina fibers have been developed with single crystal (α-alumina) and polycrystalline structures, both of which can be produced as short fibers or continuous filaments. For monocrystalline filaments, alumina is melted at a temperature slightly above its melting point, and then a capillary tube is immersed in the molten material. Next, a small seed of α-alumina is dipped in the molten alumina—which rises to the top of the capillary tube—and then slowly extracted. The single filament is grown in a draw rate of about 150 mm/min. Due to its very high degree of crystallinity and its density of about 3970 kg/m^3, α-alumina fibers have a high modulus of about 460 GPa and a high tensile strength of 2.5 GPa at room temperature, but these values drop significantly at temperatures above 900°C. For continuous polycrystalline filaments, a viscous spinning mix is prepared to produce α or γ alumina fibers with diameters of about 20 μm. For α alumina fibers, slurry is extruded through a spinneret and then the filaments are mechanically drawn as explained previously. The precursor fibers are heated in air at 1,300°C and then sintered. For γ alumina fibers, a solution is extruded by dry spinning and then the fibers are fired in air at 1,000°C. Although α alumina fibers have a higher modulus (380 GPa) than that of γ fibers (280 GPa), they have poor handling ability and rather high density. Polycrystalline filament room temperature tensile strengths are 1400 and 1800 MPa for α alumina and γ alumina, respectively. Short-staple polycrystalline alumina fibers are also produced, available in diameters of about 3 μm and lengths of several tens of millimeters.

Due to their superior structural properties and their suitability in extreme temperature applications, ceramic fibers are mainly used in metal- and ceramic-based matrices. The very high price and brittleness of the ceramic fiber, however, make it a less attractive candidate for concrete applications, in which ceramic fibers are not commonly used.

2.2.2 Natural fibers

The first composites ever made by humans were produced from natural fibers such as straw or horse hair. Traditionally, natural plant fibers were chopped into short fiber units for the production of thin roofing and cladding elements (Coutts and Warden, 1992). Short fibers can be used in a mat form as nonwoven fabrics, but these are relatively dense and less attractive for cement-based composites, in which the cement particles and aggregates in the composite do not completely penetrate between the fibers of the nonwoven fabric. Yarns can be produced from these short fibers for use in woven and knitted fabric structures, but the yarn may need some twisting to keep the fibers together. The short nature of the fibers and the requirement to twist the yarn are expected to have a negative effect on cement-based composite performance. Natural fibers can also be produced from minerals such as asbestos and basalt, from animals farmed for their wool, and from plants such as bamboo or sisal, for example (see Figure 2.1).

2.2.2.1 Asbestos

The first fibers to be commercially used in cement-based composites were made from asbestos, which is comprised of a set of six naturally occurring silicate minerals, all of which consist of long (roughly 1:20 aspect ratio), thin, microscopic fibrous crystals. Among their beneficial properties are extreme durability and resistance to fire and most chemical reactions, especially the alkalis of the cement environment, all of which promoted their use for many years in different commercial and industrial capacities, including the building and construction industries. The strength of asbestos combined with its resistance to heat and the relatively low cost of its fibers made it an ideal material for use in a variety products, including roofing shingles, floor tiles, siding, flat and corrugated sheets, and other cement compounds (Michaels and Chissick, 1979). In addition, the low-cost production methods of asbestos–cement-based components and their applicability in rapid construction techniques promoted their wide use in lightweight housing and industrial buildings.

Use of asbestos–cement composites in the building industry started in the 1880s (Woods, 2000), and by 1950, approximately 1 billion square feet of asbestos cement products had been produced for building applications. As a building product, the asbestos–cement combination exhibits many desirable material characteristics, including good fire resistance, high mechanical strength (due to the fibers' high length-to-diameter ratio), flexibility, and resistance to friction and wear, and it is also lightweight, impermeable to water, durable, tough, and resistant to rotting (The Industrial Uses of Asbestos, 1876). However, after an Environmental Protection Agency (EPA) prohibition on its use in 1973 when it became apparent that asbestos poses a serious threat to human health and safety, by the late of 1970s its use had declined sharply (National Trust for Historic Preservation, 1993). Today classified as a known human carcinogen, asbestos is strictly regulated, as exposure to this toxic mineral has been directly and scientifically linked to a number of lung and respiratory health conditions such as asbestosis, a form of lung cancer.

2.2.2.2 Basalt fibers

Basalt is made of a variety of dark gray volcanic rocks formed from molten lava that solidified in the open air (Sim et al., 2005; Singha, 2012). The mineral levels and chemical makeups of basalt formations can vary significantly from location to location. In addition, the crystal structures of basalt stones were influenced by the rate at which they cooled when the original flow reached the earth's surface, and therefore, basalt stones are found in a variety of compositions.

Basalt fiber production, for which a patent was registered in 1991, is from extremely fine basalt fibers, which comprise plagioclase, pyroxene, and olivine minerals (Singha, 2012). Similar to glass fibers, basalt fiber is produced in a continuous process. Quarried basalt rock is crushed, washed, and then transferred to melting baths in gas furnaces heated to 1,300°C–1,700°C. The molten rock is then extruded through a spinneret of small nozzles, drawn, and subsequently cooled to produce hardened, continuous filaments in the form of rovings containing varying numbers of elementary fibers (Militky and Kovacic, 1996). Compared to glass fiber processing, continuous basalt fiber production is much simpler, as the basalt fiber has a less complex composition, and it does not contain any secondary materials, a big advantage in its processing and final cost (Sim et al., 2005). Short, very cheap basalt fibers can be produced directly from crushed basalt stones, but this process results in fibers with relatively poor and uneven mechanical properties.

Basalt fibers exhibit high moduli in the range of 79.3–93.1 GPa, high strengths in the range of 3,000–4,840 MPa, high corrosion and temperature resistance, extended operating temperature ranges, and they are easy to handle (Artemenko, 2003; Cater, 2002; Militky and Kovacic, 1996). Basalt has a low density of 2800–2900 kg/m^3 much lower than that of metal (steel) and closer to those of carbon and glass fibers, but it is cheaper than carbon fiber, and its strength is higher than that of glass fiber. The moisture regain and moisture content of basalt fibers are both less than 1%. In terms of concrete applications, perhaps the most important property of basalt fibers, which can withstand pH values up to 13 or 14, is their good resistance to the alkaline environment. Accordingly, they are relatively less stable in strong acids. An extremely hard material, basalt has hardness values between 5 and 9 on the Mohs scale, which confers on it good abrasion properties. In addition, basalt fibers have excellent thermal properties compared to that of glass (E-type), and they can easily withstand continuous temperatures of 1,100°C–1,200°C for hours without suffering any physical damage. Unstressed basalt fibers and fabrics can maintain their integrity even up to 1,250°C. Basalt fibers are used to produce yarns and fabrics.

2.2.2.3 Plant fibers

Grown in a wide variety of regions around the world, plant fibers are commonly used to make ropes, carpets, backings, and bags, to name a few examples. The components of natural fibers comprise cellulose microfibrils dispersed in an amorphous matrix of lignin and hemicellulose (Saheb and Jog, 1999). Depending on the type of the natural fiber, it contains 60–80 wt% cellulose and 5–20 wt% lignin. The moisture content in natural fibers can be up to 20 wt%, a parameter that should be considered when using natural fibers in the cement product field. The properties of some selected plant fibers in use today are given in Table 2.5 (Mallick, 2007; Wambua et al., 2003).

Compared to synthetic fibers such as carbon, polypropylene, and aramid, natural plant fibers possess numerous beneficial properties for the building industry.

Table 2.5 Properties of selected natural fibers

Property	Hemp	Flax	Sisal	Jute
Density (kg/m^3)	1480	1400	1330	1460
Modulus (GPa)	70	60–80	38	10–30
Tensile strength (MPa)	550–900	800–1,500	600–700	400–800
Elongation to failure (%)	1.6	1.2–1.6	2–3	1.8

Source: Adapted from Wambua, P. et al., *Compos. Sci. Technol.*, 63, 1259, 2003.

Biodegradable, renewable, and recyclable, they do not pose any health risk, and less energy is consumed in their production. In addition, requiring a relatively low degree of industrialization, they can be produced with a small investment and at low cost, and therefore, natural plant fibers are attractive to use, particularly in developing countries, as reinforcements for cement composites.

Natural plant fibers suffer from several limitations: their tensile strengths are relatively low, they exhibit low melting points, they can absorb high moisture content, and they are sensitive to the high alkalinity of Portland-cement-based materials. In addition, sensitive to high-temperature environments, they begin to degrade at temperatures higher than 200°C, which cause severe deterioration in their mechanical properties.

Among the most widely used vegetable fibers, sisal fibers are exceeded only by cotton in terms of volume of production. Similar to cotton, they are widely used in textiles, and during the past two decades, sisal fibers have been applied as a reinforcing material in cement and polymer-based composites (Bisanda and Ansell, 1992; Toledo Filho et al., 2003). Sisal fibers are extracted from sisal plant leaves in the form of long fiber bundles with lengths of about 0.5–1.0 m (Murherjee and Satyanarayana, 1984; Oksman et al., 2002). The sisal plant leaf is reinforced by three types of fibers—structural, arch, and xylem—the first of which has significant commercial value because it almost never splits during the extraction process. The other fiber types constitute secondary reinforcement in the sisal plant leaf. Structural fiber morphology is characterized by an irregular cross section that usually has a "horseshoe" shape (Figure 2.10) with mean areas ranging from 0.04 to 0.05 mm^2 and a mean density of 900 kg/m^3.

With Young's moduli between 9 and 19 GPa and tensile strengths from 347 to 577 MPa, sisal fibers exhibit wide variability in their tensile properties. Such inconsistent behavior, a characteristic inherent to vegetable fibers, can be explained by the distribution of defects within the fiber or on the fiber surface. Notwithstanding plant characteristics, differences in test parameters/conditions and fiber cross-sectional area measurements can contribute to such variability in the fiber's tensile properties. Sisal fibers exhibit no fatigue failure below 10^6 cycles when subjected to a ratio of maximum applied fatigue stress to ultimate tensile strength of 0.5. No significant stress–strain hysteresis was observed during fatigue under monotonic tensile tests (Silva et al., 2008).

2.3 Yarn types

Comprising natural and/or synthetic fibers and used to make fabrics, yarns can be produced from either short (staple) or continuous (filament) fibers. Synthetic fibers come in three basic forms: staple, monofilament (Figure 2.11a), and multifilament (Figure 2.11b), the latter can be in a tow

(a)

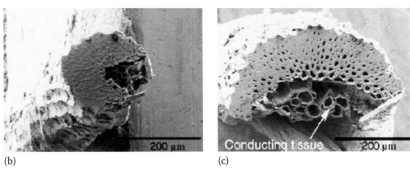

(b)   (c)

Figure 2.10 Sisal fiber: (a) the plant, (b) structural fiber with horseshoe-shaped geometry, and (c) arch fiber. (From Silva, F.A. et al., *Sci. Technol.*, 68, 3438, 2008.)

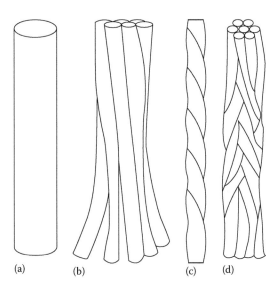

(a)   (b)   (c)   (d)

Figure 2.11 Yarns types: (a) monofilament, (b) multifilament, (c) twisted, and (d) twisted bundle.

structure that consists of many filaments assembled side-by-side. All of the basic yarn forms can be used to produce fabric structures.

Short-staple fiber yarns, known as spun yarns, are usually made by twisting (Figure 2.11c) or bonding the staple fibers into a cohesive thread. Spun yarns can contain a single type of fiber material or a variety of blends that combine several fiber materials in a single yarn. In addition, a single spun yarn can be created with only natural or synthetic fiber blends, but it is also possible and common to produce single yarns with blends of natural and synthetic fibers. Thicker yarn is produced by assembling several single spun yarns, known as plies, together by twisting the group in the direction opposite that in which the plies were twisted. In contrast, a monofilament yarn is composed of a single, thick filament (Figure 2.11a) usually with a relatively large diameter of at least 100 μm. These are usually made of low-cost materials such as PE or polypropylene, among others.

A multifilament yarn (bundle) is an assembly of parallel filaments, which are very long, fine, continuous fibers that are either twisted (Figure 2.11d) or grouped together (Figure 2.11b). One multifilament yarn consists of several hundreds to thousands of single filaments. Therefore, the fineness of the yarn, expressed in units of tex (yarn linear mass density), depends on the number of filaments, the average filament diameter, and the fiber density. For example, glass filaments have diameters in the range of 9–27 μm, and anywhere from about 400 to 2,000 filaments are used to produce a glass roving. For carbon, the filaments are usually less than 10 μm in diameter, and a bundle can contain from 1,000 up to 10,000 filaments. Aramid and polypropylene are usually produced from several hundred filaments. As for spun yarns, filament yarns can also be fabricated into assembled and nonassembled yarns. Assembled yarns are made in a two-step process that consists of twisting two or more twisted filament yarns (or plies) together in a process known as cording, whereas nonassembled yarns include a single tow of elementary fibers.

Filament yarns can also be produced in a hybrid structure that can comprise combinations of several different fiber materials in a single bundle of yarn. Such combined or hybrid yarns created by coating an AR glass or carbon filament core with a mantle of AR thermoplastics are used to produce fabric for reinforcing concrete. The thermoplastic material can be melted in an additional process to bind the yarns together to provide highly stabilized fabric structures. Other combined yarns can be manufactured by coextruding, a similar process in which the core yarn is spray coated with another polymer and additional yarns are then wound around the core yarn for reinforcement applications (Gries et al., 2006).

Composite element properties are influenced by yarn geometry and shape. For reinforcing applications, filament yarns (in either monofilament or multifilament form) made of continuous fibers are more common, as they exhibit small structural elongation and better reinforcing efficiency. For composite materials, ideally, all the filaments constituting the yarn should be parallel and drawn along the yarn's longitudinal direction, but alignments may diverge depending on the method of fabrication. For example, twisting of the filaments within the bundle, which results in filament misalignment (Figure 2.11d), may compromise reinforcing efficiency. In contrast to a yarn with no twist, in which maximum packing density in the cross section is obtained by the closely packed hexagonal arrangement of the filaments, the twisting of filament yarns significantly lowers their packing density, thus further reducing their reinforcing ability. Therefore, for most reinforcing applications, the bundles are usually made of straight yarns grouped together side-by-side without any twist.

In cement applications in particular, besides the influence of yarn alignment and orientation in the reinforcing direction, the ability of the cement particles to penetrate the spaces between the filaments in the bundle and the extent to which the yarn mechanically anchors with the cement matrix are also important considerations. Mechanical anchorage with the cement matrix can be achieved by using yarns having complex shapes. Therefore, although the reinforcing efficiency of yarn is highest when aligned as described earlier, complicated yarn shapes can result in better mechanical anchoring of the yarn with the cement-based matrix to improve cement composite performance.

In order to exploit the high mechanical potential of the elementary filaments in the bundle, sizing is applied to the yarn on the single filament surface to improve its processing behavior. The sizing material has a marked influence on the quality of the filaments, influencing properties such as the load-bearing performance. Research on the development of new yarn sizes and coatings will be discussed later in this chapter where the influence of different sizing and coating materials on the mechanical properties of fiber/yarn/fabric is discussed.

2.4 Main fabric structures

Fabrics can be manufactured with a variety of geometries that differ mainly in how the yarns are combined together to form the fabric structure. The connection method of the yarns influences fabric geometry, opening, yarn geometry, fabric properties, and handling. Fabrics are typically classified based on their production method, which may vary vastly from simple to complicated forms.

There are several advantages to using textile fabrics to reinforce a cementitious matrix. Depending on fabric type and yarn placement, the composite can be reinforced in all the required directions in one processing step. Relatively large numbers of continuous yarns can be easily incorporated in the desired directions. In addition, fabrics are well suited to reinforce curved, complex components with varying load paths since they can be placed with a reproducible accuracy. Finally, the geometry of the fabric and its yarns may have a significant influence on composite properties and performance.

The most important criterion that determines the efficacy of fabric-reinforced cement composite applications is the extent to which the fabric can be created in an open grid structure and a high displacement stability, which the fabric must possess to be incorporated in cement-based structures. The size of the grid opening should be such that the cement-based matrix—which comprises cement particles with an average size of ~10 μm and fine aggregates that are typically up to 1 mm—must be able to penetrate the spaces between the yarns of the open grid fabric structure. Furthermore, to ensure satisfactory handling, there must be no displacement of the threads.

In the following sections, the main fabric types available today and their architectural characteristics, as well as yarn orientation, yarn crimping, and yarn interlocking within each fabric structure, are addressed.

2.4.1 Woven fabric

Woven fabrics are manufactured by interlacing two sets of yarn perpendicular (0°/90°) to each other by crossing each over and under the other in alternating fashion (Figure 2.12). The yarns along the fabric length are termed warp yarns, and those along its width are termed weft or fill yarns. In woven fabrics, the yarns are connected together by friction at the joint points, so a sufficient number of connection points are needed to hold the fabric together in a stable manner. Thus, these fabrics are relatively dense so that the yarns will not come apart during service, and therefore, they cannot be in a mesh form. In woven fabric, due to the method of connection, that is, passing the warp

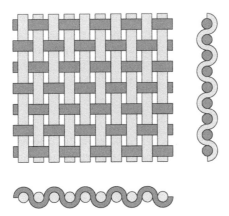

Figure 2.12 Woven fabric, plain weave structure.

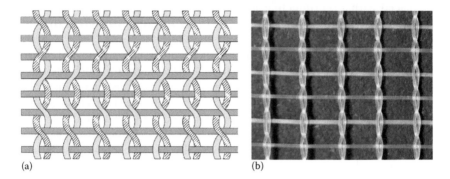

(a)   (b)

Figure 2.13 Leno fabric structure—warp yarns in eight patterns are oriented lengthwise
perpendicular to the straight weft yarns running widthwise: (a) scheme and
(b) image of a fabric.

and fill yarns under and over each other, both yarns in both directions are in crimp shapes. In addition, as the connection method is by friction alone, the filaments within the yarn in both fabric directions are not tightened, leaving the bundles relatively loose and open. Fabric density, yarn crimping, and the looseness of the bundle can influence composite properties.

Fabrics woven in an open, net design require special processing or a special weave such as leno (Figure 2.13) to facilitate connections that are strong enough to hold the yarns together while preserving the desired open fabric structure. The leno weave is a peculiar woven structure in which two warp yarns are twisted around the weft yarns to provide strong interlocking, resulting in a durable fabric with an openwork structure. Dimensional stability—with virtually no yarn slippage or displacement of the threads—is conferred on the open structure of the fabric by firmly binding pairs of twisted warp yarns to the weft yarns. Dictated by its intended application, the extent of openness in this fabric structure is determined by the size of the spacing between the paired warp yarns. The leno weave produces a strong, lightweight, and open fabric structure that is ideal for cement-based composite applications. Note that in this fabric, the weft yarns are straight.

2.4.2 Knitted fabric

The knitted fabric structure is produced by interlooping the yarns, which in the knitted fabrics is manufactured using a set of closely spaced needles that pull the yarns to form the loops (stitches) (Figure 2.14). The two main knitted fabric production methods are weft knitting (Figure 2.14a) and warp knitting (Figure 2.14b).

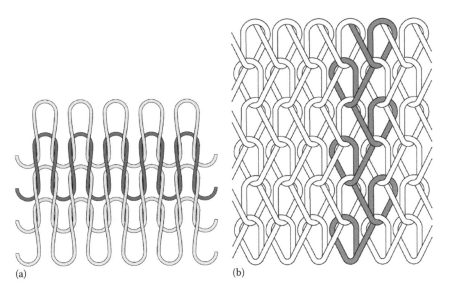

Figure 2.14 Knitted fabric schemes: (a) weft knitted and (b) warp knitted.

2.4.2.1 Weft knitted

Weft-knitted fabrics are very common in clothing applications and used less frequently for reinforcing applications, as the ultimate fabric shape ends up with high curvature. In these cases, the yarn forming the loops runs in the weft (cross-machine) direction of the fabric, passing from one loop to the next. Similar to the method used in hand knitting, the whole fabric structure can be produced from a single yarn. Moreover, because they possess low dimensional stability, weft-knitted fabrics are easily stretched.

2.4.2.2 Warp knitted

In warp-knitted fabrics, the yarns that make the loops run in the fabric's warp direction (lengthwise, in the machine direction). To produce the whole fabric structure, multiple yarns are needed. The method involves connecting a set of warp yarns together by loops that shift in a zigzag manner along the fabric length from one column to another, alternately, depending on the fabric geometry required for the intended end use. Compared to weft-knitted fabrics, these fabrics exhibit much higher structural stability. Due to the strong, loop-based connection at the fabric joints, warp-knitted fabrics can be produced in a wide range of mesh openings and mesh structures.

2.4.3 Bonded fabric

Bonded fabrics are manufactured by connecting two perpendicular sets of warp and weft (0°/90°) yarns at their junctions with glue or by heating (Figure 2.15). Due to the strong connections at the yarn junction points, mesh structures are available in a wide range of openings. Because the yarns in both directions are in straight form, bonded fabrics are suitable for cement-based composites, for which fabrics are commonly made from multifilament AR glass yarns.

2.4.4 Nonwoven fabric

Nonwoven fabrics are manufactured from webs of fibers (Figure 2.16) connected together using a variety of means, such as gluing, heating, and needle

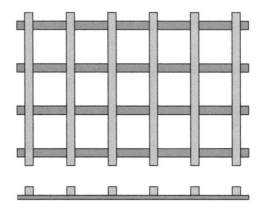

Figure 2.15 Bonded fabric.

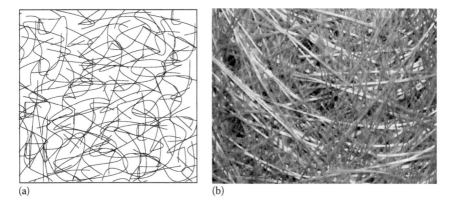

(a)   (b)

Figure 2.16 Nonwoven fabric: (a) scheme and (b) image of a fabric.

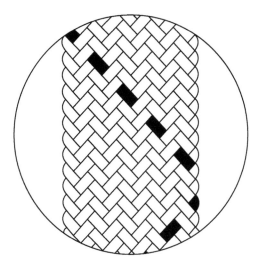

Figure 2.17 Braided fabric.

punching, among others. In contrast to all other fabric production methods, nonwoven fabric is made from fibers instead of yarns. The fibers can be short or long as well as continuous. The resulting fabric has a dense structure, in which the fibers are arranged randomly in all directions.

2.4.5 Braided fabric

Braided fabrics are manufactured by intertwining sets of continuous yarns together (Figure 2.17). Fabric width and yarn orientation can be controlled by changing the feeding speed of the yarns and the number of yarns. The production speed of these fabrics is slow compared to those of other methods. Therefore, they are used mainly for reinforcing applications such as tubular structures. Although the high density of braided fabrics precludes their production in mesh form—making them unsuitable for cement-based composites—they constitute an attractive reinforcing component for polymer-based composites.

2.5 Principal architectural characteristics of fabric and its yarns

2.5.1 General concept

For reinforcement purposes in composite materials, not only are the production methods of the fabric and its geometry important, but also the orientation and shape of the yarn within the fabric must be considered.

A wide range of fabrics have been developed especially for composite material applications. The desired properties of the final product are ensured by controlling the exact location, orientation, and geometry of the reinforcing yarns based on the requirements of the reinforcement application. Figure 2.18 presents a range of fabric structures for composite materials.

In general, the reinforcing yarns for composite materials should be as straight as possible in the reinforcing direction to achieve high reinforcing efficiency. To exploit their full reinforcing potential, the reinforcing yarns must be able to transfer the applied load along the yarn length efficiently. This depends on the extent to which they are oriented in parallel with the direction of the main stresses (Figure 2.19). The larger the yarn angle to load direction, the lower the yarn reinforcing efficiency. For example, changes in the yarn density in woven fabrics using the same yarn diameter can influence the crimping angle of the reinforcing yarn. This is clearly demonstrated in Figure 2.19, showing that the more closely bunched yarns are situated at a greater angle. In addition, most fabrics for composite

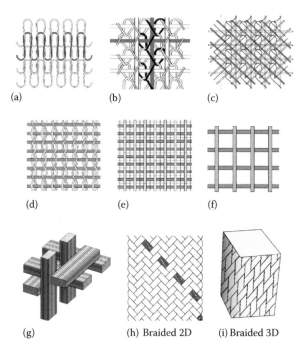

(a)

(b)

(c)

(d)

(e)

(f)

(g)

(h) Braided 2D

(i) Braided 3D

Figure 2.18 Structures of reinforcing fabrics for composite materials: (a) weft knit, (b) warp knit—laid in weft and warp, (c) warp knit multiaxial, (d) woven—leno, (e) woven—laid in weft and warp, (f) bonded, (g) 3D fabric, (h) braided 2D, and (i) braided 3D.

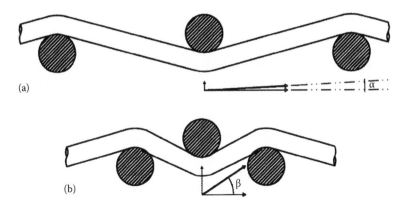

Figure 2.19 Influence of yarn geometry on stress development: (a) small angle and (b) large angle in the loading direction of the reinforcing yarn.

materials are made with high-modulus fibers that in many cases are brittle. During fabric production, the winding and bending of these brittle fibers and the crimping shape within the final fabric structure can cause the filaments to break, reducing their mechanical properties. As is the case for all types of composite materials, low yarn reinforcing efficiency and fiber breakage typically lead to an overall reduction in the mechanical properties of the composite compared to a similar composite with straight yarns in the main stress direction.

In addition to yarn alignment discussed already, the penetrability of cement paste into fabric openings and the areas between the bundle filaments is essential in providing sufficient bonding between filaments and matrix. This is an important consideration for fabric reinforcements in light of the particulate nature of the cement paste, the viscosity of which is not as low as that typical of polymer materials. In addition, the bonding of many fiber materials with the cement matrix is relatively low, as the cement matrix is hydrophilic whereas the fibers are hydrophobic. In some cases, therefore, mechanically anchoring the reinforcing yarn (which must have a complex and not necessarily straight shape and geometry) within the cement matrix can produce cement-based composites with superior mechanical properties. Thus, the design concepts of fabric architectures for cement-based composites are rather complex, taking into consideration a range of parameters such as yarn orientation, yarn shape, opening of bundle filaments, and yarn tightening effects, as well as fabric geometry and density. In the rest of Section 2.5, the basic fabric architectural properties are presented for composite materials in general and for cement-based composites in particular.

2.5.2 Reinforcing yarn orientation

Depending on the intended application of the fabric, the orientation of its component yarns can be varied to produce different architectures. These can be classified into two main categories: flat fabrics (2 dimensional, 2D) in which the reinforcing yarns are in the fabric plane only and bulky fabrics (3 dimensional, 3D) in which the reinforcing yarns are oriented in several directions, not only parallel with the fabric plane.

2.5.2.1 Flat structures (2D)

2.5.2.1.1 UNIDIRECTIONAL FABRICS

The reinforcing yarns are parallel to each other and located in only a single reinforcing direction of the fabric (Figure 2.20) to produce composite materials with enhanced mechanical properties in one direction. In this case, the only function of the yarns in the perpendicular direction is to hold the reinforcing yarns in place, and therefore, they are very thin (small diameters) and have low structural properties. Reinforcing efficiency, however, is promoted by the fine diameter of the perpendicular yarn, which allows the reinforcing yarns to be straight.

2.5.2.1.2 TWO-DIRECTIONAL FABRICS

The reinforcing yarns are located in two mutually perpendicular directions (0° and 90° orientations) relative to the fabric plane to provide a composite with improved properties in those two directions (e.g., see Figure 2.12

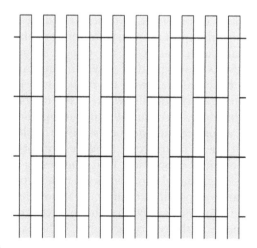

Figure 2.20 Unidirectional fabric.

Figure 2.21 Two-directional (2D) fabric for composite material comprises two sets of yarn: one set of fine diameter yarns holds a second set of reinforcing yarns in place.

[plain weave]). To obtain composite materials with high reinforcing efficiency, however, the reinforcing yarns should be straight. Therefore, flat 2D fabrics for composite materials are typically produced with two sets of perpendicular yarns in 0° and 90° orientations: reinforcing yarns with high structural properties and fine diameter yarns with low structural properties, whose main function is to hold the reinforcing yarns in place while allowing them to maintain a straight form (Figure 2.21).

2.5.2.1.3 MULTIDIRECTIONAL FABRICS

The reinforcing yarns are oriented in several directions (e.g., *x* and *y* directions as well as diagonally) relative to the fabric plane (Skelton, 1971). Multiaxial fabrics can be produced by different production methods, one of which is warp knitting (Guenther, 1983; Raz, 1987). The structure of a multiaxial warp-knitted fabric comprises several layers of reinforcing thread systems with different orientations bound together by knitting threads (loops). Any combination of angles between the thread layers with fabric weight per unit area (yarns density) is possible, and this architecture also allows for the production of open net structures. These fabrics provide composites with improved properties, typically with isotropic behavior in the direction of the fabric plane. In addition, they exhibit better dimensional stability, with higher tearing and shear resistance better than unidirectional and bidirectional fabrics. Widely used as reinforcement in composite materials, these fabrics can be produced relatively fast, which reduces their final cost. An example of a warp-knitted multiaxial fabric is presented in Figure 2.22, where the reinforcing yarns—oriented in four directions (90°, 0°, and ±45°)—are connected by stitches to form an open fabric structure.

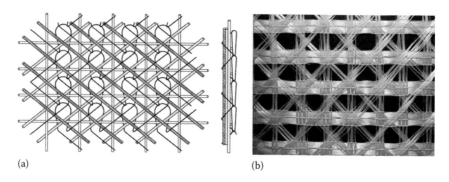

Figure 2.22 Multiaxial fabric, warp knitted: (a) scheme and (b) image of a fabric.

2.5.2.2 Bulky fabrics (3D)

Bulky fabrics contain reinforcing yarns in the fabric plane and along its thickness (Ko, 1985; Summerscales, 1987). For example, schematic depictions of three-directional and four-directional bulky fabrics are presented in Figure 2.23. Although bulky fabrics have been developed with reinforcing yarns in up to 11 different directions, the production of such multidirectional fabrics is slow and costly. Therefore, the most popular bulky fabrics are the three-directional type, in which the reinforcing yarns are located along the x, y, and z directions, with mutually perpendicular yarns along the plane and thickness of the fabric. Due to the strong connections in all three yarn directions, these fabrics resist crack development and delamination (separation between the fabric layers) within the composite. A variety of production methods such as weaving and braiding, among others, have been used to develop about twenty different three directional fabrics, with correspondingly different structures, for use in composite materials.

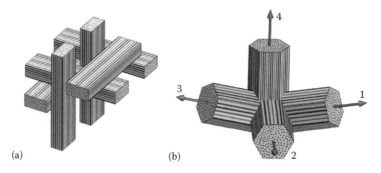

Figure 2.23 Bulky fabrics: (a) three-directional and (b) four-directional.

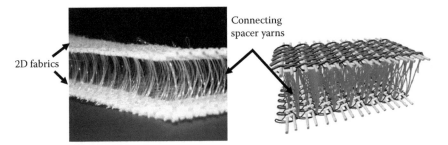

Figure 2.24 Three directional bulky spacer fabric. (From Janetzko, S. and Maier, P., Development and industrial manufacturing of innovative reinforcements for textile reinforced concrete, in B. Küppers (ed.), *Proceedings Aachen-Dresden International Textile Conference,* Aachen, Germany, November 29–30, 2007; Roye, A. and Gries, T., *Kettenwirk-Praxis,* 39(4), 20, 2004.)

Because most of these fabrics have very dense structures, they cannot be used as reinforcements in cement-based composites.

A three directional fabric possessing a relatively open structure was specially developed for cement-based composites using weft insertion double–needle bar warp-knitting technology (Figure 2.24) (Gries and Roye, 2003; Roye and Gries, 2004; Roye et al., 2004; Roey, 2007). In this fabric type, the structure is created by knitting two sets of independent, 2D knitted fabrics together with a third set of yarns along the thickness (z direction) of the fabric (Figure 2.24). Usually referred to as spacer yarns, the connecting yarns in the z direction, which should be relatively stiff and rigid to enable a stable bulky structure that stands alone and resists collapse, are made of low-modulus monofilament yarns such as polyester (PES), PE, or similar. Therefore, the resulting fabrics, whose thicknesses can be from several millimeters up to several centimeters, are known as 3D spacer fabrics. Moreover, a single, variable-thickness fabric unit can be produced to fulfill the requirements of special, composite element shapes with varying thicknesses. The technology for producing such bulky knitted fabrics is efficient, and they can be economically manufactured (Janetzko and Maier, 2007).

2.5.3 Reinforcing yarn shape

Typically, in conventional fabric production methods for clothing applications, the yarns are not in a straight geometry. Instead, their shapes are affected by the way the yarns are connected. For example, in woven fabrics the yarns in both directions are in a crimp shape, whereas the yarns in knitted fabrics are in curved shapes created by the interlooping technique used in knitting. Several fabric architectures that allow the reinforcing yarns to be in a straight geometry along the main stress direction of the composite have been developed and are available.

2.5.3.1 Woven fabrics

Woven fabrics with different geometries can be produced by changing the number of interlacing points per unit length, that is, altering the numbers of fill and warp yarns that are passed over and under each other (Bailie, 1989; Cumming, 1987; Zeweben et al., 1989). Three basic types of weave pattern are presented:

Plain weave (Figure 2.25a)—One warp yarn passes over and under one filling yarn. This type of weave has the maximum number of interlacing points between warp and fill yarns of all the weave patterns, and therefore, this woven fabric also has the highest crimping geometry. As the most densely woven fabric pattern, the plain weave has high dimensional stability in terms of slippage and fabric distortion, and it exhibits high shear resistance at the fabric plane. However, its reinforcing efficiency for composite materials is low due to the high crimping properties of its yarns in both directions. In addition, due to its high structural stability,

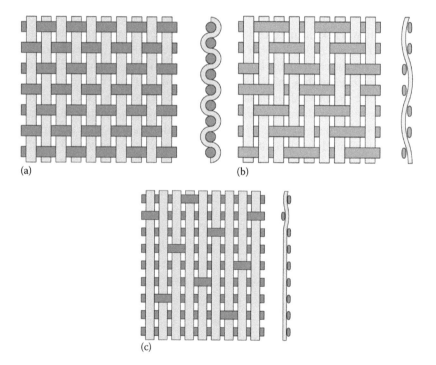

Figure 2.25 Various woven fabric architectures: (a) plain weave, (b) twill weave, and (c) satin eight-harness weave.

plain weave fabric is difficult to bend, precluding its use in composite applications that entail complex shapes.

Twill weave (Figure 2.25b)—One or more warp yarns pass over and under two or more filling yarns to produce a diagonal pattern. Twill fabrics have both high structural stability and high draping capacity, that is, they are easy to bend and therefore more suitable for use in complex composite shapes than plain weave fabrics. Yarn crimping is less severe than for the plain weave.

Satin weave (Figure 2.25c)—One warp yarn passes first over three or more fill yarns and then under only one fill yarn and so forth to produce the woven fabric with the lowest number of interlacing points between warps and fills per unit length. The greater the number of fill yarns passed by a warp yarn, the straighter their form in the fabric and the better their reinforcing efficiency; thus, satin weaves are used to produce the highest strength composites. The single warp yarn can be interlaced with a maximum of eight fill yarns. The small number of interlacing points between warps and fills confers on satin the highest drapability and bending ability among all the woven fabrics, making them easily conformable to very complex component geometries and shapes. However, the small number of interlacing points results in fabrics with low structural and dimensional stability, and thus, yarn slippage and fabric distortion are typical.

Yarn density in the woven fabric is determined by the number of interlacing points such that the higher the number of interlacing points between the fill and the warp yarns, the lower the possible number of yarns per unit length, that is, yarn density decreases with the increase in the number of interlacing points. This is due to the nature of the connection between the two yarn types in woven fabrics: interlacing the yarns by passing them under and over each other opens spaces between the yarns and reduces yarn density. Because the number of interlacing points is the smallest in satin weaves, these fabrics enable the densest woven fabric, which is used to produce composites with the greatest yarn content of the woven fabrics. The straight yarn geometry and the dense fabric structure give satin woven fabrics the highest composite properties among all the composites based on woven fabrics, making them attractive for polymer-based composite material reinforcement.

For cement-based composites, the dense structure of satin fabrics can be problematic in terms of fabric penetrability by the matrix due to yarn closeness. Although this drawback is most significant for the satin weave, the less dense twill and even plain weave structures can also suffer from poor fabric penetrability between the fabric yarns by the matrix.

2.5.3.2 Warp-knitted fabrics

2.5.3.2.1 SHORT WEFT WARP KNITTED

In warp-knitted fabrics, the yarns composing the loops run in the warp direction (lengthwise) of the fabric in a zigzag manner, alternating from one column to another (Figure 2.14b), to provide strength and stability. Due to this zigzag shifting of the warp yarn, the loop structure can be divided into two parts (Figure 2.26): one is the loop itself, referred to as the overlap, and the second, referred to as the underlap, is the part of the shifted yarn length that runs from one loop column to the other. The spaces between the loops (needles in the knitting machine) can be varied to produce longer and shorter lengths of underlapping yarn. The longer the underlap, the more it lies in the weft direction (fabric width), leading to straighter yarns along the width of the fabric. It is also possible to run the underlap yarn to a column farther from the immediately adjacent column. Knitting with a long underlap provides fabric that is more stable, thicker, and denser (Raz, 1987). By varying the underlap length, mesh structure fabrics can be produced with different openings to suit the particulate nature of the cement matrix. In short, the yarn shape of weft warp-knitted fabric is complex and strongly connected to the fabric structure.

2.5.3.2.2 WEFT INSERTION WARP KNITTED

Weft insertion warp-knitted fabric uses straight weft yarns, that is, yarns inserted along the fabric width, in the warp-knitted fabric (Figure 2.27), in which the loops run parallel to each other lengthwise along the warp direction of the fabric, without any overlapping connections (Raz, 1987). Another set of yarns, laid widthwise in the fabric's weft direction, are straight and

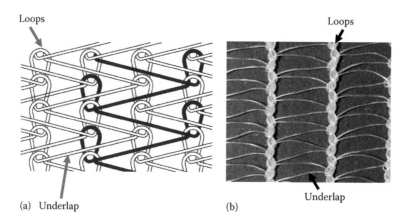

Figure 2.26 Warp-knitted fabrics: (a) scheme and (b) image of short weft.

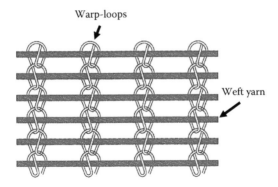

Figure 2.27 Weft insertion warp-knitted fabric with straight yarns in the weft direction.

connected to each other by the warp loops to form the fabric structure. The weft yarns can be made of one or several different yarn materials (e.g., high modulus such as carbon, glass, and aramid or low modulus such as PE, PP, and more). The weft insertion warp-knitting method produces a fabric with straight yarns along its width that exhibit virtually no crimping or lapping, a property that is advantageous for composite material applications. However, this fabric has relatively low stability, rendering it hard to handle during composite production. Moreover, the unidirectional structure of the fabric, whose reinforcing yarns run in only one direction, also limits its applicability in composites.

To improve the stability and reinforcing potential of weft insertion fabric, straight yarns can also be inserted in the warp direction to produce a fabric with reinforcing yarns in two directions: 0° and 90°. In addition, the warp and weft yarns within this fabric can be made from fibers with high structural properties for reinforcement purposes, making weft insertion warp-knitted fabric suitable for composite material structures. In addition, it is more dimensionally stable than the unidirectional weft insertion fabric described earlier. These advantageous properties of warp-knitted fabric technology make it applicable for use with brittle fibers such as glass and carbon, which are easily damaged by the severe curving and bending due to the weaving process. Instead of bending the brittle yarns, the weft insertion warp-knitting process entails their insertion into the fabric as straight fibers, thereby limiting fiber damage and breakage.

Two mechanisms are employed to bind the reinforcing warp yarn onto the 2D weft insertion fabric structure. In the first, the warp yarns are inserted into the stitches during stitch formation, surrounding the reinforcing yarn with stitches to produce what is known as chain or pillar-knit fabric (Figure 2.28a). In the second, termed tricot-knit fabric, placing of the warp yarns between the overlaps of the tricot stitches causes the

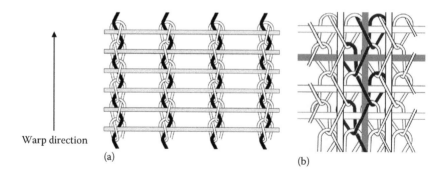

Warp direction

(a)   (b)

Figure 2.28 Schematic description of warp-knitted weft insertion fabrics: (a) chain/pillar and (b) tricot.

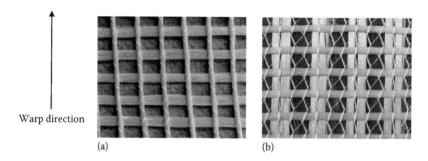

Warp direction

(a)   (b)

Figure 2.29 Images of warp-knitted weft insertion fabrics with straight yarns along the warp and weft directions: (a) chain/pillar and (b) tricot.

overlaps to hold the reinforcing warp yarns in place within the fabric structure (Figure 2.28b). In these fabrics, the loops are made of fine yarns whose only purpose is to hold the reinforcing yarns together in the warp and weft directions (Figure 2.29).

2.5.3.2.3 CHAIN-KNITTED FABRIC

The reinforcing warp yarns in these knitted fabrics are held tightly by the stitches to provide a strong and stable fabric structure (Figures 2.28a and 2.29a). The amount of stitch-tightening tension can be controlled by changing the number of the stitches per unit length such that increasing the number of stitches results in smaller stitches with a correspondingly higher tightening effect (Dolatabadi et al., 2011). From the perspective of fabric stability, although tightening is beneficial since it firmly holds the yarns together, it can also cause warp yarn misalignment and crimping (Figure 2.28a), which can limit their reinforcing efficiency. Indeed, research has empirically demonstrated that the level of tightening tension of the knitting threads has a marked influence on the ultimate properties of the chain-knitted fabric and of the composite (Schleser, 2008). An analytical model was developed to calculate the amount of tightening tension by a stitch (knot) needed to deform bundle yarn to its maximum possible packing density (in a hexagonal arrangement of circular cross sections) for a given stitch length. Greater tightening tension is required to deform the warp yarn than that required to deform the weft yarn due to differences in their geometries within the chain-knitted fabric. For AR glass roving of 2,400 tex, a tension of 0.41 N is sufficient to obtain maximum deformation in the weft direction, whereas a tension of 3.40 N is required to deform the warp yarn embedded in a 10 mm stitch length (Dolatabadi et al., 2011).

In conventional chain-knitted fabric, the knitting thread (stitch yarn) is only located on one face of the fabric. In addition, only at the fabric junctions does the knitting thread surround the warp yarn, connecting it to the weft yarns, whereas between the weft yarns (between fabric connection points) the knitting thread is concentrated on the upper surface of the fabric leaving the other side of the warp yarn free and loose. This form promotes marked deflection and widening of the reinforcing warp yarns on the side of the fabric without the knitting thread, and therefore, the portions of the reinforcing warp yarns situated between fabric junctions are widened, deflected, and wavy. This can result in an undesirable straining of the warp yarns, stress concentrations within the composite, and spalling of the concrete surface even at service load level when the fabric is used in TRC. The deflection of the warps is strongly influenced by the pattern and arrangement of the yarns within the fabric.

Fabrics for TRC applications must have open grid structures such that the long lengths of the reinforcing warp yarns are free between the connections with the weft yarns. Free warp yarn portions result in high yarn deflection between the fabric junction points (Figures 2.30a and 2.31a), a limitation that was overcome with the development of an extended stitch-bonding knitting process (Hausding and Cherif, 2010) that effectively reduces warp deflection by arranging the knitting thread evenly around the warp yarns (Figure 2.30b). Using this method, the warp yarns are fixed on both sides of the fabric (Hausding et al., 2007), and the knitting thread also runs diagonally along the backside of the fabric. Likewise, parallel reinforcing warp yarn layers can be symmetrically arranged on both faces of the fabric (Figures 2.30c and 2.31b). The symmetrical layer arrangement enables a completely straight, even, and stretched arrangement of the warp yarns within the fabric without any deflection or waviness.

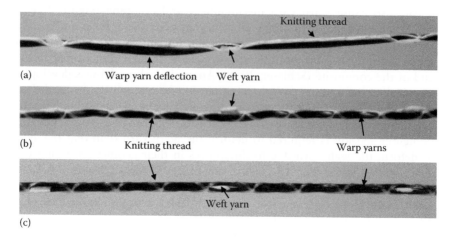

Figure 2.30 Warp yarns in three different knitting methods: (a) conventional stitch-bonded, (b) extended stitch-bonded, and (c) extended stitch-bonded with symmetrical warp yarn arrangement. (From Hausding, J. et al., Concrete reinforced with stitch-bonded multi-plies—A review, in C. Aldea (ed.), *Thin Fiber and Textile Reinforced Cementitious Systems,* American Concrete Institute, Farmington Hills, MI, ACI Symposium Publication 244, 2007, pp. 1–16.)

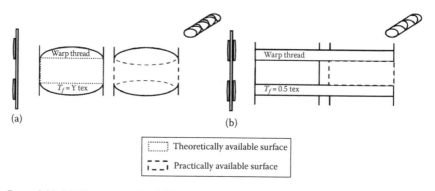

Figure 2.31 (a) Conventional and (b) extended stitch-bonding fabrics.

2.5.3.2.4 TRICOT-KNITTED FABRIC

In tricot fabric, the warp yarns are not inserted into the loops, but rather, they are connected to the fabric by the loop overlaps, that is, the yarns are situated to the sides of, and not within, the loops. This approach provides fabrics with yarns aligned along the warp and weft directions with no crimping and very little tightness tension (Figures 2.28b and 2.29b). This arrangement exploits the reinforcing ability of tricot fabrics for use in composite materials; however, their dimensional stability is relatively low due to the weak binding at fabric junction connections. The slight tightening of the fabric yarns makes tricot a relatively loose fabric that is difficult to handle and use for composite production. In contrast to warp-knitted chain fabric, as there is no tightening by stitches, both warp and weft bundles have relatively open structures in the tricot fabric, which improves cement matrix penetrability and, consequently, its reinforcing efficiency.

Advanced weft insertion warp-knitted technology also enables the production of fabrics with noncrimped yarns in diagonal ±45° directions, in addition to the straight noncrimped yarns at 0° and 90° (warp and weft), to produce multiaxial fabrics as discussed earlier (see Section 2.5.2.1.3 and Figure 2.22). This fabric is more stable than weft insertion fabric and can provide reinforcements in several directions at the composite plane to improve composite in-plane and shear properties.

Weft insertion technology can also be used to produce 3D spacer fabric using double–needle bar warp-knitting technology (see Section 2.5.2.2) with 2D warp (chain)-knitted fabrics at its surfaces. High-performance reinforcing yarns can be either brittle or ductile since they are not bent during the production phase. It is also possible to insert high-performance yarns along the z direction to incorporate structural reinforcing yarns in all dimensions. However, because some bending of these yarns is necessary during fabric production, it is difficult to use very brittle fibers such as glass, and more ductile fibers such as aramid should be employed (Figure 2.32).

In summary, the weft insertion warp-knitting technique is an effective method to integrate high-performance brittle fibers into a fabric structure without the potential for buckling or crimping that is typical in woven fabrics. This knitting technology prevents fiber damage during the reinforcing process and enhances the reinforcing efficiency of the fabric in composite materials. Furthermore, with its relatively low resistance to deformation in comparison with conventional woven fabrics, weft insertion warp-knitted

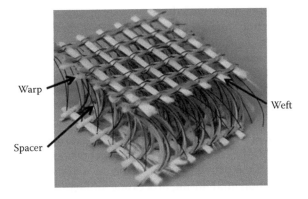

Figure 2.32 Three-dimensional (3D) warp-knitted fabric with AR glass in the weft and warp directions and aramid in the fabric thickness direction (z direction).

fabric can be easily deformed without the fabric buckling under complex deformation (Hu and Jiang, 2002). In addition, this knitting technology is amenable to fabric structures with a wide range of densities, up to very open mesh structures, which promote fabric penetrability by the particulate cementitious matrix, making these fabrics very attractive for cement-based composites. Weft insertion warp-knitted fabrics are thus the most popular reinforcement fabrics for concrete applications, particularly when the inclusion of high-performance fibers is desired.

2.6 Mechanical properties and coating of textiles

The geometry and mechanical properties of the fabric play vital roles in the composite's performance. The mechanical properties of the fabric strongly depend on the yarn and fiber properties as well as textile processing as discussed in the following sections.

Brittle fiber/filament and roving tensile strengths are commonly improved by sizing and coating the textile. Fibers, bundles, and fabric surfaces are coated for several reasons (Mäder et al., 2000; Zinck et al., 2001): (1) preservation of fiber performance, especially in terms of strength and abrasion resistance, within the textile fabric (Glowania and Gries, 2010); (2) binding of the filaments within a bundle to improve their reinforcement abilities as the filaments within the fabric structure carry the loads as a single unit; (3) improved interphase formation during composite production for better adhesion of the fabric structure with the surrounding cementitious matrix (Gao et al., 2004, 2007); (4) structural stabilization of the fabrics to ensure the yarns are aligned in the reinforcing direction in the cementitious matrix; and (5) fiber and bundle protection against chemical attack, which improves the fabric's reinforcement durability within the concrete.

Many resins/polymers are available in the market for fiber/bundle/fabric coating applications with varied viscosity properties. The higher the viscosity of the coating compound, the lower its penetration of the bundle filaments. Textile coating compounds for TRC applications have been examined in a range of studies that experimentally evaluated the effects of aqueous dispersion polymers, including PVA, ethyl acetate (PVAc) (Weichold, 2010), poly(ethylene oxide) monomethyl ether (PEO-MME), acrylate dispersions, nonionic polyurethane (Glowania and Gries, 2010; Glowania et al., 2011), styrene–butadiene copolymers, and various epoxy resins (Butler et al., 2006; Kruger et al., 2003; Raupach and Orlowsky, 2007; Raupach et al., 2006; Schleser, 2008; Schorn et al., 2003). Sizing is usually applied on the filament/fiber surface during the spinning process to limit the damage potential during processing and handling while increasing fiber tensile performance. Later, coating is applied on top of the sizing to the filament, bundle, and fabric surfaces.

2.6.1 Fibers

As discussed earlier, the tensile strength of a fiber is inversely dependent on its diameter (see Section 2.2.1.1 and Figure 2.5). This behavior was clearly observed for different diameters (13, 17, and 19 μm) of AR glass filaments of the same type (Figure 2.33), with reported roving strength values of 427, 353, and 304 MPa, respectively (Scheffler et al., 2009). The lower strengths of the thicker filaments were explained by the higher content of surface defects (Ehrenstein and Spaude, 1983). The strength of a given fiber depends on the critical flaw, which is generally located at the fiber surface where such flaws concentrate stresses at crack tips, leading to failure at very low stress levels. Consequently, the measured tensile strengths of glass and other brittle filaments with large-diameter fibers can be significantly lower than their theoretical values (Schmitz and Metcalfe, 1967). By protecting the fiber surface from mechanical damage during handling, using surface sizing and coatings, its mechanical properties can be significantly improved. The sizing improves tensile strength by partially filling the flaws on the fiber surface and increasing the crack tip radius (Zinck et al., 2001), thereby leading to a mechanical "healing" effect. Improved tensile strength of sized AR glass fibers due to the healing is obvious in Figure 2.33, in which they are compared to virgin, unsized fibers of various diameters (Scheffler et al., 2009).

Filament tensile properties can be further improved by adding nanoparticles to polymer sizing, which reduces the stress concentration at the flaw tip (Tyagi et al., 2004) due to the self-healing of the flaws as the nanoparticles

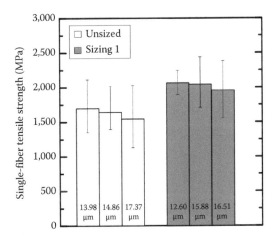

Figure 2.33 Comparison of single-fiber tensile strengths of unsized fibers and fibers with sizing (N-propyltrimethoxysilane) for different filament diameters. (From Scheffler, C. et al., *Compos. Sci. Technol.*, 69, 531, 2009.)

migrate into the filament crack/flaw (Gupta et al., 2006). Indeed, coating of the fiber surface with a nanocomposite may yield a defect-free surface due to nanodefect repair. In addition, coating of the fiber surface leads to enhanced environmental durability and results in improved mechanical properties of the fibers and, consequently, of the composites. AR glass fibers coated with hybrid styrene–butadiene copolymer containing a low fraction (0.2 wt% in the sizing) of multiwalled carbon nanotubes exhibited significantly improved tensile strength (up to 70%) due to a high healing efficiency factor. The incorporation of an organoclay at 1 wt% in the sizing increased AR glass fiber strength by 25% and was also shown to improve the alkali resistance of the roving (Figure 2.34) (Gao et al., 2007). Taken together, the specific polymer coating used, the "bridging" effect of the nanotubes, and interface debonding/plastic deformation around the crack tip redistribute stress and arrest further growth of the crack within the fiber. A nanometer-scale hybrid coating layer based on styrene–butadiene copolymer with single- or multiwalled carbon nanotubes and/or nanoclays was also found to heal the surface flaws of glass fibers and improve their alkali resistance for TRC applications (Gao et al., 2008).

Brittle fibers tend to suffer dramatic reductions in their tensile strength during both the yarn and the fabric production processes (Colombo et al., 2013, Curbach and Jesse, 1999; Glowania and Gries, 2010, Harting et al., 2011). For example, the tensile strength of a single 14 μm AR glass filament is about 1,700 MPa, that of AR glass roving comprising 160 of the same 14 μm filaments of 2,400 tex is about 900 MPa, and finally, the tensile

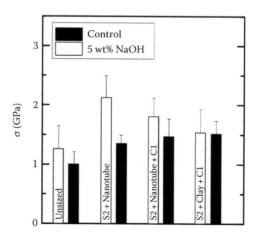

Figure 2.34 Tensile strengths of single AR glass fibers coated with nanoparticles before and after alkaline treatment (7 days in NaOH aqueous solution at ambient environment). (From Gao, S.L. et al., *Acta Materialia*, 55, 1043, 2007; Gao, S.L. et al., *Compos. Sci. Technol.*, 68, 2892, 2008.)

strength of 2D warp-knitted fabric constructed of this roving is about 600 MPa. Hence, the tensile strength of the roving is about half that of the filament, and the tensile strength of the fabric is only one-third that of the roving (Glowania and Gries, 2010). Harting et al. (2011) reported that the reduction in AR glass filament tensile strength relative to its yarn (bundle) is also dependent on filament diameter. Thus, a 40% reduction for 13.5 µm filaments and a 70% reduction for greater diameter filaments of 16 µm were recorded (reduction from 2,300 to 1,400 MPa and from 2,100 to 600 MPa of filament vs. yarn for the 13.5 and 16 µm diameter filaments, respectively). Further tensile strength reductions of those filaments were reported by Colombo et al. (2013) when they were part of a fabric structure. Possible reasons for such reductions can be attributed to a nonuniform distribution of the load between the filaments inside the yarn (Curbach and Jesse, 1999), the presence of defects, and damage suffered in the yarn and fabric manufacturing processes during spinning, twisting, weaving, and/or knitting.

2.6.2 Bundles

Coating is also applied to the bundle/roving materials used to produce the fabric and its effect on yarn tensile strength is highly dependent on the weight content of the coating (Mäder et al., 2004). AR glass bundles with styrene–butadiene coatings of 0%, 2.4%, 5.7%, 7.0%, and 14.3% by weight exhibited tensile strengths of 1,351, 1,367, 2,029, 2,002, and 2,145 MPa, respectively. A similar trend is observed in carbon yarns, in which coatings of 0%, 3.5%, 8.4%, and 12.95% by weight result in tensile strengths of 1,567, 2,131, 2,365, and 2,682 MPa, respectively. The type of coating also influences the tensile strengths of AR glass and carbon bundles. The improvements in roving tensile properties of AR glass fibers with carboxylated (ARG C1) or thermoplastic (ARG C2) styrene–butadiene copolymer coatings compared to uncoated AR glass bundles (ARG) are evident in Figure 2.35 (Scheffler et al., 2009). Carbon bundles with high molecular epoxy resin dispersions (C3–C5) and epoxy ester (C6) benefit from similar improvements in their tensile properties. Similar findings of the important contribution of coatings to tensile strength were provided by Glowania and Gries (2010) and Dilthey et al. (2007), who compared the tensile behavior of an AR glass bundle coated with epoxy and acrylic dispersion relative to that of uncoated bundle (Figure 2.36). The research consistently shows that coated yarn bundles are stronger than those without coating, while the coating type plays an important role in the ultimate properties of the bundle. In the case of bundle yarns, the coating improves filament mechanical properties, fills the bundle spacing, and improves the tensile strength of the bundle as the filaments are effectively combined in a single unit to carry the load.

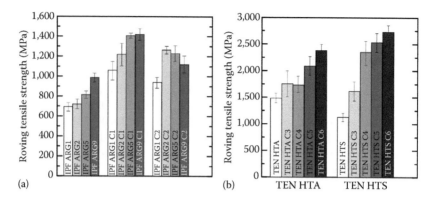

Figure 2.35 Comparison of roving tensile strengths with various coatings: (a) AR glass [filaments sized during spinning with silane coupling agent, c-aminopropyltriethoxy silane (APS), and N-propyltrimethoxysilane (PTMO), in conjunction with film formers and nanoparticles] and (b) carbon (fiber sizing is based on epoxy resin [HTA] or polyurethane [HTS]). (From Scheffler, C. et al., *Compos. Sci. Technol.*, 69, 531, 2009.)

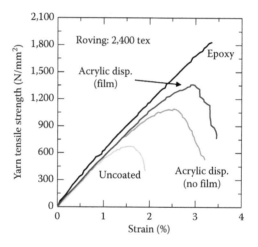

Figure 2.36 Tensile strength of 2,400 tex AR glass bundle with various polymer coating compounds. (From Glowania, M. and Gries, T., Coating of high performance fibers and textiles for textiles reinforced concrete, in *The Eighth fib PhD Symposium in Kgs*, Lyngby, Denmark, June 20–23, 2010, pp. 1–6.)

In all the coatings described earlier, the bundles are impregnated with water-insoluble polymers, which effectively transforms it into a fiber-polymer composite rod. However, with this approach, the individual filaments are not in direct contact with the cementitious matrix—a setup that does not fully exploit the potential improvement in properties that could be achieved if every individual filament was immersed in the cement matrix. Weichold and Möller (2007) developed a different coating approach based on the use of water-soluble polymers that contain reactive, unhydrated, fine cement particles, that is, hybrid cement-in-polymer dispersions. They investigated the use of water and/or an alkali-soluble polymer such as PVA, poly(ethylene-co-vinyl acetate [EVA]), or PVAc. The reactive, nonhydrated cement particles in the cement-in-PVA dispersion penetrate the spaces between the individual filaments before the roving comes in contact with the aqueous cement paste. Subsequent embedment of the hybrid-coated roving in fresh concrete and its exposure to the aqueous environment dissolves the water-soluble polymer, but the reactive cement particles remain between the bundle spaces. Their subsequent reaction with the cement paste water activates the hydration process between the filaments of the bundle to produce a roving with hydration products continuously distributed inside the roving (Glowania et al., 2009).

Low–molecular weight poly(ethylene oxide)s with melt viscosities lower by 3–5 orders of magnitude than that of PVA were studied for cement-in-polymer dispersions by Hojczyk et al. (2010). The high solubility of these polymers in water enables the use of a high volume fraction of cement of about 62%.

2.6.3 Fabrics

Although a coating can also be applied on the fabric surface, the spaces between its yarns are left open to allow cement matrix penetration. Coating the fabric improves its mechanical performance and promotes better handling properties.

Two principal methods, off-line and on-line coating, are used to coat fabric-reinforced structures. The off-line coating process involves coating the fabric after it is manufactured and is considered to be the inferior method, since yarn tension during the coating process is not constant, which results in some misalignment of the yarns within the coated fabric. On-line coating coats the fabric during its production where constant yarn tension is ensured as the control of the coating roll drive is directly tied to the warp-knitting machine control system. Yarn tension can therefore be kept high, even during the production process, and airflows detrimental to the final coat structure can be avoided. Furthermore, the amount of polymer coating applied to the textile fabric is controllable by adjusting the circumferential speed of the roll coater and the transportation speed of the warp-knitting machine and the polymer content in the water-based polymer dispersion bath (Koeckritz et al., 2010). Disadvantages of the on-line coating process include the slower speed of the fabric manufacturing machine—six times slower (2.5 m/min) than that of the coating machine (15 m/min)—and the complexity of combining the manufacturing and coating processes in a parallel operation (Glowania et al., 2009).

For concrete applications in which an off-line process is used, the standard method for fabric reinforcement is dip coating, in which the textile is immersed in a liquid coating bath, after which it is squeezed through a pair of rollers. Fabric penetrability by the coating material may be limited using this process due to the extremely short processing dwell and the potential for air to become trapped inside the bundle. To prevent air bubbles from remaining inside the fabric, the sponge-suck effect can be exploited to increase coating material penetration of the spaces between the yarns (Glowania et al., 2011) by squeezing the textile while it is still immersed in the coating compound. The squeezing rollers, which must also be fully dipped in the bath, press against each other at a predetermined, constant pressure (varied for different fabric materials and types) to compress the yarns and force out any air locked between the filaments. After the compression and while the fabric is still immersed in the coating bath, the yarn expands in the process pulling the coating compound into the voids between the filaments. In addition, off-line coating may also include the use of heated rollers—which melt the coating material and increase its penetration into the bundle of the fabric—through which the coated fabric is drawn at the end of the coating process.

The improvement in fabric mechanical performance gained by applying a polymer coating is presented in Figure 2.37, which shows the increase in tensile strength of warp-knitted fabric as functions of both the coefficient

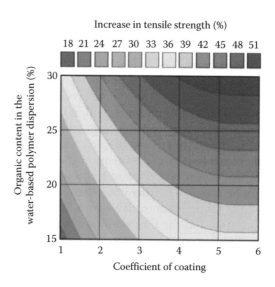

Figure 2.37 Relationship between the coating coefficient, polymer organic content, and warp-knitted fabric tensile strength. (From Koeckritz, U. et al., *J. Ind. Textil.*, 40(2), 157, October 2010.)

of coating and the organic content in the water-based polymer dispersion (Koeckritz et al., 2010). In the fabric used for Figure 2.37, the coating was applied on-line during fabric production. The coefficient of coating is defined as the relative movement between the circumferential speed of the roll coater and the running speed of the warp-knitting machine. Fabric tensile strength is highly dependent on the coating coefficient and content such that it increases as the organic content and coating coefficient are increased. Similar behavior was reported by Glowania et al. (2011), who showed improved tensile behavior of polyurethane and epoxy-coated warp-knitted fabrics, between which the latter exhibited better performance compared to noncoated fabric (Figure 2.38).

The tensile properties of the fabrics also depend on the fabric structure, that is, chain or tricot, and the direction, 0° versus 90° (Figure 2.38). In general, the tensile strength of chain-knitted fabric, in which the warp yarns are inserted in the loops (Figures 2.28a and 2.29a), is lower than that of tricot, where the warp yarns are located outside the loops (Figures 2.28b and 2.29b). This is due to the lower tightening effect of the stitches on the warp yarn of the tricot knit, which facilitates deeper polymer penetration into the bundle of the roving and results in the higher tensile strengths of tricot versus chain-knitted fabric (Glowania et al., 2009, 2011). In addition, the higher warp yarn tightening of chain-knitted fabrics causes reinforcing

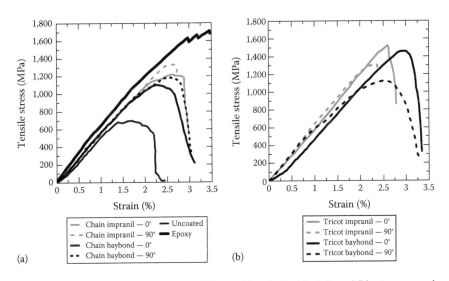

Figure 2.38 Tensile behavior of yarns of different fabric knits: (a) chain and (b) tricot, coated with polyurethane and epoxy. (From Glowania, M.H. et al., Coating of AR-glass fibers with polyurethane for textile-reinforced concrete, in *Ninth International Symposium on High Performance Concrete—Design, Verification & Utilization*, Energy Events Centre, Rotorua, New Zealand, August 2011.)

yarn misalignment due to its location within the stitches (see Figure 2.28a), further reducing its tensile properties.

A comparison of tensile strengths in the 0° and the 90° directions of roving for chain fabric with different polymer coatings shows approximately 10% greater strength in the latter direction regardless of the polymer (Figure 2.38a). This outcome is caused by compaction of the roving in the 0° direction by the fabric's loops and, consequently, limited bundle penetrability by the polymer. Equivalent comparison of the 0/90° tensile strengths of tricot fabric reveals the opposite trend, namely, the roving in the 0° direction exhibits higher tensile strength than that in the 90° direction of the fabric. In its 0° direction, tricot fabric is nearly noncompacted compared to that in the 90° direction, which, in this case, is entrapped by the loops (Figure 2.38b).

Complete coating of the bundles within the fabric with a polymer compound that uniformly penetrates the multifilament yarns binds the single filaments of each strand together, thus enabling them to carry loads as a single structural unit. Tensile failure, therefore, represents all the filaments at once. Such a coating is also beneficial in cementitious applications, in which forces can be transferred from the cement matrix onto all the filaments, thus increasing the total strength of the composite. Due to the requirement that the coating material fully penetrates the yarns, the polymer should be with low viscosity.

Another parameter that influences the tensile performance of chain-knitted fabric is the length of the loops/stitches around the warp yarn. Different loop sizes and tensions of the knitting (loop) yarn during fabric production were studied for their effects on the cross-sectional areas and perimeters of the yarns of virgin weft- and warp-knitted AR glass rovings without any coating (Dolatabadi et al., 2014). The cross-sectional area and bundle perimeter of the weft yarns were larger than those of the warp yarns for all loop sizes (Figure 2.39a) due to the greater lateral contraction of the knitting (stitch) yarns over the warp yarn in comparison to the weft yarns. In addition, smaller stitch lengths resulted in high bundle filament packing densities—resulting in smaller bundle cross sections and yarn perimeters—in both warp and weft yarns. Furthermore, knitting tension did not show a marked effect on yarn cross-sectional area or perimeter for all stitch lengths. Accordingly, it can be deduced that stitch length is a dominant factor that controls bundle filament packing density, that is, bundle tightening. The tensile strength of uncoated chain-knitted fabric in its warp direction is dependent on the lengths of the stitches surrounding the warp yarns (Figures 2.39b and 2.40). Reducing the stitch length increases the tensile strength of the warp yarns in the warp direction of a chain fabric and increases the packing density of the bundle filaments. Consequently, the tightening effect of smaller stitching on the warp yarn improves the friction between the filaments within the bundle and improves fabric tensile performance in its warp direction.

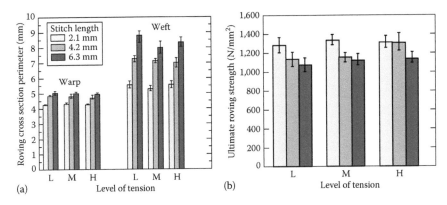

Figure 2.39 Effects of knitting yarn tension and stitch length on (a) warp and weft roving cross-sectional perimeter and (b) tensile strength of warp yarn in a chain-knitted fabric structure. (From Dolatabadi, M.K. et al., *J. Text. Inst.*, 105(7), 711, 2014.)

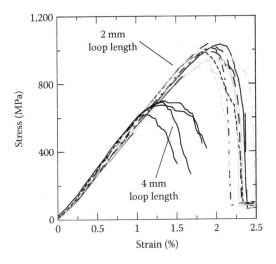

Figure 2.40 Effects by loop size (2 and 4 mm) on the tensile behavior of chain warp-knitted fabrics tested in the warp direction. (From Peled, A. et al., *Cement Concr. Compos.*, 30(3), 174, 2008a.)

Insofar as the use of relatively small stitches around the warp bundles (and the resulting strong connection between the warp and weft yarns of a knitted fabric) firmly holds the yarns together in chain fabric form, it is beneficial from the point of view of fabric stability (Figure 2.41a). However, this technique can also cause warp yarn misalignment and

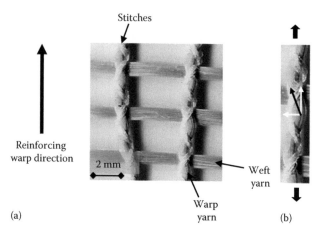

Figure 2.41 (a) Chain warp-knitted fabric and (b) higher magnification of the warp yarn showing crimping. (From Peled, A. et al., *Cement Concr. Compos.*, 30(3), 174, 2008a.)

crimping (Figures 2.28a and 2.41b), which negatively affects the fabric and composite strengths. The influences of crimping are evident when comparing the tensile properties of a virgin (noncoated) chain-knitted fabric in its weft and warp directions (Figure 2.42). Despite the use of identical AR glass fibers in both directions, tensile strength in the warp direction is lower than

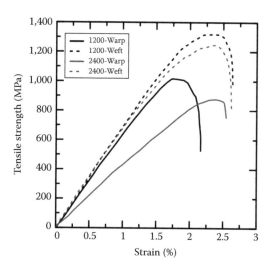

Figure 2.42 Tensile behavior of weft insertion warp-knitted fabrics in the weft and warp fabric directions for two tex values. (From Peled, A. et al., *Cement Concr. Compos.*, 30(3), 174, 2008a.)

that in the weft direction (Peled et al., 2008a) due to crimping, which is more severe the higher the yarn tex value (i.e., thicker yarn) for yarns under similar tensions. Note that a similar trend of improved strength in the weft direction was also observed for the coated chain fabrics discussed earlier (Figure 2.38). Despite the greater strength of coated chain fabrics in the weft direction, from a practical point of view, the stress on fabrics used for composites should be in the warp direction because fabric length, restricted in the weft direction by knitting machine width, is effectively unlimited in the warp direction.

The compression behaviors of 3D spacer fabrics were investigated by Roye and Gries (2005, 2007), who evaluated the effects of spacer yarn crossing patterns and sizes having different angles (Figure 2.43). It was reported that the compression resistance of the 3D fabrics decreases from structure 3A to structure 3E, that is, from low to high spacer yarn angles. Linear behavior was observed for the relation between force-compression and spacer yarn angle.

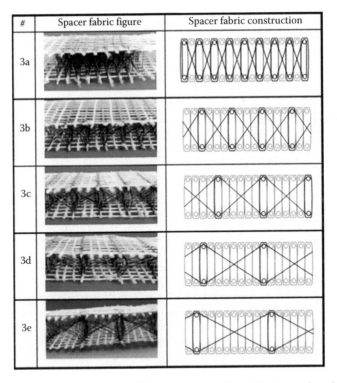

Figure 2.43 3D spacer fabrics with different patterns of varied sizes and angles of the z yarns oriented in the fabric thickness direction. (From Roye, A. and Gries, T., *J. Ind. Textil.*, 37(2), 163, 2007.)

2.7 Summary

To summarize, the performance and ultimate properties of a fabric are strongly dependent on the properties of the fibers and yarns used in its manufacture and on the method of textile processing. The number and arrangement of the fibers and yarns within the fabric have strong influences on fabric behavior, and therefore, they affect the composite performance significantly. Among the various textile fabrication processes, only a few can be realistically considered for use for cement matrix reinforcements, in which the most important criterion is that open structures with high displacement stability be created to enable complete penetration of the fabric by the cement matrix. In the manufacture of reinforcing fabrics, the leno weave and warp-knitting weft insertion fabrics, including multiaxial and 3D spacer fabrics, are the most commonly used, as these fabrics can be produced with relatively stable structures across a wide range of fabric openings. However, other fabric types can also be used for specific requirements.

Sizing and coating have significant influences on fiber, yarn, and fabric properties, which can be evaluated based on the quality of the adhesion between the filaments and on filament load-bearing performance. Furthermore, effective coatings form a protective layer against chemical attack of the fabric in the alkaline concrete environment, and in so doing, they preserve the fabric's properties during the service life.

References

ACI Committee 544, State of the art report on synthetic fiber-reinforced concrete, Draft Document, American Concrete Institute, MI, 2005.

Artemenko, S. E., Polymer composite materials made from carbon, basalt and glass fibers—Structure and properties, *Fiber Chemistry*, 35(3), 226–229, 2003.

Bailie, J. A., Woven fabric aerospace structures, in C. T. Herakovich and Y. M. Tranopoloskii (eds.), *Handbook of Composites Structures and Design*, Vol. 2, 1989, pp. 353–391.

Bentur, A., Mindess, S., and Vondran, G., Bonding in polypropylene fiber reinforced concrete, *International Journal of Cement Composites and Lightweight Concrete*, 11, 153–158, 1989.

Bisanda, E. T. N. and Ansell, M. P., Properties of sisal–CNSL composites, *Journal of Materials Science*, 27, 1690–1700, 1992.

Bunsell, A. R., *Fiber Reinforcements for Composite Materials*, Composite Materials Series, R. P. Pipes (Series ed.), Elsevier, Amsterdam, the Netherlands, Vol. 2, 1988.

Bunsell, A. R. and Simon, G., Ceramic fibers in fiber reinforcements for composite materials, Chapter 9, in A. R. Bunsell (ed.), *Fibre Reinforcements for Composite Materials*, Composite Materials Series, Elsevier, Amsterdam, the Netherlands, Vol. 2, 1988, pp. 427–477.

Butler, M., Hempel, R., and Schorn, H., Bond behavior of polymer impregnated AR-glass textile reinforcement in concrete, in *ISPIC International Symposium Polymers in Concrete*, University of Minho, Guimaraes, Portugal, 2006, pp. 173–183.

Calundann, G., Jaffe, M., Jones, R. S., and Yoon, H., High performance organic fibers for composite reinforcement, Chapter 5, in A. R. Bunsell (ed.), *Fiber Reinforcements for Composite Materials*, Composite Materials Series, R. P. Pipes (Series ed.), Elsevier, Amsterdam, the Netherlands, Vol. 2, 1988, pp. 211–248.

Cater, S., Editorial, *International Composites News*, 40, 559–568, March 2002.

Coleman, D., Man-made fibers before 1945, in D. Jenkins (ed.), *The Cambridge History of Western Textiles*, Cambridge University Press, Cambridge, U.K., 2003, pp. 933–947.

Colombo, I. G., Magri, A., Zani, G., Colombo, M., and di Prisco, M., Textile reinforced concrete: Experimental investigation on design parameters, *Materials and Structures*, 46, 1933–1951, 2013.

Coutts, R. S. P. and Warden, P. G., Sisal pulp reinforced cement mortar, *Cement and Concrete Composites*, 14(1), 17–21, 1992.

Cumming, W. D., Unidirectional and two-directional fabrics, in T. J. Reinhart (ed.), *Engineering Materials Handbook Composites*, ASM International, Materials Park, OH, Vol. 1, 1987, pp. 125–128.

Curbach, M. and Jesse, F., Lecture No. 513: Basic tensile tests on strain specimens of textile-reinforced concrete, in *Techtextil Symposium*, Frankfurt, Germany, 1999.

Dilthey, U., Schleser, M., Chudoba, R., and Konrad, M., Experimental investigation and micro-mechanical simulation of concrete reinforces with epoxy impregnated fabrics, in *Proceedings of Eighth International Symposium on Fiber Reinforced Polymer Reinforcement for Concrete Structures (FRPRCS-8)*, University of Patras, Patras, Greece, July 16–18, 2007, pp. 456–457.

Dolatabadi, M. K., Janetzko, S., and Gries, T., Deformation of AR glass roving embedded in the warp knitted structure, *The Journal of the Textile Institute*, 102(4), 308–314, April 2011.

Dolatabadi, M. K., Janetzko, S., and Gries, T., Geometrical and mechanical properties of a non-crimp fabric applicable for textile reinforced concrete, *The Journal of the Textile Institute*, 105(7), 711–716, 2014.

Edie, D. D., The effect of processing on the structure and properties of carbon fibers, *Carbon*, 36(4), 345–362, 1998.

Ehrenstein, G. W. and Spaude, R., Rissbildung in Elementar glasfasern, *Z Werkstofftech*, 14, 73–81, 1983.

Fitzer, E. and Heine, M., Carbon fibers and surface treatment in fiber reinforcements for composite materials, Chapter 3, in A. R. Bunsell (ed.), *Fibre Reinforcements for Composite Materials*, Composite Materials Series, Elsevier, Amsterdam, the Netherlands, Vol. 2, 1988, pp. 74–148.

Franklin, W. and Warner, J. R., Boron and silicon carbide/carbon fibers, Chapter 8 in A. R. Bunsell (ed.), *Fiber Reinforcements for Composite Materials*, Composite Materials Series, Elsevier, Amsterdam, the Netherlands, Vol. 2, 1988, pp. 371–425.

Gao, S. L., Mäder, E., and Plonka, R., Coatings for glass fibers in a cementitious matrix, *Acta Materialia*, 52, 4745–4755, 2004.

Gao, S. L., Mäder, E., and Plonka, R., Nanostructured coatings of glass fibers: Improvement of alkali resistance and mechanical properties, *Acta Materialia*, 55, 1043–1052, 2007.

Gao, S. L., Mäder, E., and Plonka, R., Nanocomposite coatings for healing surface defects of glass fibers and improving interfacial adhesion, *Composites Science and Technology*, 68, 2892–2901, 2008.

Geil, P. H., *Polymer Single Crystal*, Interscience, New York, 1963.

Glass industry, January 1987, p. 27.

Glowania, M. and Gries, T., Coating of high performance fibers and textiles for textiles reinforced concrete, in *The Eighth fib PhD Symposium in Kgs*, Lyngby, Denmark, June 20–23, 2010, pp. 1–6.

Glowania, M., Weichold, O., Hojczyk, M., Seide, G., and Gries, T., Neue Beschichtungsverfahren für PVA-Zement-Composite in textilbewehrtem Beton ("New coating techniques for PVA-Cement-Composite"), in *Textilbeton. Theorie und Praxis. Tagungsband zum 4. Kolloquium zu textilbewehrten Tragwerken (CTRS4) und zur 1*, Anwendertagung, Dresden, Germany, June 2009, pp. 75–86.

Glowania, M. H., Linke, M., and Gries, T., Coating of AR-glass fibers with polyurethane for textile-reinforced concrete, in *Ninth International Symposium on High Performance Concrete—Design, Verification & Utilization*, Energy Events Centre, Rotorua, New Zealand, August 2011.

Gries, T., Offermann, P., Engler, T., and Peled, A., in W. Brameshuber (ed.), Textiles *in Textile Reinforced Concrete: State of the art Report, RILEM TC 201-TRC*, RILEM, Paris, France, 2006, pp. 11–27.

Gries, T. and Roye, A., Three dimensional structures for thin walled concrete elements, in M. Curbach (ed.), *Textile Reinforced Structures: Proceedings of the Second Colloquium on Textile Reinforced Structures (CTRS2)*, Sonderforschungsbereich 528, Technische, Dresden, Germany, 2003, pp. 513–524.

Griffith, A. A., *The Phenomena of Rupture and Flow in Solids*, The Royal Society, London, U.K., 1920.

Guenther, W., Knitted fabric composites, *Reinforced Plastics*, 7(27), 213, July 1983.

Gupta, P. K., Glass fibres for composite materials, Chapter 2, in A. R. Bunsell (ed.), *Fibre Reinforcements for Composite Materials*, Composite Materials Series, Elsevier, Amsterdam, the Netherlands, Vol. 2, 1988, pp. 20–71.

Gupta, S., Zhang, Q. L., Emarick, T., Balazs, A. C., and Russell, T. P., Entropy-driven segregation of nanoparticles to cracks: A route to self-healing systems, *Nature Materials*, 5, 229–233, 2006.

Hannant, D. J. and Zonsveld, J. J., Polyolefin fibrous networks in cement matrices for low cost sheeting, *Philosophical Transactions of the Royal Society of London, Series A*, 294, 591–597, 1980.

Harting, J., Jesse, F., Schicktanz, K., and Haussler-Combe, U., Influence of experimental setups on the apparent uniaxial tensile load-bearing capacity of textile reinforced concrete specimens, *Material and Structure*, 45, 433–446, 2011.

Hausding, J. and Cherif, C., Improvements in the warp-knitting process and new patterning techniques for stitch-bonded textiles, *The Journal of the Textile Institute*, 101(3), 187–196, March 2010.

Hausding, J., Engler, T., Franzke, G., Köckritz, U., and Offermann, P., Concrete reinforced with stitch-bonded multi-plies—A review, in C. Aldea (ed.), *Thin Fiber and Textile Reinforced Cementitious Systems*, American Concrete Institute, Farmington Hills, MI, ACI Symposium Publication 244, 2007, pp. 1–16.

Hayashi, T. and Fumiaki, A., A chromosomal aberration study of fibrillated PVA fiber in cultured mammalian cells, *Environmental Mutagen Research*, 24, 23–27, 2002.

Hojczyk, M., Weichold, O., Glowania, M., and Gries, T., Cement-in-polymer dispersions as coatings for glass rovings and textiles, in *Proceedings of the International RILEM Conference on Material Science—Second ICTRC 2010—Textile Reinforced Concrete—Theme 1*, 2010, Aachen, Germany, pp. 79–89.

Hu, J. and Jiang, Y., Modeling formability of multiaxial warp knitted fabrics on a hemisphere, *Composites: Part A*, 33, 725–734, 2002.

Hull, D., *An Introduction to Composite Materials*, Cambridge Solid State Science Series, Cambridge University Press, Cambridge, U.K., 1981.

Janetzko, S. and Maier, P., Development and industrial manufacturing of innovative reinforcements for textile reinforced concrete, in B. Küppers (ed.), *Proceedings Aachen-Dresden International Textile Conference*, Aachen, Germany, November 29–30, 2007.

Johnson, D. J., Microstructure of various carbon fibers, in *Proceedings of the First International Conference on Carbon Fibers, Their Composites and Applications*, Plastic Institute, London, U.K., 1971, pp. 52–56.

Ko, F. K., Recent advances in textile structure composites, in *Proceedings of Advanced Composites*, Dearborn, MI, December 1985, pp. 83–84.

Koeckritz, U., Cherif, C. H., Weiland, S., and Curbach, M., Reinforced concrete application in-situ polymer coating of open grid warp knitted fabrics for textile, *Journal of Industrial Textiles*, 40(2), 157–169, October 2010.

Kruger, M., Ozbolt, J., and Reinhardt, H. W., A new 3D discrete bond model to study the influence of bond on the structural performance of thin reinforced and prestressed concrete plates, in A. E. Naaman and H. W. Reinhardt (eds.), *Fourth International Workshop on High Performance Fiber Reinforced Cement Composites (HPFRCC 4)*, Ann Arbor, MI, RILEM Publications, Paris, France, 2003, pp. 49–63.

Li, V. C., Horikoshi, T., Ogawa, A., Torigoe, S., and Saito, T., Micromechanics-based durability study of polyvinyl alcohol-engineered cementitious composite (PVA-ECC), *ACI Materials Journal*, 101(3), 242–248, 2004.

Li, V. C., Wu, C., Wang, S., Ogawa, A., and Saito, T., Interface tailoring for strain-hardening PVA-ECC, *ACI Materials Journal*, 99(5), 463–472, 2002.

Lowenstein, K. L., *The Manufacturing Technology of Continuous Glass Fibers*, Elsevier, Amsterdam, the Netherlands, 1983, p. 38.

Mäder, E., Mai, K., and Pisanova, E., Interphase characterization on polymer composites—Monitoring of interphasial behavior in dependence on the mode of loading, *Composite Interfaces*, 7, 133–147, 2000.

Mäder, E., Plonka, R., Schiekel, M., and Hempel, R., Coatings on alkali-resistant glass fibres for the improvement of concrete, *Journal of Industrial Textiles*, 33(3), 191–207, January 2004.

Majumdar, A. J. and Nurse, R. W., Glass fiber reinforced cement, Building Research Establishment Current Paper, CP79/74, Building Research Establishment, Hertfordshire, England, 1974.

Mallick, P. K., *Fiber Reinforced Composites Materials, Manufacturing, and Design*, 3rd edn., CRC Press, Taylor & Francis Group, Boca Raton, FL, 2007.

Matovich, M. A. and Pearson, J. R. A., *Industrial & Engineering Chemistry Fundamentals*, ACS Publications, Washington, DC, Vol. 8, Chapter 4, 1969.

Michaels, L. and Chissick, S. S., *Asbestos: Properties, Applications, and Hazards*, John Wiley & Sons, New York, 1979.

Mileiko, S. T. and Tikhonovich, S., *Metals and Ceramic Based Composites*, Composite Materials Series, Elsevier, Amsterdam, the Netherlands, Vol. 12, 1997.

Militky, J. K. and Kovacic, V., Ultimate mechanical properties of basalt filaments, *Textile Research Journal*, 66(4), 225–229, 1996.

Mobasher, B., *Mechanics of Fiber and Textile Reinforced Cement Composites*, CRC Press, Taylor & Francis Group, Boca Raton, FL, 2011.

Moncrieff, R. W., *Man-Made Fibers*, 6th edn., Butterworth, London, U.K., 1979.

Morgan, R. J. and Allred, R. E., Aramid fiber reinforcements, in S. M. Lee (ed.), *Reference Book for Composites Technology*, Technomic Publishing Company, Lancaster, PA, Vol. 1, 1989.

Murherjee, P. S. and Satyanarayana, K. G., Structure properties of some vegetable fibers. Part 1: Sisal fibre, *Journal of Materials Science*, 19, 3925–3934, 1984.

Naaman, A. E., *Ferrocement and Laminated Cementitious Composites*, Techno Press 3000, Ann Arbor, MI, 2000, p. 372.

Naaman, A. E. and Shah, S. P., Tensile test of ferrocement, *Journal of the American Concrete Institute*, 68, 693–698, 1971.

National Trust for Historic Preservation, *Coping with Contamination: A Primer for Preservationists*, Information Bulletin No. 70, Washington, DC, 1993, p. 12.

Nishioka, K., Yamakawa, S., and Shirakawa, K., Properties and applications of carbon fiber reinforced cement composites, in R. N. Swamy, R. L. Wagstaffe, and D. R. Oakley (eds.), *Proceedings of the Third RILEM Symposium on Developments in Fibre Reinforced Cement and Concrete*, Sheffield, U.K., RILEM Technical Committee 49-FTR, Lancaster, England, Paper 2.2, 1986.

Oksman, K., Wallstrom, L., and Toledo, F. R. D., Morphology and mechanical properties of unidirectional sisal–epoxy composites, *Journal of Applied Polymer Science*, 84(13), 2358–2365, 2002.

Peled, A., Cohen, Z., Pasder, Y., Roye, A., and Gries, T., Influences of textile characteristics on the tensile properties of warp knitted cement based composites, *Cement and Concrete Composites*, 30(3), 174–183, 2008a.

Peled, A., Zaguri, E., and Marom, G., Bonding characteristics of multifilament polymer yarns and cement matrices, *Composites Part A*, 39(6), 930–939, 2008b.

Peterlin, A., in A. Cifferri and I. M. Ward (eds.), *Ultra High Modulus Polymers*, Applied Science Publishers, Barking, U.K., 1979.

Raupach, M. and Orlowsky, J., *Textilbeton—Ein neuer Verbundwerkstoff ("TRC—A New Composite Material")*, 10, Internationales Aachener Schweißtechnik Kolloquium, Aachen, Germany October 2007, pp. 24–25.

Raupach, M., Orlowsky, J., Büttner, T., Dilthey, U., and Schleser, M., Epoxy-impregnated textiles in concrete—Load bearing capacity and durability, in *First International RILEM Conference on Textile Reinforced Concrete*, Aachen, Germany, September 6–7, 2006, pp. 77–88.

Raz, S., *Warp Knitting Production*, Melliand, Heidelberg, Germany, 1987.

Roye, A., Hochleistungsdoppelraschelprozessfür Textilbetonanwendungen, Doctoral dissertation, Institut fürTextiltechnik der RWTH Aachen University, Aachen, Germany, 2007.

Roye, A. and Gries, T., Design by application—Customized warp knitted 3-D textiles for concrete applications, *Kettenwirk-Praxis*, 39(4), 20–21, 2004.

Roye, A. and Gries, T., Technical spacer fabrics: Textiles with advanced possibilities, in *13th International Techtextil-Symposium "Focusing on Innovation,"* Frankfurt am Main, Germany, Paper: V5_ROYE, June 6/9, 2005.

Roye, A. and Gries, T., 3-D textiles for advanced cement based matrix reinforcement, *Journal of Industrial Textiles*, 37(2), 163–173, 2007.

Roye, A., Gries, T., and Peled, A., Spacer fabric for thin walled concrete elements, in M. Di Prisco, R. Felicetti, and G. A. Plizzari (eds.), *Fiber Reinforced Concrete—BEFIB*, PRO 39, RILEM, Bagneux, France, 2004, pp. 1505–1514.

Saheb, D. N. and Jog, J. P., Natural fiber polymer composites: A review, *Advances in Polymer Technology*, 18(4), 351–363, 1999.

Scheffler, C., Gao, S. L., Plonka, R., Mäder, E., Hempel, S., Butler, M., and Mechtcherine, V., Interphase modification of alkali-resistant glass fibres and carbon fibres for textile reinforced concrete I: Fibre properties and durability, *Composites Science and Technology*, 69, 531–538, 2009.

Schleser, M., Einsatz polymerimprägnierter, alkaliresistenter glastextilien zur bewehrung zementgebundener matrices ("Use of polymer-impregnated, alkali-resistant glass fabrics for reinforcement of cement-based), Doctoral dissertation, Institut für Schweißtechnikund Fügetechnik der RWTH Aachen University, Aachen, Germany, 2008.

Schmitz, G. K. and Metcalfe, A. G., Testing of fibers, *Materials Research and Standards*, 7(4), 146–152, 1967.

Schorn, H., Hempel, R., and Butler, M., Mechanismen des Verbundes von textilbewehrtem Beton ("Mechanism of TRC composite"), Tagungsband 15. Internationalen Baustofftagung ibausil, Bauhaus-Universität Weimar, Germany, September 2003, pp. 24–27.

Silva, F. A., Chawla, N., and Toledo Filho, R. D., Tensile behavior of high performance natural (sisal) fibers composites, *Science and Technology*, 68, 3438–3443, 2008.

Sim, J., Park, C., and Moon, D. Y., Characteristics of basalt fiber as a strengthening material for concrete structures, *Composites Part B: Engineering*, 36(6–7), 504–512, 2005.

Singha, K., A short review on basalt fiber, *International Journal of Textile Science*, 1(4), 19–28, 2012.

Skelton, J., Triaxially woven fabrics: Their structures and properties, *Textile Research Journal*, 8(41), 637–647, August 1971.

Smith, P. and Lemstra, P. J., Ultra-high strength polyethylene filaments by solution spinning/drawing. 3. Influence of drawing temperature, *Polymer*, 21, 1341–1343, 1980.

Summerscales, J., High performance reinforced fabrics, *Progress in Rubber and Plastics Technology*, 3(3), 20–31, 1987.

The Industrial Uses of Asbestos, *Scientific American*, 258–259, April 22, 1876.

Toledo Filho, R. D., Ghavami, K., England, G. L., and Scrivener, K., Development of vegetable fiber–mortar composites of improved durability, *Cement and Concrete Composites*, 25, 185–196, 2003.

Tyagi, S., Lee, J. Y., Buxton, G. A., and Balazs, A. C., Using nanocomposite coating to heal surface defects, *Macromolecules*, 37, 9160–9168, 2004.

Wambua, P., Ivens, J., and Verpoest, I., Natural fibres: Can they replace glass in fibre reinforced plastics? *Composites Science and Technology*, 63, 1259, 2003.

Ward, I. M., *Structure and Properties of Polymers*, Wiley, New York, 1975.

Weichold, O., Preparation and properties of hybrid cement-in-polymer coatings used for the improvement of fiber-matrix adhesion in textile reinforced concrete, *Journal of Applied Polymer Science*, 116, 3303–3309, 2010.

Weichold, O. and Möller, M., A cement-in-poly(vinylalcohol) dispersion for improved fiber-matrix adhesion in continuous glass-fiber reinforced concrete, *Advanced Engineering Materials*, 9(8), 712–715, 2007.

Wild, J. P., Introduction, in D. Jenkins (ed.), *The Cambridge History of Western Textiles, I*, Cambridge University Press, Cambridge, U.K., 2003, pp. 9–25.

Woods, A. L., *Keeping a Lid on It: Asbestos–Cement Building Materials*, Technical Preservation Services, National Park Service, Philadelphia, PA, 2000. http://www.nps.gov/history/nps/tps/recentpast/index.htm. Accessed June 30, 2008.

Wunderlich, B., *Macromolecular Physics*, Academic Press, New York, Vols. 1–3, 1973–1980.

Yang, H. H., Aramid fibers in fiber reinforcements for composite materials, Chapter 6, in A. R. Bunsell (ed.), *Fiber Reinforcements for Composite Materials*, Composite Materials Series, Elsevier, Amsterdam, the Netherlands, Vol. 2, 1988, pp. 249–330.

Zeweben, C., Hahn, H. T., Chou, T., and Tsu-Wei, C., *Mechanical Behavior and Properties of Composite Materials*, Delaware Composites Design Encyclopedia, Technomic, Lancaster, U.K., Vol. 1, 1989, pp. 3–45.

Zinck, P., Mäder, E., and Gerard, J. F., Role of silane coupling agent and film former for tailoring glass fiber sizings from tensile strength measurements, *Journal of Materials Science*, 36, 5245–5252, 2001.

Chapter 3

Fabrication of TRC

3.1 Introduction

Production technique for manufacturing composite-based components is a key factor for achieving reliable, high-quality products that are also cost effective. A characteristic of a production process is the way of bringing together the matrix and reinforcing materials to the final required product, which may consist of chopped discrete fibers, continuous fiber systems (yarns or rovings), fabrics, or combinations of them. Generally, in the composite materials industry, the early manufacturing method for elements was based on a hand lay-up technique. Although hand lay-up is a reliable process, it is by its nature very slow and labor intensive. Mechanized and automated methods were developed that could support mass production, such as compression molding, extrusion, pultrusion, and filament winding. The type of production method, controls the shape and quality of the element, and can also greatly affect the reinforcement content and the overall properties of the final product (Mallick, 2007).

For the production of cement-based composites where short fibers are used as the reinforcement, several techniques were developed and commercially used. The simplest is the conventional casting technique that has traditionally been used for the development of cement-based composites in which short fibers are added into the fresh mix during mixing. This technique results in composites with a relatively low-volume fraction of short discrete fibers distributed in a random two- or three-dimensional orientation. In this method, only a limited quantity of fibers can be added (commonly below 2% by volume); exceeding this quantity of fiber can cause workability difficulties during mixing and placing. As a result, this method is used for concrete where the fibers are mainly employed to control cracking, that is, post-cracking response (Mobasher et al., 1990), but less often to significantly improve the concrete tensile strength. For addition of greater content of fibers, special mixers were developed with high-energy shear to eliminate the agglomeration and fiber balling effects,

(Garlinghouse and Garlinghouse, 1972) and reducing fiber damage normally observed in conventional drum mixers (Haupt, 1997; Mobasher and Li, 1994, 1996, 2003).

The production of high-performance cement-based composites that exhibit a substantial strain-hardening type of response with improved tensile strength, ductility, and toughness requires a large content of fibers with high-quality composite matrix. For the production of such high-performance products, special techniques were developed including extrusion, spray-suction technique, and slurry infiltration fiber concrete (SIFCON and SIMCON). In the extrusion process, a highly viscous, plastic-like mixture is forced through a die, which is a rigid opening having the geometry of a desired cross section (Akkaya et al., 2000; Peled and Shah, 2003; Peled et al., 2000; Shao and Shah, 1997). Materials are formed in this process under high shear and high compressive forces, and the fiber volume fraction can reach about 3%–4%. In the spray-suction technique, chopped fibers are fed into a compressed air gun and sprayed on the required surface. The content of fibers can be about 5% by volume. In the slurry infiltration fiber concrete, steel fibers are generally used, and the fibers content can range from 5% to 20% by volume (Shah and Balaguru, 1992). In this technique, fibers are placed in the mold and then infiltrated with cement-based slurry.

It is known that the processing method can significantly affect the properties of the composite even when keeping the same matrix and fibers. Igarashi et al. (1996) found that increasing the processing time of the fresh mixture influences the fiber–matrix bond strength due to changes in the interfacial microstructure when using the same processing method, material, and fiber. Delvasto et al. (1986) investigated the effects of applied pressure after casting on the flexural response of cement composites. They found that the performance of the composite depended on the applied pressure. Pressed composites showed an increase in flexural strength but reduction in the post-cracking response. Peled and Shah (2003) compared the properties of cast and extruded composites with similar matrix and fiber. They also found significant effects of the processing method. Following these findings, choosing the optimal manufacturing method for composite component production is essential to meet the final product requirements.

The technologies outlined above briefly reviewed the manufacture of cement-based products made of short discrete fibers; yet, there is a need for new processing and manufacturing techniques to produce cement-based products reinforced with textile fabrics. For textile-reinforced concrete (TRC), the reinforcement is incorporated in a lamination manner having several layers of fabrics in a single composite, in which the fabrics and the matrix can be made from different raw materials. Textile cement-based composites exhibit significant strength and ductility improvement and therefore can be the ideal choice for a range of applications: cement boards including wall panels, exterior siding, roofing and flooring tiles; structural components; and strengthening, repairing, and retrofitting of structural concrete elements exposed to static as well as severe loading conditions. Note that in the nonmetallic TRC, no significant concrete cover is required to protect the reinforcement against corrosion, as is the case for steel-reinforced concrete, which allows the production of durable, economical, thin, lightweight TRC elements.

Cement-based composite laminates prepared with continuous reinforcement by fabrics enables the full potential of fibers to be materialized because the manufacturing technique is fully controlled, where all the fibers can be positioned in the composite at any desired direction depending on yarn orientation in the fabric. Thus, the composite laminates can be designed for the specific service loads they may encounter. Furthermore, an important advantage of textile reinforcement is its formability, which allows it to easily adapt to complex free-form geometries and produce components with complex shapes and geometries (Hegger et al., 2004; Weiland et al., 2008). In cement-based composites made of textile fabrics, the cement matrix is particulate in nature, which requires an open fabric structure in order to form a continuous phase; therefore, the rheology of the fresh mix and the open texture of the reinforcement fabric are key factors that should be strongly taken into consideration during fabrication to produce high-quality, high-performance components. Depending on the mix design, form and size of the element and the desired application, different innovative manufacturing methods with computational and control power are needed to develop economic and versatile products. Some available processes are spraying method or laminating technique, including pultrusion (Peled and Mobasher, 2004), filament winding (Mobasher and Pivacek, 1998), cross-ply and sandwich lamination techniques (Mobasher et al., 1997), and complex curve–shaped elements (Schneider and Bergmann, 2008). This chapter provides a brief description of several developments in the production and manufacture of various TRC components.

3.2 Hand lay-up

The hand lay-up technique, or hand laminating technique, is one of the oldest and simplest manufacturing methods to produce composite materials in the polymer-based composite industry. In this method, layers of fabrics are continuously positioned on top of each other until the required composite thickness and number of layers are reached, while the fabrics are either previously impregnated with the matrix or sprayed with the matrix during production, creating laminated composites. The advantages of this method are its simplicity and the possibility of manufacturing large and complex-shaped products. The disadvantages are the low production rate and high labor intensity, as well as the required skills to keep the product quality within acceptable limits of variation.

An efficient way to use the simple casting technique is for the production of composites reinforced with three-dimensional (3D) fabrics. Here two fabric layers are connected in advance during fabric production (see Chapter 2), while in the course of the composite preparation, the cementitious matrix is poured into the fabric opening until the total 3D fabric is completely filled and covered (Adiel and Peled, 2015). In this way, one simple step is required to prepare the entire composite. This method provides a two-ply fabric composite with through-thickness connecting yarns, unless several 3D fabrics are placed on top of each other in advance and then the matrix is poured to produce the composite. It should be noted that using 3D fabric makes it difficult to manufacture complex-shaped products, which are more easily prepared with hand lay-up of 2D fabric layers.

3.3 Filament winding

Filament winding technique was developed to manufacture cross-ply composites made of continuous roving (Mobasher et al., 1997). This computer-controlled process can be used to produce pipes, laminated composite sheets, and cross-ply and angle-ply composites. In this process, continuous yarn is pulled from a single spool and passed through a wetting tank; after it is partially drained, it travels through a cement-based slurry bath in a tube shape. The end of this tube is located on a sliding table that moves transverse to the roving travel direction and aligns the yarn before it is wound around the mold. The continuous cement-impregnated yarn is then wrapped around a rotating mandrel, producing a laminated composite. Figure 3.1 presents a schematic of this process. The orientation of the yarn on the mandrel can be varied, providing 0°, 90°, and ±45° laminated composites, depending on the final product requirements (Pivacek and Mobasher, 1997). PP and AR glass yarns were successfully used to produce cross-ply laminated composites with this method using different ply orientations, [0°/±45°/90°], [0°/±45°], and [±45°]. Tensile strength of 50 MPa was reported for the AR glass composites made with the filament winding process. The weak point of these composites is the ply interface layer. The object produced in this method is a unidirectional aligned composite lamina. Based on this process, the pultrusion process was developed, in which fabrics are used to produce the composite rather than individual yarns (described in Section 3.4).

3.4 Pultrusion

Pultrusion is a continuous process for producing fabric–cement sheets of straight and constant cross-sectional area. In this process, fabrics are passed through a slurry infiltration chamber similar to the filament winding process

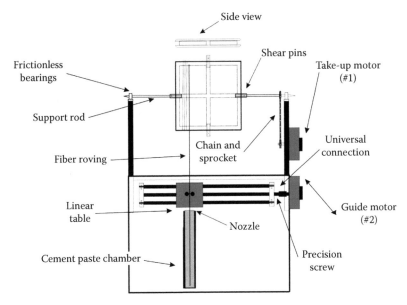

Figure 3.1 Scheme of the filament winding process, top view. (From Mobasher, B. and Pivacek, A., *Cement Concr. Compos.*, 20, 405, 1998.)

described earlier. The mixture should be sufficiently fluid to enable the fabric to transfer through the cement slurry without causing any damage to the fabric, but dense enough so that it will adhere to the fabric when leaving the cement bath. Furthermore, the penetration of the cement paste between the openings of the fabrics, as well as in between the filaments of the bundle, is a controlling factor in the ability of the system to develop and maintain bonding. Such penetration depends on the size of the bundle and fabric opening, but also on the rheology of the matrix. The use of additives such as fly ash, silica fume, polymer, and short fibers can influence the viscosity of the fresh cement mixture and its penetration ability. For that purpose, during cement bath impregnation, the fabric is pulled through a series of rollers to evenly push the matrix between the fabric opening and the bundle filament spaces. The impregnated fabric is then pulled through another series of rollers to squeeze out the excess matrix (located at the matrix bath exit), which also further improves matrix penetration into the bundles. Fabric–cement composite laminate sheets are then formed on a rotating plate-shaped mandrel. The pultrusion setup is presented in Figure 3.2a. A picture of the preliminary setup of the equipment while preparing specimens is shown in Figure 3.2b.

Sheets with various widths, lengths, and thickness of composites can be produced using the pultrusion technique. After forming the samples, pressure can be applied on top of the fabric–cement laminates to improve penetration of the matrix between the openings of the fabrics and the bundle filaments.

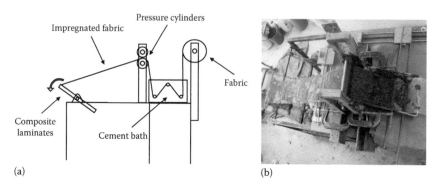

Figure 3.2 (a) Pultrusion setup and (b) during processing. (a: From Peled, A. and Mobasher, B., *ACI Mater. J.*, 102(1), 15, 2004; b: From Peled, A. and Mobasher, B., *ASCE J. Mater. Civil Eng.*, 19(4), 340, 2007.)

The intensity of this applied pressure is limited and cannot reach high levels, as the matrix is still fresh at this stage, where elevated pressures can remove most of the matrix between the fabric laminates leaving insufficient matrix to bind the fabrics to a uniform complete board. The number of fabric layers can be varied depending on composite requirements. Fabric–cement boards with 20 layers were successfully produced (Haim and Peled, 2011), and specimens with up to 10% by volume reinforcements have been made (Peled and Mobasher, 2004). This method can use any type of fabric or fiber material; however, due to the forces applied on the fabric during pulling and squeezing, it should retain sufficient dimensional stability. Bonded, knitted, and plain weave fabric types made of varied fiber materials including PP, PE, AR glass, carbon, aramid, PVA, as well as hybrid fabrics combining several fiber types were used in this process (Cohen and Peled, 2012; Mobasher et al., 2014; Peled and Mobasher, 2007).

The pultrusion assures uniform production of fabric–cement composites that exhibit significant improved performance in which their tensile behavior is characterized by multiple cracking. Several parameters related to the processing effects have been addressed, which include matrix formulation, processing direction, fabric types, and fiber material types (Mobasher et al., 2006; Peled and Mobasher, 2004, 2006; Peled et al., 2006). These influences are discussed in more detail in subsequent chapters.

3.5 Prestressed technique

Prestressed concrete is a traditional method to overcome the weakness of concrete in tension. In the traditional way, steel cable or rods are prestressed under tension, and when the tension is released against a hardened matrix, it generates a compressive stress that balances the tensile stress that the concrete member would otherwise experience during loading. The advantages of prestressed concrete include crack control and more durable concrete. A similar method was applied to textile reinforcement in which fabrics or continuous yarns were first stressed under tension prior to composite production (Krüger and Reinhardt, 2004; Kruger et al., 2003; Meyer and Vilkner, 2003; Reinhardt and Krüger, 2003; Reinhardt et al., 2003; Xu et al., 2004).

Fine, aggregate fresh concrete mixture was cast into a mold where the prestressed fabrics/yarns are placed. The prestress was removed 1 day after casting. Several fabric types were successfully used with this method, such as aramid, carbon, and glass. Special clamping devices were developed in order to anchor the fabrics and apply the required prestress. Different types of frames and prestressing setups were introduced. Figure 3.3 presents a laboratory square frame that was developed for such purposes using ten hydraulic pistons at each side with about 4,000 N force at individual pistons (Krüger et al., 2002; Reinhardt and Krüger, 2003; Reinhardt et al., 2002). In general, the clamping devices can lead to fracture of the yarn filaments in the anchoring area; in order to avoid this damage and to simplify the prestress method, either the ends of the rovings were blocked with epoxy or the entire rovings were impregnated in resin. A prestress of about 1.5 MPa was applied for glass fabric, 3 MPa for the carbon fabric (Xu et al., 2004), and 14 MPa for the aramid fabric (Meyer and Vilkner, 2003).

The prestressed textile reinforcements greatly improved the structural member mechanical properties with higher bond strength between fabric (yarn) and concrete, enhanced durability due to delay in cracking (Meyer and

Figure 3.3 Prestressed frame setup. (From Reinhardt, H.W. and Krüger, M., Prestressed concrete plates with high strength fabrics, in Sixth RILEM Symposium on Fiber-Reinforced Concrete (FRC), BEFIB 2004, Verona, Italy, 2003, pp. 188–196.)

Vilkner, 2003), and reduced crack width (Krüger et al., 2002). Furthermore, especially when woven types of fabrics are used, the curved shape of their yarns causes the concrete to spall off the element face; when the fabric is first pretensioned, such curvature effects become negligible, with no concrete spalling (Curbach et al., 2001; Meyer and Vilkner, 2003). More details on the influence of this technique on the mechanical performance of concrete beams are provided in Chapter 5.

3.6 Sandwich panels

Manufacture of sandwich panels made with laminated TRC was developed using discontinuous production technique. Such panels can be made of two thin TRC panels at the surfaces with rigid foam core in between (Ehlig et al., 2012; Shams et al., 2014). The advantages of these panels are their low weight with high structural capacity, thus fulfilling structural and physical demands. For example, TRC-based sandwich panels were manufactured with polyurethane rigid foams (hc = 150 mm) as a core, attached to two TRC shells at the facings (each of 15 mm thickness) either by gluing or by pressing a notched core into the fresh concrete layer. The notches were oriented perpendicular to the beam axis and established a mechanical connection that enables an efficient load transfer between core and facings. The TRC shells were produced by placing fine-grained concrete with short glass fibers and two layers of fabric alternately in a lamination process (Hegger et al., 2008).

Dey et al. (2015) developed two types of sandwich composites using a strong skin, hollow core elements. Both the systems used TRC as a ductile skin. One of the systems used an autoclaved aerated concrete (AAC) as a brittle core and the other system used a polymeric Fiber-reinforced aerated concrete (FRAC) as the ductile core material. Skin layer consisted of two layers of Alkali Resistant Glass (ARG) textiles. Performance of was evaluated in terms of flexural stiffness, strength, and energy absorption capacity.

3.7 Complex-geometry-shaped elements

One significant advantage of textile reinforcement is its availability in sheets with high formability, which allows them to be readily adapted to complicated shapes and geometries. Therefore, when textile reinforcement is used in combination with cementitious matrices of high fluidity, elements with complex geometry can be produced. These cement-embedded textiles can be used for production of different structural profiles and curve-shaped components, including strengthening layers on profiles or shell-shaped components. The fibers within the textile can be positioned in the desired direction of the element to fully exploit their load-bearing capacity.

Figure 3.4 Thin, curved barrel shell made of TRC for roof construction. (From Hegger, J. and Voss, S., *Eng. Struct.*, 30, 2050, 2008.)

Furthermore, these elements can be used to design lightweight structures that require only a minimal amount of materials.

An example of complex-curved barrel shell element for roof construction made by thin TRC is given in Figure 3.4 (Hegger and Voss, 2008). This element was manufactured by a simple application of shotcrete where layers of concrete with a thickness of 3–5 mm and AR glass fabric were applied alternately. The span of this barrel shell is 7 and 1.5 m long cantilevers on both sides, with a thickness of 2.5–6 cm at the most tensioned longitudinal edges, in which up to ten layers of the AR glass fabric were placed.

A quasicontinuous process was developed to produce U-shaped profiles, (Brameshuber, 2006; Brameshuber and Brockmann, 2001; Brockmann and Brameshuber, 2005; Weck and Kozler, 2003). In this process, a textile is rolled first by rollers into a mold and positioned in such a manner that it will be aligned in the middle of the profile. Then fine concrete with high flow ability is pumped and injected into the mold. A fine-grained concrete matrix was developed with suitable rheological properties for the pumping and injection process while offering complete penetration of the matrix into the textile reinforcement. In order to demold the element shortly after casting, pressure is applied on the fresh concrete to remove excess water and harden it. The pressed U-profile is then passed through an oven and shifted for the next component to be pressed. Pressure up to 2 MPa can be applied with this process.

An example of U-shaped profile TRC elements, each 2.50 m in width, 900 mm length, and a thickness of only 30 mm, is presented in Figure 3.5. It was used to build a footbridge over the Döllnitz River with a span of 9 m

Figure 3.5 U-profile segments of a pedestrian bridge at the University of Dresden. (From Schneider, H.N. and Bergmann, I., The application potential of textile-reinforced concrete, in A. Dubey (ed.), *ACI Fall Convention 2005 on Textile-Reinforced Concrete*, SP-250-1, 2008, pp. 7–22; Curbach, M. et al., Marktpotential Brückenbau—die erste textilbewehrte Segmentbrücke von der Idee bis zum Bauwerk, in FBF Betondienst GmbH (ed.), Tagungsband zu 50, BetonTage, Ernst & Sohn a Wiley Brand GmbH & Co. KG, Berlin, Germany, 2006, pp. 54–56.)

(Weiland et al., 2007). This bridge was made with textiles and prestressed steel; due to the textile reinforcement, it reaches only 20% of the weight of a similar steel-reinforced structure.

The ability to produce strong, light, complex-shaped elements with textile-reinforced fine-grained concrete also allows the production of complex rhombic filigree lattice structures as presented in Figure 3.6. Such a

(a)   (b)

Figure 3.6 Prototype (a) and detail (b) of the rhombic framework of TRC. (From Hegger, J. and Voss, S., *Eng. Struct.*, 30, 2050, 2008.)

structure is too complex to put into practice with conventional concrete construction. The framework consists of rhombic basic elements with outer dimensions of 1,000 × 600 × 160 mm, a wall thickness of 25 mm, and a weight of about 23 kg. A demonstration building was erected by assembling three parallel rows of arcs with 12 rhombi, resulting in a span of 10 m, vault rising of 3 m, and width of 1.8 m. The dimensions of the rhombic elements were determined considering the supported glass roofing.

The flexibility of textiles enabled the production of complex hypar surfaces (De Bolster et al., 2009; Scholzen et al., 2015). The basic section of the free-form structures can be made of sandwich panels with a rigid foamed core (such as polyurethane) and TRC at the faces, shaped in the form of hyperbolic paraboloids (hypars) (De Bolster et al., 2009). The hypars' single elements can be produced on a mold made of a cable net (Mollaert and Hebbelinck, 2002) or of technical textiles. The cable net system allows for the production of a great variety of shapes and spans due to its high flexibility, but it requires a lot of preparation time and labor cost. A mold made of a textile can be used for the production of several hypar shells on the same mold (Van Itterbeeck et al., 2009). A pavilion made of TRC to produce a load-bearing structure was produced, composed of four TRC hypar shells supported by a steel-reinforced concrete column in the center. Nonimpregnated fabrics were used to enable the complex shape as they can be easily adapted to the shell geometry. Precast shells were produced by shotcrete technology (Figure 3.7a). The overall plan dimensions of the structure were 14 × 14 m and 4 m in height (Figure 3.7b) (Scholzen et al., 2015). This shape was designed by the Spanish architect Félix Candela (1910–1997), who created in the fifties and sixties of the twentieth century many buildings in Mexico that are based on variations of such hypar shells (Cassinello et al., 2010).

(a)   (b)

Figure 3.7 Hypar shell roof structure made of TRC: (a) shell production using shotcrete and (b) the final roof structure. (Photo: bauko 2, RWTH Aachen University, Aachen, Germany.) (From Scholzen, A. et al., *Struct. Concr.*, 16(1), 106, 2015.)

Figure 3.8 Strengthening of a barrel-shaped roof made of reinforced concrete by a layer of TRC. (From Lieboldt, M. and Mechtcherine, V., *Cement Concr. Res.*, 52, 53, 2013.)

Another example of using the deformability of textile can be for repair and strengthening of structural concrete components having either flat or curved surfaces (Brückner et al., 2006; Schladitz et al., 2009; Weiland et al., 2008). Figure 3.8 shows an example of a reinforced roof structure with carbon fabrics and fine-grained concrete during processing, where a carbon fabric is applied by hand on top of a thin fresh layer of fine-grained concrete and then covered by hand. The manufactured roof has a barrel shell form 16 m in length and 7 m in width. A thin TRC layer of only 10–15 mm total thickness is applied on the roof (Butler et al., 2009; Lieboldt and Mechtcherine, 2013).

3.8 Summary

Textile cement-based composites are the ideal choice for a range of applications: cement boards including wall panels, exterior siding, roofing, and flooring tiles; structural components; strengthening, repairing, and retrofitting of structural concrete; and complex free-form elements for architectural and structural applications. Several production techniques and manufacturing methods were developed for the production of such TRC components, and some full-scale elements are already applied successfully. This area is rapidly growing, offering new applications and opportunities, which will require further developments of new processing and manufacturing techniques.

References

Adiel, S. E. and Peled, A., 3D fabrics as reinforcement for cement-based composites, *Composites Part A*, 74, 153–165, 2015.

Akkaya, Y., Peled, A., and Shah, S. P., Parameters related to fiber length and processing in cement composites, *Materials and Structures*, 33, 515–524, 2000.

Brameshuber, W., RILEM TC 201-TRC: Textile Reinforced Concrete. State-of-the-Art Report of RILEM Technical Committee 201-TRC, Bagneux: RILEM, Report 36, 2006.

Brameshuber, W. and Brockmann, T., Development and optimisation of cementitious matrices for textile reinforced elements, in *Proceedings of the 12th International Congress of the International Glass Fibre Reinforced Concrete Association*, Dublin, Republic of Ireland, May 14–16, 2001, Concrete Society, London, U.K., 2001, pp. 237–249.

Brockmann, T. and Brameshuber, W., Matrix development for the production of textile reinforced concrete (TRC) structural elements, in P. Hamelin, D. Bigaud et al. (eds.), *The Third International Conference on Composites in Constructions*, EPFL tous droits réservés, Lyon, France, Vol. 2, 2005, pp. 1165–1172.

Brückner, A., Ortlepp, R., and Curbach, M., Textile reinforced concrete for strengthening in bending and shear, *Materials and Structures*, 39, 741–748, 2006.

Butler, M., Mechtcherine, V., and Hempel, S., Experimental investigations on the durability of fibre-matrix interfaces in textile-reinforced concrete, *Cement and Concrete Composites*, 31, 221–231, 2009.

Cassinello, P., Schlaich, M., and Torroja, J. A., Félix Candela. In memorian (1910–1997). From thin concrete shells to the 21th century light weight structures, *Informes de la Construcctión*, 62(519), 5–26, 2010.

Cohen, Z. and Peled, A., Effect of nanofillers and production methods to control the interfacial characteristics of glass bundles in textile fabric cement-based composites, *Composites: Part A*, 43, 962–972, 2012.

Curbach, M., Baumann, L., Jesse, F., and Martius, A., Textilbewehrter Beton für die Verstärkung von Bauwerken, *Beton*, 51(8), 430–434, 2001.

Curbach, M., Jesse, D., and Weiland, S., Marktpotential Brückenbau—die erste textilbewehrte Segmentbrücke von der Idee bis zum Bauwerk, in FBF Betondienst GmbH (ed.), *Tagungsband zu 50*, BetonTage, Ernst & Sohn a Wiley Brand GmbH & Co. KG, Berlin, Germany, pp. 54–56, 2006.

De Bolster, E., Cuypers, H., Van Itterbeeck, P., Wastiels, J., and De Wilde, W. P., Use of hypar-shell structures with textile reinforced cement matrix composites in lightweight constructions, *Composites Science and Technology*, 69, 1341–1347, 2009.

Delvasto, S., Naaman, A. E., and Throne, J. L., Effect of pressure after casting on high strength fiber reinforced mortar, *Journal of Cement Composites and Lightweight Concrete*, 8(3), 181–190, 1986.

Dey, V., Zani, G., Colombo, M., Di Prisco, M., and Mobasher, B., Flexural impact response of textile-reinforced aerated concrete sandwich panels, *Materials & Design*, 86, 187–197, 2015.

Ehlig, D., Schladitz, F., Frenzel, M., and Curbach, M., Textilbeton–Praxisprojekte im Überblick (Textile concrete—An overview of executed projects). *Beton- und Stahlbetonbau*, 107(11), 777–785, 2012 (in German).

Garlinghouse, L. H. and Garlinghouse, R. E., The Omni mixer—A new approach to mixing concrete, *ACI Journal*, 69(4), 220–223, 1972.

Haim, E. and Peled, A., Impact behavior of fabric-cement hybrid composites, *ACI Materials Journal*, 108-M25, 235–243, May–June 2011.

Haupt, G. J., Study of cement based composites manufactured by extrusion, compression molding and filament winding, MS thesis, Arizona State University, Tempe, AZ, May 1997.

Hegger, J., Horstmann, M., and Scholzen, A., Sandwich panels with thin-walled textile-reinforced concrete facings, in *ACI Fall Convention 2007 on Design and Applications of Textile-Reinforced Concrete*, SP 251-73, 2008, pp. 109–123.

Hegger, J. and Voss, S., Investigations on the bearing behavior and application potential of textile reinforced concrete, *Engineering Structures*, 30, 2050–2056, 2008.

Hegger, J., Will, N., Schneider, H. N., and Kolzer, P., New structural elements made of textile reinforced concrete—Applications and examples, *Beton – und Stahlbetonbau*, 99(6), 482–487, 2004.

Igarashi, S., Bentur, A., and Mindess, S., The effect of processing on the bond and interfaces in steel fiber reinforced cement composites, *Cement and Concrete Composites*, 18, 313–322, 1996.

Krenchel, H. and Stang, H., Stable microcracking in cementitious materials, in A. M. Brandt, V. C. Li, and I. H. Marshall (eds.), *Brittle Matrix Composites 2*, Elsevier Applied Science, Warsaw, Poland, 1994, pp. 20–33.

Krüger, M., Ožbolt, J., and Reinhardt, H. W., A discrete bond model for 3D analysis of textile reinforced and prestressed concrete elements, *Otto-Graf-Journal*, 13, 111–128, 2002.

Kruger, M., Ozbolt, J., and Reinhardt, H. W., A new 3D discrete bond model to study the influence of bond on structural performance of thin reinforced and prestressed concrete plates, in H. W. Reinhardt and A. E. Naaman (eds.), *High Performance Fiber Reinforced Cement Composites (HPFRCC4)*, RILEM, Ann Arbor, MI, 2003, pp. 49–63.

Krüger, M. and Reinhardt, H. W., Prestressed textile reinforced elements, *Beton- und Stahlbetonbau*, 99(6), 472–475, June 2004.

Lieboldt, M. and Mechtcherine, V., Capillary transport of water through textile-reinforced concrete applied in repairing and/or strengthening cracked RC structures, *Cement and Concrete Research*, 52, 53–62, 2013.

Mallick, P. K., *Fiber Reinforced Composites Materials, Manufacturing, and Design*, 3rd edn. CRC Press, Taylor & Francis Group, LLC, New York, 2007.

Meyer, C. and Vilkner, G., Glass concrete thin sheet prestressed with aramid fiber mesh, in A. E. Naaman and H. W. Reinhardt (eds.), *Fourth International Workshop on High Performance Fiber Reinforced Cement Composites (HPFRCC4)*, RILEM, Ann Arbor, MI, 2003, pp. 325–336.

Mobasher, B., Castro-Montero, A., and Shah, S. P., A study of fracture in fiber reinforced cement-based composites using laser holographic interferometry, *Experimental Mechanics*, 30(90), 286–294, 1990.

Mobasher, B., Dey, V., Peled, A., and Cohen, Z., Correlation of constitutive response of hybrid textile reinforced concrete from tensile and flexural tests, *Cement and Concrete Composites*, 53, 148–161, 2014.

Mobasher, B. and Li, C. Y., Tensile fracture of carbon whisker reinforced cement based composites, in K. Basham (ed.), *ASCE Materials Engineering Conference, Infrastructure: New Materials and Methods for Repair*, 1994, pp. 551–554, November 13–16, 1994.

Mobasher, B. and Li, C. Y., Mechanical properties of hybrid cement based composites, *ACI Materials Journal*, 93(3), 284–293, May–June 1996.

Mobasher, B. and Li, C. Y., Fracture of whisker reinforced cement based composites, in A. M. Brandt, V. C. Li, and I. H. Marshall (eds.), *Proceedings of the International Symposium on Brittle Matrix Composites 4*, Warsaw, Poland, September 2003, pp. 116–124.

Mobasher, B., Peled, A., and Pahilajani, J., Distributed cracking and stiffness degradation in fabric–cement composites, *Materials and Structures*, 39(3), 317–331, 2006.

Mobasher, B. and Pivacek, A., A filament winding technique for manufacturing cement based cross-ply laminates, *Cement and Concrete Composites*, 20, 405–415, 1998.

Mobasher, B., Pivacek, A., and Haupt, G. J., Cement based cross-ply laminates, *Journal of Advanced Cement Based Materials*, 6, 144–152, 1997.

Mollaert, M. and Hebbelinck, S., ADAPTEN, a generating system for cable net structures, in *Fifth International Conference on Space Structures*, Guildford, Surrey, 2002, August 19–21, 2002.

Peled, A., Cyr, M. F., and Shah, S. P., High content of flyash class F in extruded cementitious composites, *ACI Materials Journal*, 97(5), 509–517, 2000.

Peled, A. and Mobasher, B., Pultruded fabric–cement composites, *ACI Materials Journal*, 102(1), 15–23, 2004.

Peled, A. and Mobasher, B., Properties of fabric–cement composites made by pultrusion, *Materials and Structures*, 39(8), 787–797, 2006.

Peled, A. and Mobasher, B., Tensile behavior of fabric cement-based composites: Pultruded and cast, *ASCE Journal of Materials in Civil Engineering*, 19(4), 340–348, 2007.

Peled, A. and Shah, S. P., Processing effects in cementitious composites: Extrusion and casting, *Journal of Materials in Civil Engineering*, 15, 192–199, 2003.

Peled, A., Sueki, S., and Mobasher, B., Bonding in fabric–cement systems: Effects of fabrication methods, *Cement and Concrete Research*, 36(9), 1661–1671, 2006.

Pivacek, A. and Mobasher, B., Development of cement-based cross-ply laminates by filament winding technique, *Journal of Materials in Civil Engineering*, 6, 55–57, May 1997.

Reinhardt, H. W. and Krüger, M., Prestressed concrete plates with high strength fabrics, in *Sixth RILEM Symposium on Fiber-Reinforced Concrete (FRC)*, BEFIB 2004, Verona, Italy, 2003, pp. 188–196.

Reinhardt, H. W., Krüger, M., and Grosse, C. U., Thin plates prestressed with textile reinforcement. Concrete: Materials Science to Application. A Tribute to Surendra P. Shah. ACI, Farmington Hills, MI, SP-206, 2002, pp. 355–372.

Reinhardt, H. W., Krüger, M., and Grosse, U., Concrete prestressed with textile fabric, *Advances in Concrete Technology*, 1(3), 231–239, 2003.

Schladitz, F., Lorenz, E., Jesse, F., and Curbach, M., Verstärkung einer denkmalgeschützten Tonnenschale mit Textilbeton, *Beton-Stahlbetonbau*, 104, 432–437, 2009.

Schneider, H. N. and Bergmann, I., The application potential of textile-reinforced concrete, in A. Dubey (ed.), *ACI Fall Convention 2005 on Textile-Reinforced Concrete*, SP-250-1, 2008, pp. 7–22.

Scholzen, A., Chudoba, R., and Hegger, J., Thin-walled shell structures made of textile-reinforced concrete: Part I: Structural design and construction, *Structural Concrete*, 16(1), 106–114, March 1, 2015.

Shah, S. P. and Balaguru, P. N., *Fiber-Reinforced Cement Composites*, McGraw-Hill, New York, 1992.

Shams, A., Horstmann, M., and Hegger, J., Experimental investigations on Textile-Reinforced Concrete (TRC) sandwich sections, *Composite Structures*, 118, 643–653, 2014.

Shao, Y. and Shah, S. P., Mechanical properties of PVA fiber reinforced cement composites fabricated by extrusion processing, *ACI Material Journal*, 94(6), 555–564, 1997.

Van Itterbeeck, P., Wastiels, J., and De Wilde, W. P., Use of hypar-shell structures with textile reinforced cement matrix composites in lightweight constructions, *Composites Science and Technology*, 69, 1341–1347, 2009.

Weck, M. and Kozler, P., Development of a production machine for textile reinforced continuous concrete profiles, in M. Curbach (ed.), *Proceedings of the Second Colloquium on Textile Reinforced Structures*, Dresden, Germany, 2003, pp. 467–480.

Weiland, S., Ortlepp, R., Brückner, A., and Curbach, M., Strengthening of RC structures with textile reinforced concrete (TRC), in C. M. Aldea (ed.), *ACI Spring 2005 Convention New York on Thin Fiber and Textile Reinforced Cementitious Systems*, SP-244-10, 2007, pp. 157–172.

Weiland, S., Ortlepp, R., Hauptenbuchner, B., and Curbach, M., Textile reinforced concrete for flexural strengthening of RC-structures—Part 2: Application on a concrete shell, ACI SP, *Design & Applications of Textile-Reinforced Concrete*, 251(3), 41–58, 2008.

Xu, S., Krüger, M., Reinhardt, H. W., and Ožbolt, J., Bond characteristics of carbon, alkali resistant glass, and aramid textiles in mortar, *Journal of Materials in Civil Engineering*, 16, 356–364, 2004.

Chapter 4

Micromechanics and microstructure

4.1 Introduction

The fundamentals of the mechanics and micromechanics of brittle matrix reinforced with fibers/yarns were laid down in the ACK model (Aveston et al., 1971). The model was first developed for continuous fibers using simplified assumptions enabling analytical solutions. These basic concepts are valid for textile-reinforced concretes (TRC), which are essentially brittle matrix composites reinforced with continuous yarns. Therefore, it will serve as a basis for the overview in this chapter.

The concepts of the ACK model are summarized graphically in Figures 4.1 and 4.2. It is assumed that both matrix and reinforcement are linearly elastic and brittle, with the reinforcing yarns having much higher ultimate tensile strain. A detailed account for the model is given in several references (e.g., Bentur and Mindess, 2007), and only a concise overview is provided here. The characteristic output of the model is a stress–strain curve consisting of three stages with a multiple cracking zone in which the matrix is successively cracked.

Up to the first cracking stress, the reinforcing yarns and the matrix are within their linear elastic range, and the overall behavior of the composite is therefore elastic–linear. The modulus of elasticity is the weighted average of the matrix and reinforcement, that is, the rule of mixtures is effective at this range:

$$E_c = E_m V_m + E_f V_f \qquad (4.1)$$

where
E_c, E_m, E_f are the moduli of the composite, matrix, and fiber/yarn, respectively
V_m, V_f are the relative volume contents of the matrix and fiber/yarn, respectively, $V_m + V_f = 1$

At the cracking strain of the matrix, as the first crack is being formed, the reinforcing yarns bridge over the cracks and take upon themselves the entire

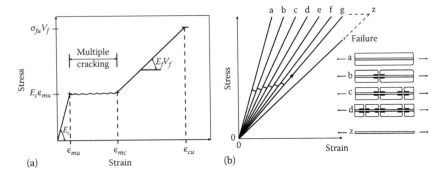

Figure 4.1 Concepts of the ACK model: (a) stress–strain curve and (b) multiple cracking mechanism. (After Allen, H.G., The purpose and methods of fibre reinforcement, in *Proceedings of the International Building Exhibition Conference on Prospects of Fibre Reinforced Construction Materials*, London, England, Building Research Station, 1971, pp. 3–14.)

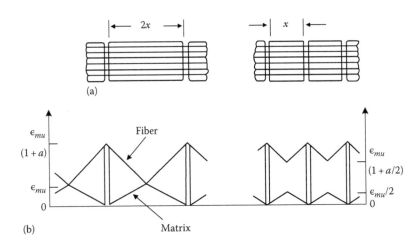

Figure 4.2 (a) Multiple cracking and (b) stress distribution in the matrix and reinforcing yarn according to the ACK model (Aveston et al., 1971), assuming linear (frictional) stress transfer and elastic brittle behavior of the matrix.

load that was carried by the matrix. Thus, at the crack surface, the stress in the matrix is zero and the load is gradually being transferred from the reinforcing yarns to the matrix. For a constant frictional shear bond at the interface, there is a linear buildup of stress in the matrix until it reaches the matrix strength and additional crack is formed. This process goes on to lead to the successive cracking of the matrix, and the multiple cracking zone is obtained, in which the stress on the composite is roughly constant and the matrix is subdivided by cracks having spacing value, in between the upper and lower bound values of x and $2x$, with x being

$$x = \frac{V_m}{V_f} \frac{\sigma'_{mu} r}{2\tau_{fu}}$$

(4.2)

where
$\quad$ V_m, V_f is the relative volume content of matrix and fiber/yarn, respectively, $V_m + V_f = 1$
$\quad$ σ'_{mu} is the stress of matrix at its ultimate strain
$\quad$ r is the radius of the fiber
$\quad$ τ_{fu} is the fiber–matrix shear bond strength

The stresses within each "block" of cracked matrix are presented in Figure 4.2, with α defined as:

$$\alpha = \frac{E_m V_m}{E_f V_f}$$

The strain at the end of the multiple cracking zone ε_{mc} is between an upper and a lower limit:

$$\varepsilon_{mu}\left(1 + \frac{\alpha}{2}\right) < \varepsilon_{mc} < \varepsilon_{mu}\left(1 + \frac{3\alpha}{4}\right)$$

(4.3)

where ε_{mu} is the matrix ultimate strain.

Following the end of the multiple cracking zone, additional stresses imposed on the composite are carried by the reinforcing yarns that have a higher ultimate strain than the matrix. The ultimate strain of the composite, ε_{cu}, is calculated as

$$\left(\frac{\varepsilon_{fu} - \alpha\varepsilon_{mu}}{2}\right) < \varepsilon_{cu} < \left(\frac{\varepsilon_{fu} - \alpha\varepsilon_{mu}}{4}\right)$$

(4.4)

where ε_{fu} is the ultimate strain of fiber/yarn.

The parameter ε_{mu} is obtained using the ACK approach (Aveston et al., 1971), which predicts the strength of the matrix phase in the presence of fibers as a function of bulk parameters and interfacial ones, as follows:

$$\varepsilon_{mu} = \left[\frac{12\tau\gamma_m E_f V_f^2}{E_c E_m^2 r V_m}\right]^{\frac{1}{3}}$$

(4.5)

In this approach, γ is the fracture toughness and r is the fiber radius. This approach has been verified to be applicable for cement-based materials as it has been clearly shown that the strength of the matrix is increased in the presence of fibers (Mobasher and Shah, 1990).

The bonding mechanism, its stiffness and strength, as well as the modulus of elasticity of the reinforcing yarns, have a profound effect on the overall mechanical behavior, especially in the multiple cracking zone, as shown schematically in Figure 4.3. The differences in behavior are quite dramatic, and therefore, control of the bond is quite important: for low-modulus yarns, a sufficiently high bond can result in an elastic–plastic composite, while for a high modulus the bond may affect the overall postcracking zone to yield a strain-hardening composite without a marked "quasiyield" point.

As demonstrated, the bond has a marked impact on the shape of the stress–strain curve. In-depth studies of the bond effect have been carried out for yarns consisting of individual fibers, where the shape is well defined and the circumference is completely surrounded by the cementitious matrix. In this case, bond models based on resolving interfacial shear stresses were developed and applied within the framework of the ACK model. These models assume either a frictional shear mode transfer or a mixed mode of adhesion and shear (Figure 4.4). Shear–lag mechanisms that are nonlinear in nature were resolved for such reinforcements and introduced into the ACK model.

Additional steps were taken to model interfacial interactions induced by fibers/yarns with complex geometry that generate anchoring mechanisms. The need to resort to such systems is often sparked by the low interfacial bond between reinforcing single fibers and the particulate cementitious matrix. These characteristics were mainly addressed with the view of enhancing the efficiency of short fiber reinforcements.

An extensive overview of these models on the micromechanics scale (bonding interactions with matrix) and macromechanics scale (behavior of

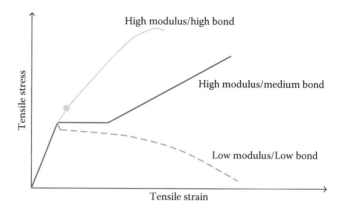

Figure 4.3 Schematic stress–strain behavior of cement composites with fibers of different moduli and bond strength.

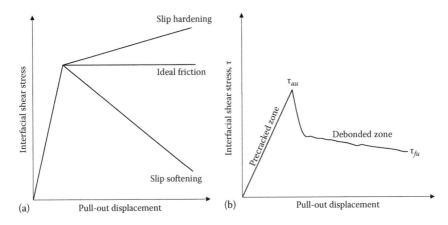

Figure 4.4 Range of interfacial shear stress–slip displacement bonding curves considered for modeling the behavior of reinforcing individual fibers embedded in a cementitious matrix, showing strain hardening and softening behavior (a), and a sharp transition from the adhesional bond to the frictional one (b). (After Bentur, A. and Mindess, S., *Fibre Reinforced Cementitious Composites*, Taylor & Francis, 2007.)

the composite) can be found in various general references (e.g., Bentur and Mindess, 2007). The discussion in this chapter will focus on the issues where the bonding in textiles has special characteristics.

4.2 Bond and pullout

A significant amount of experimental and analytical investigations have characterized the interface in man-made fiber–cement matrix systems. Tests for resolving the nature of the bond of cementitious composites have been performed by many researchers on steel, glass, and polymeric fibers. Naaman and Najm (1991) state that there are four main factors that influence the bond adhesion: (1) physical and chemical adhesion; (2) mechanical component of bond such as deformed, crimped, and hooked end fibers; (3) fiber-to-fiber interlock; and (4) friction. Peled and Bentur (2003b) investigated the pullout behavior of straight and crimped polyethylene (PE) yarns. Increasing the crimp density serves to enhance the mechanical anchoring and the equivalent adhesional bond strength increasing from 1 to 1.84 MPa (10 mm fiber embedded length). The bond properties of carbon fibers were investigated by Katz et al. (1995). Kim et al. (2007) investigated steel hooked end and toroid-shaped fibers and Markovich et al. (2001) studied the hooked end steel fibers.

Several fiber pullout models have been proposed to characterize bond properties of a single fiber and a cement matrix (Banholzer et al., 2005;

Naaman and Najm, 1991; Stang and Shah, 1986; Sueki et al., 2007). Several models (Mobasher and Li, 1995; Naaman et al., 1991a,b) proposed an analytical solution assuming that the fiber and the interface behave elastically in the bonded region and are characterized by a constant residual shear strength in the debonded region. More sophisticated bond strength models (Naaman et al., 1991a; Stang et al., 1990) that better describe bonding in pre- and postpeak response were also proposed; however, additional details of bond strength model significantly increase the complexity of analytical solutions.

Several studies dealt with the characterization of the bond and pullout of the fabric–cement system with the view that these characteristics can be quite different than those of the fibers that make up the fabric. Sueki et al. (2007) modified Naaman's model to analyze pullout test results and quantified the equivalent bond properties of several fabrics. In their analysis, the role of fill yarns in providing anchorage points along the length were modeled in a discrete approach. Banholzer et al. (2005) proposed another analytical model to simulate the pullout response of a fiber–matrix system in which an N-piecewise linear bond stress versus slip relation is adopted, simulating the behavior of a typical roving in a fabric, where the roving is made up of bundled filaments.

Test setups similar to the one in Figure 4.5 have been used. In some of them, imaging techniques were applied to monitor the nature of slip and breakage of fibers as the pullout progresses.

Typical results for glass fabric are shown in Figure 4.6a, demonstrating the effect of the matrix (addition of fly ash enhances the bond mechanism) as well as the special nature of the fabric, which in this case shows up as a dual peak response. This response is associated with the failure of the junction bonds and transfer of the load down the length of the fiber. Sophisticated modeling of curves of this kind using techniques such as nonlinear finite differences mode (e.g., Soranakom and Mobasher, 2009) enables to resolve several bonding parameters such as the stiffness of the interface (Figure 4.6b).

Currently, there is no fabric pullout model available (Peled and Bentur, 2003b; Peled et al., 2006; Zastrau et al., 2003), although attempts have been made to resolve and quantify some specific mechanisms that occur in fabric pullout, such as those shown in Figure 4.7. Most bond properties of the fabric structures are obtained from a straight fiber pullout or treated fabric (grid reinforcement) and are quantified as an equivalent smooth longitudinal fiber. Therefore, the equivalent bond properties will include the effect of mechanical anchorage of transverse yarn and slack existing in fibers.

The use of a model based on straight fiber pullout represents properties of a fabric structure in terms of quasistraight yarns, enabling a comparison with available fiber composites. Furthermore, it simplifies the correlation

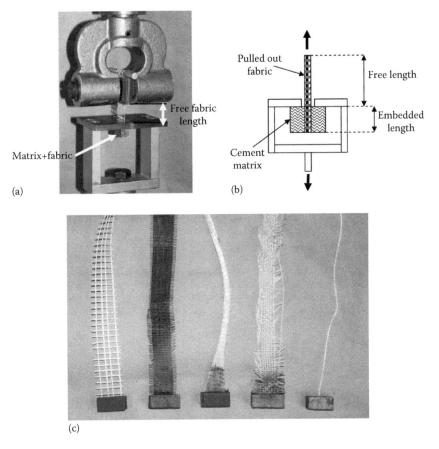

Figure 4.5 Setup for testing of pullout of fabrics from cementitious matrix: (a) image, (b) scheme, and (c) examples of pulled fabrics.

of the geometrical interlock of a fabric system into apparent chemical and mechanical bond characteristics.

The two unique characteristics of the textile that affect its bonding and pullout response are the nature of the yarns that make up the textile fabric and the overall geometry of the textile and the makeup of the yarns:

1. The textile is made up of yarns that could consist of individual fibers or bundled filaments. The latter structure is more typical of high-performance reinforcement such as glass and carbon, where the bundled yarns are assemblies of several hundreds or thousands of filaments with diameters of the order of magnitude of 10 μm. The filaments in the bundle are held together in the textile fabric, and upon production of the

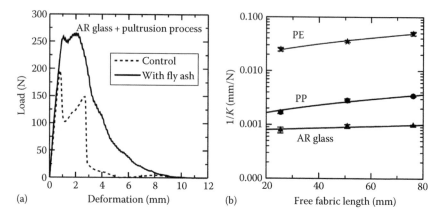

(a) Deformation (mm) (b) Free fabric length (mm)

Figure 4.6 The pullout response of an AR glass fabric using control and fly ash–blended matrix (a) and the effective stiffness of the interface transition zone obtained from the pullout experiments (b). (From Mobasher, B. and Soranakom, C., Effect of transverse yarns in the pullout response and toughening mechanisms of textile reinforced cement composites, in *Fracture and Damage of Advanced Fibre-Reinforced Cement-Based Materials, Proceedings of the 18th European Conference on Fracture*, ECF18, Dresden, Germany, Aedificatio Publishers, 2010, pp. 59–66.)

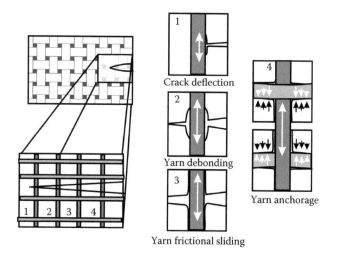

Figure 4.7 Various stages of contribution of the fabric to resisting the growth of the cracks. (After modeling concepts of Mobasher, B. and Soranakom, C., Effect of transverse yarns in the pullout response and toughening mechanisms of textile reinforced cement composites, in *Fracture and Damage of Advanced Fibre-Reinforced Cement-Based Materials, Proceedings of the 18th European Conference on Fracture*, ECF18, Dresden, Germany, Aedificatio Publishers, 2010, pp. 59–66.)

composite, they do not tend to open up and they retain the bundle structure. As a result, cement grains that are 10–30 μm in diameter and more cannot penetrate into the spaces in the bundle. Only hydration products can deposit in these small spaces, but to a very limited extent and at a very slow rate. As a result, the reinforcing unit is not of a well-defined geometry but rather an assembly of filaments, with the external ones (sleeve filaments) being in intimate contact with the cementitious matrix and the internal ones (core filaments) having limited points of contact, through hydration products, which may have penetrated inward. The bonding mechanisms of this unit are quite complex, and there is a need to take into account that many of the filaments are poorly bonded to the matrix, or none at all. Thus, the efficiency of the reinforcement is less than would be expected on the basis of the total surface area of the filaments.

2. The overall geometry of the textile fabric is characterized by intersections of weft and warp yarns, which interact in a variety of modes to create complex geometries. These geometrical characteristics may lead to anchoring effects that enhance the bond, or just the opposite, compromising impregnation of the textile by the matrix, leading to a reduction in bonding.

These two will be discussed in Sections 4.3 and 4.4.

4.3 Microstructure and bonding of multifilament yarns

4.3.1 Nature of the reinforcement and matrix

Most of the textile fabrics used in TRC are made of AR (alkali-resistant) glass filaments in the form of bundled yarns. Thus, much of the attention of the binding processes has been given to this type of reinforcement. The principles, however, can serve for the evaluation of other types of bundled systems where the filaments are made of high-modulus material with elastic–brittle behavior such as carbon. The moduli of both are much higher than that of the cementitious matrix, being 70 and 210 GPa for glass and carbon, respectively, compared to 20–30 GPa for the matrix.

The matrix is made of mortar (sometimes labeled as fine aggregate concrete), with the maximum sand particles size less than 1 mm and the water/binder ratio in the range of 0.40–0.45. The typical composition is presented in Table 4.1.

The compressive strength is in the range of 82–100 MPa for 10–40 mm cubes, and the flexural strength can be as high as 30 MPa.

The individual glass filaments have a modulus of about 70 GPa and a tensile strength of about 1,400 MPa. However, the properties of the

Table 4.1 Typical composition of fine aggregate concrete used as the matrix in TRC

Cement CEM I 52.5 (c)	490 kg/m³
Fly ash (f)	175 kg/m³
Silica fume (s)	35 kg/m³
Water (w)	280 kg/m³
w/bᵃ	0.40
Plasticizer	1.00% by mass of binder
Siliceous fines (0–0.25 mm)	500 kg/m³
Siliceous sand (0.2–0.6 mm)	714 kg/m³

Source: Banholzer, B. et al., *Mater. Struct.*, 39, 749, 2006.

ᵃ $w/b = w/(c + f + s)$.

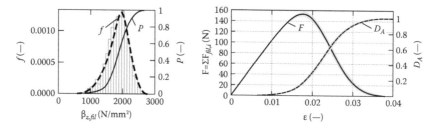

Figure 4.8 Filament strength distribution and stress–strain relation of a multifilament yarn. (From Zastrau, Z. et al., *Technische Mechanik*, Band 28 Heft 1, 53, 2008.)

multifilament yarn are much lower in strength, about 700 GPa (Banholzer et al., 2006). This difference is due to statistical spread as well as damage that may have been caused to the filaments upon handling. In the case of a bundle test, the weaker filaments would have a strong effect on the overall strength. Therefore, when considering the mechanics of the yarn, statistical effects should be taken into account (Chudoba et al., 2006; Vorechovsky and Chudoba, 2006).

The spread in the properties of the filament changes the nature of tensile behavior of the multifilament yarn from linear elastic for a single filament to a more ductile one in the yarn (Figure 4.8) (Zastrau et al., 2008).

4.3.2 Microstructure and bonding processes

Bonding processes associated with the special structure of a multifilament yarn in a cement matrix were resolved qualitatively in the context of studies of glass–fiber-reinforced cement composites that were produced by the spray method. The composites are manufactured by chopping the yarn and spraying it on a mold along with a mortar matrix. The outcome is a ductile composite in which the chopped yarns remain in a bundled form, engulfed by the cementitious matrix. The aging of the composites is associated with embrittlement, which could not be always correlated with alkali attack as the glass was AR. It was suggested that this change in the mechanical properties is associated with the deposition of hydration products within the inner filaments of the yarn. Intensive studies confirmed that such a change indeed takes place, changing the micromechanical bonding mechanisms. Before aging, the yarn can bridge over cracks due to the flexible nature of the yarn (external filament [sleeve] bonded tightly while the internal [core] ones can bend across an inclined crack—Figure 4.9), to one which is brittle and cannot accommodate the locally large deformation in the oriented crack due to the filling in of the spaces by hydration products.

These characteristics of a flexible reinforcing yarn were analyzed by Stucke and Majumdar (1976) in terms of fracture mechanics considerations and will be discussed to a greater extent in Chapter 9, in conjunction with the durability of TRC.

Limited penetration of hydration products should show up in pullout tests of multifilament yarns, resulting in lower bond strength than in monofilament testing. This was indeed confirmed by several investigators as shown in Table 4.2. The values in this table represent the effective bonding of the bundle, calculated as the pullout load divided by the surface area of all the filaments, relative to the bond obtained in a monofilament testing.

Various quantitative treatments of the contact between the filaments in the bundle and the matrix were reported, and they can provide insight into the values in Table 4.2. They are based on the feasibility of impregnation of the spaces between the filaments in the bundle by cement grains

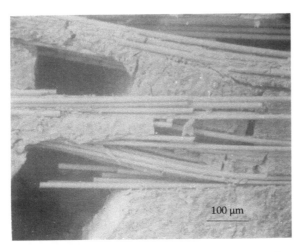

Figure 4.9 Local bending in a multifilament glass yarn bridging across a crack as observed by in situ SEM study. (After Bentur, A. and Diamond, S., *Cement Concr. Res.*, 14, 31, 1984.)

Table 4.2 Effect of bundle structure on the bond values in a bundled yarn

Fiber	Filament diameter (μm)	Number of filaments in bundle	τ_{single} (MPa)	T_{bundle} (MPa)	$\tau_{bundle}/\tau_{single}$	Reference
Nylon	27	220	0.16	0.051	0.321	Li and Wang (1990)
Kevlar	12	1,000	4.50	0.198	0.044	
Polyethylene (Spectra)	38	20	1.02	0.328	0.322	
		40	1.02	0.502	0.492	
		57	1.02	0.505	0.495	
		118	1.02	0.352	0.352	
Glass[a]	12.5	204	1.1	0.38	0.350	Oakley and Proctor (1975)

Source: Adapted from Bentur, A. and Mindess, S., *Fibre Reinforced Cementitious Composites*, Taylor & Francis, 2007; Based on Li, V.C. and Wang, Y., *Composites*, 21, 132, 1990; Oakley, D.R. and Proctor, B.A., Tensile stress-strain behavior of glass fiber reinforced cement composites, in A. Neville (ed.), *Fibre Reinforced Cement and Concrete*, The Construction Press, Lancaster, England, 1975, pp. 347–359.

[a] Based on microscopic observations.

(Dolatabadi et al., 2010, 2011) and the exposed surface area of the filaments in the bundle that could form contact with the matrix (Li and Wang, 1990).

Dolatabadi et al. (2011) considered the size of the spaces between the filaments assuming a hexagonal filament packing, with a space d_g between the filaments (Figure 4.10), as well as the particle size gradation of the cement grains (Figure 4.11) to resolve the portion of the grains that can potentially penetrate into these spaces. These approaches present the physical limitations with regard to the potential penetration of cement particles into the bundle, and they do not take into account other

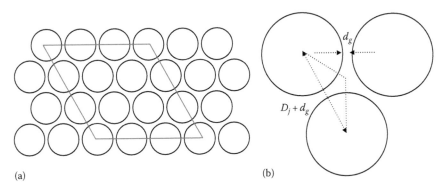

(a)   (b)

Figure 4.10 Modeling parameters of the packing density of a bundle assuming hexagonal packing (a) and spacing d_g between filaments of diameter D_f (b). (After Dolatabadi, N. et al., An analytical investigation of cement penetration within bundle of fibers, in *International RILEM Conference on Material Science*, Aachen, Germany, Vol. 1, 2010, pp. 69–78.)

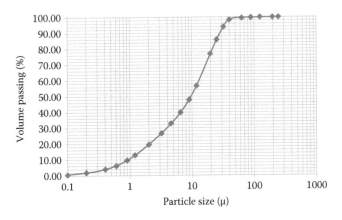

Figure 4.11 Particle size gradation of cement used for production of TRC composite behavior. (After Dolatabadi, M.K. et al., *Mater. Struct.*, 44, 245, 2011.)

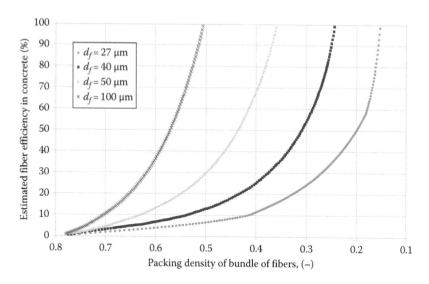

Figure 4.12 Calculated efficiency of filaments in cement matrix as a function of density of 2,400 tex yarn with different filament diameters. (After Dolatabadi, M.K. et al., *Mater. Struct.*, 44, 245, 2011.)

parameters such as capillary forces and the opening up of the bundle during the production process.

The packing density of the filaments could be calculated in terms of the geometrical parameters given in Figure 4.10, and the bundle efficiency (i.e., the space penetrated relative to the total space between the filaments) could be calculated by considering the particle size distribution of the cement. The relation between these two parameters is shown in Figure 4.12 for filaments in the size range of 27–100 μm. For a typical bundle packing density in the range of 0.4–0.8 and filament diameter of 27 μm, the efficiency is expected to be lower than 10%.

A second approach for calculating the efficiency of the filaments in the bundle was reported by Li and Wang (1990), based on calculation of the exposed surface of the filaments relative to the total one. This relative surface decreases with the increase in bundle size (number of filaments making up the bundle) (Figure 4.13).

The efficiency values expected from the model of Dolatabadi et al. (2011) (Figure 4.12) and Li and Wang (1990) (Figure 4.13) are less than 10%, much smaller than the ones predicted from the relative bond in Table 4.2. This suggests that the bundle probably opened up during the production process, allowing more surface area to become exposed relative to the calculated one.

In the actual bundle, one would expect a gradient of filling up of the spaces between the filaments, as the impregnation takes place by penetration from

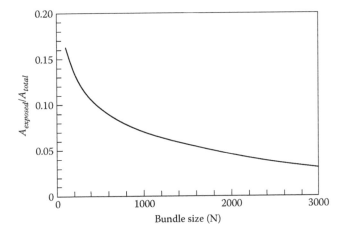

Figure 4.13 The calculated exposed area relative to the total area of the filaments as a function of the bundle size (number of filaments making up the bundle). (From Li, V.C. and Wang, Y., *Composites*, 21, 132, 1990.)

the periphery of the sleeve filaments inward, toward the core filaments. This is expected to be the case in the flow of particles of sufficiently small cement grains during the production of the composites. Additional impregnation by deposition of hydration products over time may have a different nature if some of these are nucleated in the spaces between the filaments.

The presence of such a gradient was confirmed by Zhu and Bartos (1997). They quantified the bundle microstructure using a special in situ push-in test arrangement, shown schematically in Figure 4.14.

The resistance to the push-in of the filaments is an indication of the bonding of the filaments and, in turn, this can serve as an estimate of the presence of matrix material around them. In an unaged composite, the resistance in the core filaments was small, while that in the sleeve was high; after aging, the resistance at the core filaments became very high (Figure 4.15a). In metakaolin-modified matrix, embrittlement upon aging was eliminated, and this was accompanied by maintaining low resistance at the inner filaments, relative to the external ones (Figure 4.15b). This is a quantitative manifestation of the special microstructure of a cementitious matrix reinforced with glass-bundled filaments: the external sleeve filaments were well bonded to the matrix, while the inner core ones were largely free, resulting in a bonding mechanism that provided ductility. If the core filaments were becoming well bonded due to deposition of hydration products, embrittlement occurred.

These observations of Zhu and Bartos were consistent with the concept of a telescopic pullout mechanism in a bundled reinforcement, where the external filaments formed a sleeve in which the filaments are tightly bonded to the matrix and fail, while the inner core filaments, which are not well

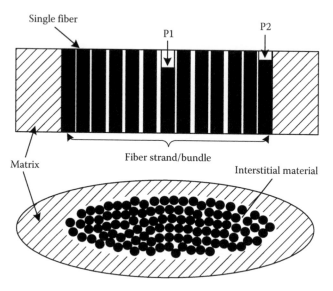

Figure 4.14 Schematic description of the setup for push-in strength testing of filaments within a yarn. (After Zhu, W. and Bartos, P.J.M., *Cement Concr. Res.*, 27(11), 1701, 1997.)

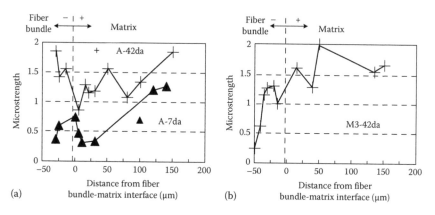

Figure 4.15 Push-in resistance of filaments in a bundle in unaged and aged (embrittled) Portland cement composite, 7 (A-7da) and 42 days (A-42da) of accelerated aging (a) and in aged composite with metakaolin matrix that did not show embrittlement upon aging for 42 days of accelerated aging (M3-42da) (b). (From Zhu, W. and Bartos, P.J.M., *Cement Concr. Res.*, 27(11), 1701, 1997.)

bonded, can engage in slip and provide toughness to the composite, as shown in Figure 4.16 (Bartos, 1987).

The understanding of the microstructure of the bundle in relation to its bonding was investigated in-depth by Banholzer (2004), Banholzer and Brameshuber (2004), and Banholzer et al. (2006), who developed a novel

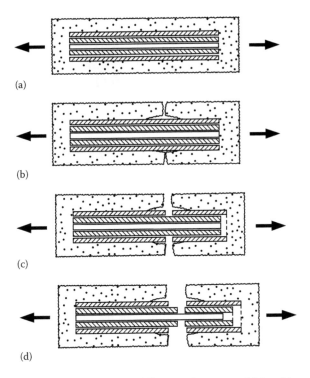

(a)

(b)

(c)

(d)

Figure 4.16 Telescopic failure in a bundled fiber reinforcement. (a) Load is transmitted to the fibers. Interfaces are intact, (b) matrix cracks, partial debonding occurs, (c) outer fibers fracture, core fibers pull out, and (d) core fibers pull out. (After Bartos, P., Brittle matrix composites reinforced with bundles of fibers, in Maso, J.C. (ed.), *From Materials Science to Construction Materials, Proceedings, RILEM Symposium*, Chapman & Hall, the Netherlands, 1987, pp. 539–546.)

technique that enables monitoring individual filaments in situ during a pullout test. This technique can detect at each stage of the loading the individual filaments that fail, to quantify their number and location. The test developed for this purpose, the FILT method, consisted of transmitting light through the glass filaments and capturing it on the other side with a camera.

This test method provides the means to quantify the number of filaments that are active (i.e., the filaments that transmit light and their proportion relative to all the original filaments in the yarn) during each stage of pullout loading. It confirms that in a typical glass yarn, filaments start breaking at the circumference and, thereafter, layer after layer of filaments continue to fracture from the periphery toward the center of the yarn, as shown in Figure 4.17. The telescopic mode of pullout can be clearly observed showing the external filaments breaking first, immediately at and after the peak load,

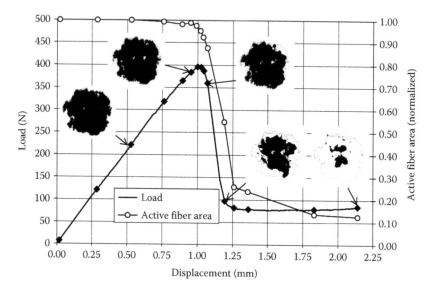

Figure 4.17 Pullout curve and images of the active filaments at each stage of loading; data published by Bentur et al. (2010), using the FILT test method developed by Banholzer et al. (2006).

with residual bond remaining along with a core of active filaments after about 1.1 mm slip displacement.

A good match between the portion of active filaments and the pullout resistance at different levels of load can be observed for systems showing a brittle failure and a ductile telescopic failure (Figure 4.18).

The colored images superimposed one over the other represent the active filament portion at each stage of slip marked with the same color on the curve. The yellow area shows the core of active filaments at the termination of the test; the red marks the area of active filaments (i.e., the whole yarn) at the onset of filament fracture and close to the tip of the pullout curve.

Separation of sleeve and core has been observed also in systems other than glass filaments yarns. Ackers and Garrett (1983a,b) have reported it for asbestos fibers, in which bundles of fibrils are often observed, providing bridging across cracks combined with sheath and core separation.

Sleeve and core mode of failure has also been reported for monoyarn reinforcement of fibrillated polypropylene (PP) (Ohno and Hannant, 1994). Microscopic observations of the interfaces after failure have indicated the occurrence of slip lines within the fiber, which is thought to affect the mechanical performance of the system. Thus, the fiber could be considered as consisting of sleeve and core (Figure 4.19), with the shear strength between the sleeve and core being smaller than that at the interface between the external face of the fiber and the matrix. Ohno and Hannant suggested that this is a mode of failure similar to the one taking place in glass yarn pullout. Based on microscopic observations of failure in glass yarn–cement system, it was suggested by Ohno and Hannant (1994) that a similar phenomenon occurs there too, justifying the modeling of such systems in terms of core and sleeve analysis of the reinforcement's interaction with the matrix.

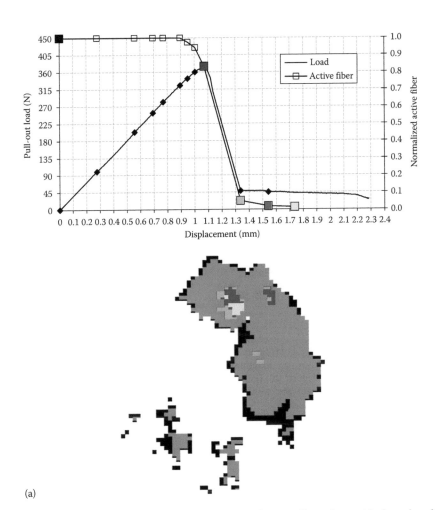

(a)

Figure 4.18 Pullout load and active filament curves during pullout, along with the colored images of the active fibers, for bundled yarn glass systems exhibiting brittle behavior (a), and ductile telescopic failure (b). (Bentur et al., 2010; Yardimci et al., 2011), using the FILT test method developed by Banholzer et al. (2006).

(Continued)

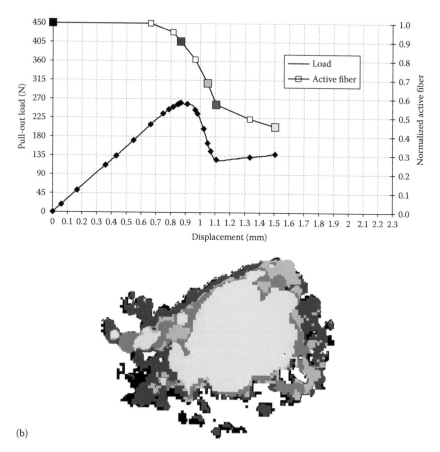

(b)

Figure 4.18 (Continued)   Pullout load and active filament curves during pullout, along with the colored images of the active fibers, for bundled yarn glass systems exhibiting brittle behavior (a), and ductile telescopic failure (b). (Bentur et al., 2010; Yardimci et al., 2011), using the FILT test method developed by Banholzer et al. (2006).

4.3.3 Quantification of the pullout in bundled reinforcement

The special nature of the reinforcing bundle and the nonuniform microstructure require quantitative modeling that is quite different than the conventional ones, which are solely based on analysis of interfacial stresses between the matrix and the sleeve periphery of the reinforceing unit. In the current case, there is a need also to consider the processes occuring within the reinforcing unit, that is, the bundled filaments that are not uniformly bonded.

The overview of the internal structure of the bundled yarn outlined in Section 4.3.2 suggests that it can be described in terms of layers of

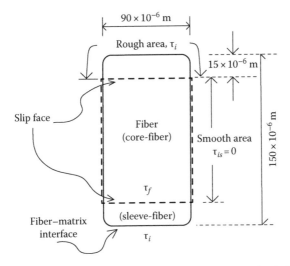

Figure 4.19 Simplified cross section of the PP fiber and simulation of its behavior in terms of a sleeve and core. (After Ohno, S. and Hannant, D.J., *ACI Mater. J.*, 91(3), 306, 1994.)

filaments with solids deposited in-between in a discontinuous mode. These solids form local connections between filaments, described as "adhesive cross links" in Figure 4.20 (Hegger et al., 2006). The density of these "adhesive cross links" decreases toward the core of the bundle, while the external sleeve filaments are continuously embedded in the cementitious matrix.

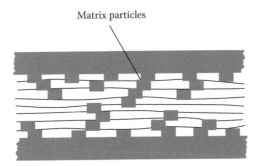

Figure 4.20 Description of the internal structure of a bundled reinforcement in cement matrix in terms of matrix particles deposited discontinuously in between the filaments, forming "adhesive cross links." (After Hegger, J. et al., *Mater. Struct.*, 39, 765, 2006.)

Models that describe the behavior of this special reinforcement assembly can be based on several assumptions that facilitate analytical or numerical treatments, with the concepts outlined as follows:

- The bundle can be described in terms of layers, each characterized by a different interfacial bond–slip relation, with the bond strength decreasing towards the core of the bundle, as shown schematically in Figure 4.21 (from Hegger et al., 2004).
- The model may be based on a simplified two-layer description of sleeve and core elements or as a multilayer structure with varying number of layers.
- The bond–slip relation in each layer can be described in terms of a smeared bond relation that is effective along the whole length of the filament, as suggested by the "bond layer model" (Hegger et al., 2004), with the bond quality decreasing towards the center, as shown in Figures 4.21 and 4.22.
- Alternately, the bonding in each layer can be described in terms of a filament that is partly engulfed with matrix particles, exhibiting at this range 100% bond quality, while the other part, is bridging over the crack, remaining free, as seen in Figure 4.23 (Banholzer, 2004). This is essentially an idealization in which all the local "adhesional cross links" of length $L_{v,i}$ are summed up together to a fully embedded zone with a length L_v, with a free length filament portion of length ψ_v.
- The quantification of the models can be based on analytical treatment or numerical simulation of a two-layer structure (Figure 4.24) or a multilayer structure (Figure 4.25).

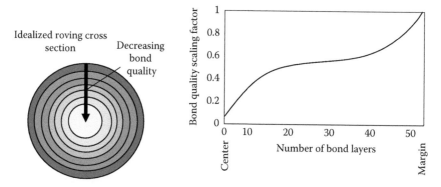

Figure 4.21 The layered structure of the bundled yarn with bond decreasing toward the center. (From Hegger, J. et al., A smeared bond-slip relation for multi-filament yarns embedded in fine concrete, in *Sixth RILEM Symposium of Fiber-Reinforced Concretes (FRC)—BEFIB*, Varenna, Italy, September 20–22, 2004.)

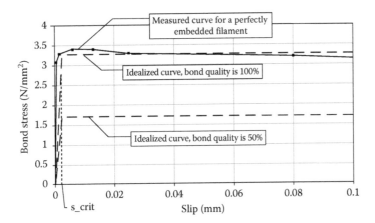

Figure 4.22 Bond–slip curves for single internal filaments with reduced bond quality of internal filaments in the bundle. (After Brameshuber, W. and Banholzer, B., Bond characteristics of filaments embedded in the fine grained concrete, in *Textile Reinforced Structures, Proceedings of the Second Colloquium on Textile Reinforced Structures (CTRS2)*, Dresden, Germany, 2003, pp. 63–76.)

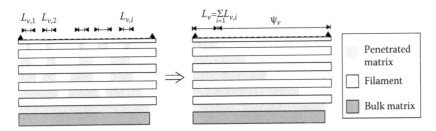

Figure 4.23 Idealization of the bonding around filaments in terms of a fully bonded portion L_v and a free portion ψ_v. (After Banholzer, B., Bond behavior of multi-filament yarn embedded in a cementitious matrix, PhD dissertation, RWTH Aachen University, Aachen, Germany, 2004.)

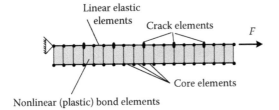

Figure 4.24 Finite element model simulation in terms of two-layer, sleeve and core bundle. (After Hegger, J. et al., A smeared bond-slip relation for multi-filament yarns embedded in fine concrete, in *Sixth RILEM Symposium of Fiber-Reinforced Concretes (FRC)—BEFIB*, Varenna, Italy, September 20–22, 2004.)

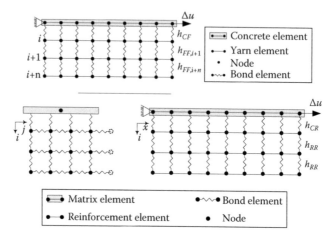

Figure 4.25 Finite element model simulation in terms of a multiple layer bundle. (After Haussler-Combe, U. and Hartig, J., *Cement Concr. Compos.*, 29, 279, 2007; Hartig, J. and Haussler-Combe, U., A model for Textile Reinforced Concrete under imposed uniaxial deformations, *CMM-2011, Computer Methods in Mechanics*, Warsaw, Poland, May 9–12, 2011.)

4.3.3.1 Sleeve and core layer modeling

4.3.3.1.1 ANALYTICAL TREATMENT

The sleeve and core two-layer quantitative analytical model was first developed by Ohno and Hannant (1994) for PP-reinforced film, which exhibited shear failure at the reinforcement–matrix interface and within the reinforcement itself (Figure 4.19).

The equilibrium equation at the reinforcement (sleeve surface) matrix in the region between 0 and x_i is as follows:

$$P_i \tau_{max} \frac{V_f}{A_f} x_i - E_m V_m \varepsilon_{fsi} = 0 \qquad (4.6)$$

where
  x_i is the transfer length
  P_i is the perimeter of the fiber
  τ_{max} is the bond stress at the sleeve–matrix interface
  V_f is the fiber volume
  A_f is the fiber cross-sectional area
  V_f/A_f are the number of fibers
  E_m is the modulus of matrix
  V_m is the matrix volume
  ε_{fsi} is the strain

The equilibrium equation at the core–sleeve, from 0 to x_i, is as follows:

$$P_i \frac{V_f}{A_f} \tau_f x_f - (1-k) E_f V_f \left(\varepsilon_{fco} - \varepsilon_{mu} \right) = 0 \tag{4.7}$$

where
   P_i is the perimeter of the core–slip surface
   τ_f is the bond stress at the core–sleeve interface
   x_f is the transfer length
   $(1-k)$ is the area fraction of the core reinforcement
   $\varepsilon_{fco}-\varepsilon_{mu}$ is the strain in the core, which is the extra strain of the bridging
      core over the crack, over the matrix ultimate strain

Upon first cracking of the matrix, the stress distribution in the matrix, sleeve and core, as presented in Figure 4.26, shows that the strain in the matrix at a distance greater than x_f from the crack is ε_{mu}, implying that another crack may form in this zone. Thus, cracking continues to develop in a mode similar to the multiple cracking according to the ACK model, until the matrix

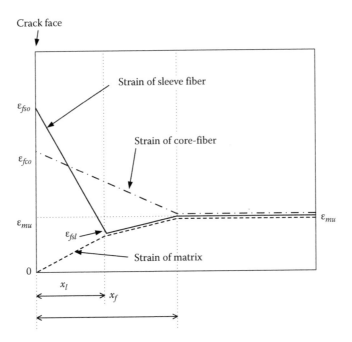

Figure 4.26 Strain distribution after first cracking. (After Ohno, S. and Hannant, D.J., ACI Mater. J., 91(3), 306, 1994.)

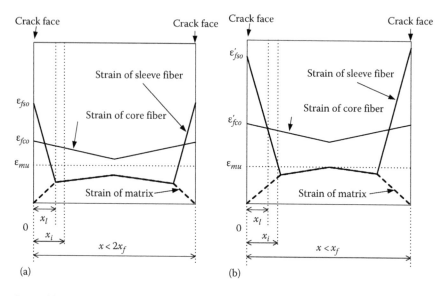

Figure 4.27 Strain distribution during multiple cracking: (a) cracks forming at spacing $x < 2x_f$ and (b) after further loading (crack spacing $x < x_f$). (After Ohno, S. and Hannant, D.J., ACI Mater. J., 91(3), 306, 1994.)

is cracked into a series of blocks in the size range of $x_f - 2x_f$, and the stress distribution is shown in Figure 4.27a. Since the strain in the matrix within blocks is smaller than the ultimate matrix strain, further loading is possible, resulting eventually in a stress distribution, as shown in Figure 4.27b, with additional matrix cracking, as the stress at the center increased to the ultimate strain. This is the result of the stress transfer into the matrix via the action of the sleeve and core and is exhibited by an increase in the transfer length from x_i to x_i' (compare (a) and (b) in Figure 4.27).

The final stress distribution after this additional cracking is presented in Figure 4.28, and the stress–strain curve of the composite is different than that of the ACK model, showing two cracking zones (Figure 4.29) labeled the first and the second cracking zones, characterized by strains ε_a and ε_b in Figure 4.29. Beyond point B on the curve, the additional load is carried by extension of the reinforcement.

The characteristic values of the stress–strain curve calculated on the basis of these considerations are as follows.

Strain at which the stress moves up after the first multiple cracking zone, ε_a:

$$\varepsilon_a = \left(1 + 0.659\Gamma\alpha\right)\varepsilon_{mu} \tag{4.8}$$

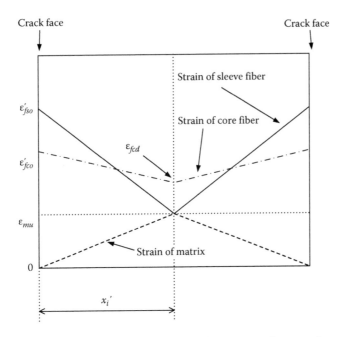

Figure 4.28 Strain distributions of fiber and matrix between cracks, at crack spacing $2x_i'$.
(After Ohno, S. and Hannant, D.J., *ACI Mater. J.*, 91 (3), 306, 1994.)

where

$$\alpha = \frac{E_m V_m}{E_f V_f}$$

$$\Gamma = \frac{\alpha\beta - \sqrt{k\beta(1+\alpha)(\alpha - \alpha\beta + k)(1-k)}}{\alpha\beta - k(1+\alpha)}$$

$$\beta = \frac{P_f \tau_f}{P_i \tau_i}$$

Strain at the end of multiple cracking, ε_b:

$$\varepsilon_b = \left(1 + 0.659\frac{\alpha(1-\beta)}{k}\right)\varepsilon_{mu} \qquad (4.9)$$

where

$$k = \frac{V_{fs}}{V_f}$$

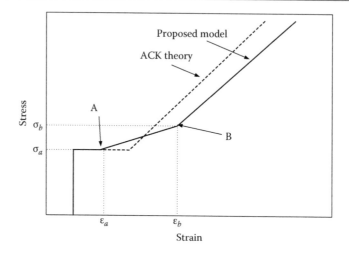

Figure 4.29 Schematic description of the stress–strain curve predicted by considering a sleeve and core reinforcement. (After Ohno, S. and Hannant, D.J., *ACI Mater. J.*, 91(3), 306, 1994.)

Stress at the end of multiple cracking, σ_b:

$$\sigma_b = E_f V_f \varepsilon_{mu} \left(1 + \Psi\alpha\right)$$

(4.10)

where

$$\Psi = \frac{(1-\beta)}{2k} + \frac{1}{2} = \frac{1-\beta+k}{2k}$$

From this model, it could be concluded that if the matrix is strong or the reinforcement is likely to slip internally, the secondary multiple cracking region, AB in Figure 4.29, expands.

When the sleeve portion fails, it can be treated as part of the matrix, that is, a reduction in the content of the reinforcement.

The model prediction was compared with experimental results for systems with fibrillated PP reinforcement having 4.8%, 6.2%, and 10.8% fiber content. The dimensions of the reinforcing yarn and the location of the internal slip plane are shown in Figure 4.19, and the shear bond values were taken as 5 and 2 MPa for the reinforcement–matrix and for the internal slip plane, respectively. Calculations at the end of the multiple cracking zone gave a reinforcement stress of 410 MPa, which is about its strength, justifying taking the sleeve portion as part of the matrix after the multiple cracking zone. Comparison of the experimental and the calculated curves based on various models and assumptions showed good agreement.

4.3.3.1.2 NUMERICAL TREATMENT

Numerical treatment of the multifilament reinforcement in terms of the two-layer, sleeve and core model was presented by Hegger et al. (2004, 2006). They considered a multilayer structure of the bundle, having a smeared bond between the filaments, with quality reducing toward the inner core filaments (bond layer model). It was suggested that a numerical simulation based on a two-layer model along the concepts of Ohno and Hannant (1994), differentiating the sleeve and core, would provide reasonable representation and facilitate a numerical solution involving a small number of parameters, as shown in Figure 4.24. The verification of the model was based on an experimental determination of the pullout curves as well as tensile test of a yarn impregnated in a matrix.

The linear elastic elements of one layer represent the uncracked concrete, and in it special elements are introduced in a smeared manner to represent cracking (Figure 4.24). The parallel chain represents the core filaments. The two chains are linked with nonlinear bond elements. The number and spacing of cracks are taken from the results of tensile test of the impregnated roving, as this test is intended to determine the bond properties of the core and not the overall behavior of the composite. Several assumptions are made:

- The slope of the linear part of the pullout curve reflects the relation between crack opening and crack bridging in the composite, which is facilitated by the sleeve element.
- Failure of the tensile specimen occurs when the force in the sleeve exceeds its maximum pullout force.
- The crack spacing is larger than the transfer length of the sleeve filament.
- The core filaments transfer load directly to the matrix.
- The smeared core–matrix bond law is bilinear elastic–plastic (Figure 4.30).
- The filament–filament bond is negligibly small.
- The matrix law follows linear elastic behavior and linear strain softening (Figure 4.31).

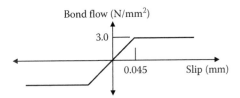

Figure 4.30 Smeared bond law matrix–core filaments. (After Hegger, J. et al., A smeared bond-slip relation for multi-filament yarns embedded in fine concrete, in *Sixth RILEM Symposium of Fiber-Reinforced Concretes (FRC)—BEFIB*, Varenna, Italy, September 20–22, 2004.)

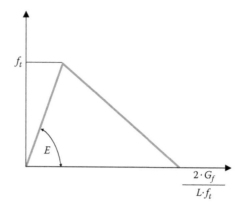

Figure 4.31 Matrix material law. (After Hegger, J. et al., A smeared bond-slip relation for multi-filament yarns embedded in fine concrete, in *Sixth RILEM Symposium of Fiber-Reinforced Concretes (FRC)—BEFIB*, Varenna, Italy, September 20–22, 2004.)

The matching between the experimental data of the tensile test and the numerical simulation was excellent. Calculations assuming a model with and without core filaments gave similar results in the elastic range and the main difference was in the postcracking range, where the reinforcement with the core filaments showed a greater strain hardening that simulated the actual experimental behavior.

The distribution of loads carried by the sleeve and core filaments is presented in Figure 4.32, showing that the core picks up the load over a distance of about 100 mm.

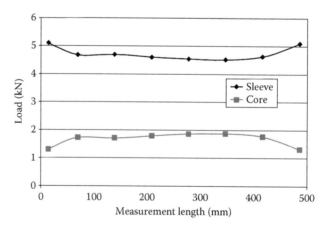

Figure 4.32 Distribution of core and sleeve load over the measurement length. (After Hegger, J. et al., A smeared bond-slip relation for multi-filament yarns embedded in fine concrete, in *Sixth RILEM Symposium of Fiber-Reinforced Concretes (FRC)—BEFIB*, Varenna, Italy, September 20–22, 2004.)

4.3.3.2 Multilayer modeling

Analysis of the bond transfer considering the multilayer nature of the bundled filaments (not just a sleeve and core) was carried out by (1) dividing each of them into bonded and free parts (Figure 4.23) (Banholzer, 2004) or by (2) smeared bonding of a multilayer model using finite element analysis (Figure 4.25) (Hartig and Haussler-Combe, 2011; Hartig et al., 2008; Haussler-Combe and Hartig, 2007).

4.3.3.2.1 INDIVIDUAL FILAMENTS WITH BONDED AND FREE ZONES

This treatment is based on the concepts shown in Figure 4.23, described in several publications (Banholzer, 2004; Banholzer and Brameshuber, 2004; Banholzer et al., 2006). Several assumptions were made with regard to the quantification of the model shown in Figure 4.23:

- The filaments carry only normal loads.
- The matrix penetrating in between the filaments carries only shear.
- The bond law between the matrix and the bonded portion in the single filaments is BSR $\tau(s)$.
- The bond law, BSR $\tau(s)$, is identical to all the filaments in the various layers of the bundle.
- The bundle structure can be idealized as a layered system made up of m layers with $o(v)$ filaments per layer (Figure 4.33).

The number of filaments of the whole bundle is $N_{F,m}$.
The number of filaments per layer:

$$o(v) = N_{F,v} - N_{F,v-1} \quad N_{F,v} \sum_{j=1}^{v} o(j) \tag{4.11}$$

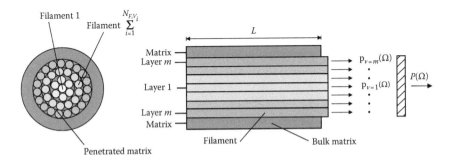

Figure 4.33 Idealized structure of the bundled filaments. (After Banholzer, B., Bond behavior of multi-filament yarn embedded in a cementitious matrix, PhD dissertation, RWTH Aachen University, Aachen, Germany, 2004.)

If the pullout (p_i)–slip (Ω) response of each filament is described by the $p_i(\Omega)$ function, then the total response of the bundle is:

$$P(\Omega) = \sum_{i=1}^{N_{F,m}} P_i(\Omega) \tag{4.12}$$

The pullout–slip response in each layer, $p_v(\Omega)$:

$$p_v(\Omega) = \frac{\sum_{i=N}^{N_{F,v}} pi(\Omega)}{o(v)} \cdot o(v) = p_{v,a}(\Omega) \cdot o(v) = \tag{4.13}$$

Hence, Equation 4.11 can be rewritten as

$$P(\Omega) = \sum_{v=1}^{m} p_v(\Omega) \tag{4.14}$$

Equation 4.12 implies that if all the layers' response is known, that is, $p_v(\Omega)$ for $1 < v < m$, then the overall pullout–slip response of the bundle, $P(\Omega)$, can be calculated.

In the bundled structure, the bond increases as one moves from the core to the sleeve, and, therefore, successive failure of filaments will occur from the external toward the internal, as shown experimentally in Figure 4.15. This can be expressed as:

$$P_v(\Omega) > p_{v-1}(\Omega) \tag{4.15}$$

Assuming that the shear response in the free portion of the filament is negligible, two conclusions can be drawn:

1. The average pullout load force p_v for each filament is dependent only on the part that is embedded in the matrix L_v.
2. The slip Ω of a layer v is a function of two components: the slip w in the bonded portion and the elongation Δ of the free filament portion:

$$\Omega = w + \Delta \tag{4.16}$$

The elongation of the free portion can be calculated by considering the elastic nature of the free filament part, and the slip w can be calculated based on the embedded length:

$$\Omega = \omega + \frac{p_V(\omega)}{E_F A_F \cdot o(V)} \Psi_V = \omega + \frac{p_{V,a}(\omega)}{E_F A_F} \Psi_V = \omega + \frac{p_V(\omega)}{E_F A_F} \Psi_V \tag{4.17}$$

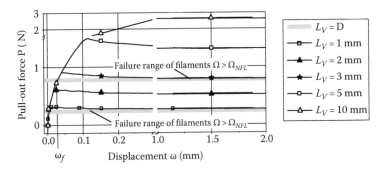

Figure 4.34 Simulated response of a filament in a single-filament pullout test with different embedded length L_v based on data of Banholzer (2004).

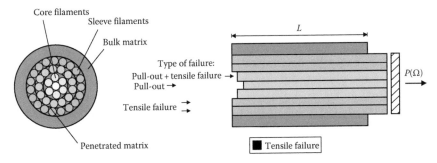

Figure 4.35 Successive failure of filaments in a bundle with the external ones having higher embedded length failing earlier and closer to the crack surface. (After Banholzer, B., Bond behavior of multi-filament yarn embedded in a cementitious matrix, PhD dissertation, RWTH Aachen University, Aachen, Germany, 2004.)

If the embedded values of L_v are known, w can be calculated based on simulated single-filament pullout tests as shown in Figure 4.34 for various embedded length values of L_v.

When the filament's embedded length is sufficiently long for the filament to reach its tensile strength, fracture of the filament will take place. In view of the increase in embedded length toward the sleeve (periphery) of the bundle, a gradual breakage of the filaments will occur upon pullout loading in successive steps from the outer filaments inward, as shown schematically in Figure 4.35. This is consistent with results of the FILT test represented in terms of reduction in the active filaments portion upon loading (Figure 4.18).

It was shown (Banholzer, 2004) that the model can be matched in stepwise calculation to simulate the reduction in the active filament–slip curve (Figure 4.36). The number of steps corresponds to the number of layers assumed.

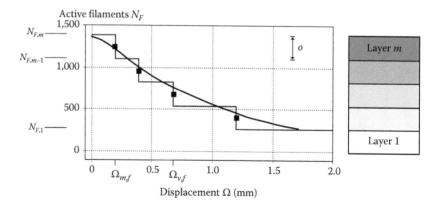

Figure 4.36 Stepwise simulation to match the active filament–slip curve. (After Banholzer, B., Bond behavior of multi-filament yarn embedded in a cementitious matrix, PhD dissertation, RWTH Aachen University, Aachen, Germany, 2004.)

On the basis of these concepts, simulations of the whole pullout load–slip curves could be determined for a system with parameters shown in Table 4.3 and pullout–slip displacement relation for a single embedded filament shown in Figure 4.37.

A similar approach for treating individual filaments by modeling them in terms of two portions each, one engulfed with matrix and bonded and the other free, bridging over a crack, was proposed also by Lepenies et al. (2007), Zastrau et al. (2008), and Richter et al. (2002), as shown in Figure 4.38a. However, instead of matching the model to observations of active filaments, they made an assumption with regard to the free length of the filaments (Figure 4.38b).

The modeling of pullout is based on the concepts illustrated in Figure 4.39, showing that as the crack is opening up under tensile loading, the external filaments break since the stress buildup in them is higher than in the core.

Table 4.3 Material properties and parameters for the model simulations in Figure 4.36

Parameter	Value
Tensile strength of filament (MPa)	1,200
Modulus of elasticity of filament (GPa)	53.3
Bond–slip function of embedded filament, BSR $\top$ (s)	Figure 4.37
Critical embedded length of filament (mm)	2.8
Number of filaments in the bundle	1,400
Number of layers	5,100
Contact perimeter (mm)	8.0

Source: After Banholzer, B., Bond behavior of multi-filament yarn embedded in a cementitious matrix, PhD dissertation, RWTH Aachen University, Aachen, Germany, 2004.

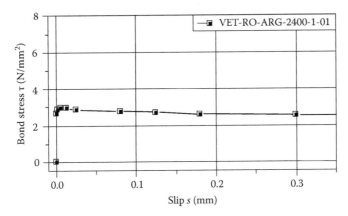

Figure 4.37 Pullout load–slip relations for modeling. (After Banholzer, B., Bond behavior of multi-filament yarn embedded in a cementitious matrix, PhD dissertation, RWTH Aachen University, Aachen, Germany, 2004.)

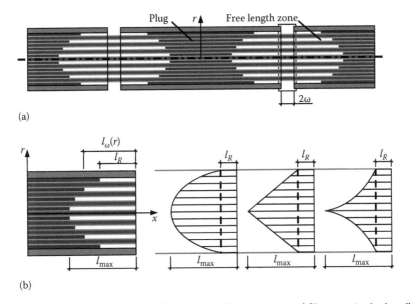

(a)

(b)

Figure 4.38 Idealized distribution of bonded and free portions of filaments in the bundle bridging the crack (a) and possible distributions of the free length portions (b). (From Lepenies, I. et al., *ACI SP*, 244-7, 109, 2007.)

This is the result of the higher rate of buildup of stress in the external filaments due to their higher bond, and therefore their strength level is approached at an earlier stage. The parameters in the free and bonded length distribution are the maximum free length in the middle core filament, l_{max}, and the free length portion of the external filament, l_R.

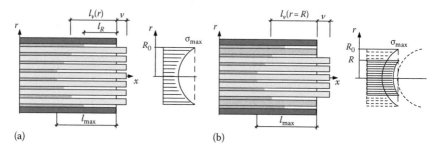

Figure 4.39 Stress distribution in the bundle: (a) before loading with no broken filaments and (b) after loading with external filaments broken. (From Lepenies, I. et al., *ACI SP*, 244-7, 109, 2007; Zastrau, Z. et al., *Technische Mechanik*, Band 28 Heft 1, 53, 2008.)

Based on these concepts, the load carried by the yarn at different stages of loading (i.e., crack opening), F(w), was calculated for a circular yarn with a radius of R and a distribution of free length of $l_w(r)$. For this calculation, it is assumed that the strain in each filament during tensile loading is half the crack width divided by the free length w/l_w, and the stress to which it is subjected, $\sigma(r)$, is equal to its modulus multiplied by the strain:

$$F(\omega) = \int \sigma(r)dA = \int_{r=0}^{R} E\frac{w}{l_w(r)}dA \tag{4.18}$$

In the special case of a roving with circular cross section and quadratic distribution of the free filament lengths along the radius, Equation 4.18 leads with

$$dA = 2\pi r dr$$

to

$$F(w) = 2\pi E w \int_{r=0}^{R} \frac{r}{l_w(r)}dr \tag{4.19}$$

$$F(w) = 2\pi E w \int_{r=0}^{R} \frac{r}{l_{max} - (\Delta l_w/R_o^2)r^2}dr \tag{4.20}$$

$$F(w) = w\frac{\pi R_o^2 E}{\Delta l_w}\ln\left\{\frac{l_{max}}{l_{max} - \Delta l_w (R/R_0)^2}\right\} \tag{4.21}$$

where

$$\Delta l_w = l_{max} - l_R$$

Introducing a failure strain of ε_F for the filament, it is possible to calculate for a given crack width of w the active portion of the bundle, namely, the portion of filaments that are not broken and can transfer stress. This portion is the residual bundle of radius R, and from that, the force that can be supported by the active filaments as a function of the residual bundle radius R, $F(R)$, can be calculated:

$$w(R) = \left\{ l_{\max} - \Delta l_w \left(\frac{R}{R_o} \right)^2 \right\} \varepsilon_F = l_R(R)\varepsilon_F \qquad (4.22)$$

$$F(R) = \pi E \frac{R_o^2}{\Delta l_w} \ln \left\{ \frac{l_{\max}}{l_R(R)} \right\} l_R(R)\varepsilon_F \qquad (4.23)$$

From a practical point of view, one would like to have a bonded strand in which the load-carrying capacity increases after the failure of the external filaments. The described model predicts that this would happen when the ratio of $l_{\max}/l_{R(Ro)} > e$. When the ratio is 1, all the filaments will fail at the same time when the crack width reaches a critical value of w_R. The results of a parametric calculation of the transition from ductile to brittle behavior for bundles of different shapes is shown in Figure 4.40, for two free length distribution profiles and for circular and increasingly elliptical-shaped bundles. For a quasiductile behavior (i.e., telescopic failure), the ratio of $l_{\max}/l_R$ should

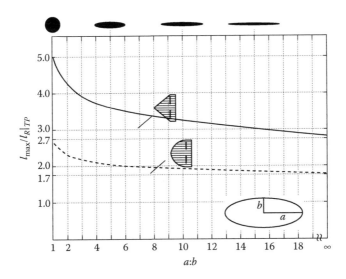

Figure 4.40 The ductile–brittle transition in the bundle as a function of its shape. (After Hartig, J. et al., *Cement Concr. Compos.*, 30, 898, 2008.)

be above the lines in Figure 4.40. The change in the shape of the bundle could be the result of the production process of the composite that can flatten the circular cross section. This becomes more marked when the number of filaments in the bundle is bigger.

4.3.3.2.2 SMEARED BOND APPROACH

Multilayer modeling with smeared bond simulation between the filaments was advanced by Haussler-Combe and Hartig (2007), Hartig et al. (2008), and Hartig and Haussler-Combe (2011), along the principles outlined in Figure 4.25.

The concepts and assumption of the treatment are as follows:

- Two-bar elements representing the matrix and the bundled reinforcement.
- Bond elements representing filament–matrix and filament–filament interactions.
- Bond elements are of zero thickness following two types of bond law, with the bond between filaments being the same for all filaments and having frictional behavior.
- The core filaments are not linked directly to the matrix but through the sleeve (different than other models where the core filaments are modeled as being directly linked to the matrix).
- Matrix modeled as a one-bar chain, assuming that shear gradients are small due to small specimen thickness and small distances between reinforcing yarns.
- Poisson transverse deformations are neglected.
- The bond surface areas, S, are also determined based on the layer model.
- Boundary conditions are given with prescribed displacements at the end nodes of the bar element chains.
- Loading can be applied in prescribed displacements, forces, and imposed strains.
- In the finite element calculation, the incremental steps make equilibrium iteration necessary at each increment.

The results of a parametric study based on a finite element calculation for matrix properties of 28.5 GPa modulus and 6.5 MPa tensile strength and reinforcement content of 1.9% by volume consisting of bundles of 800 filaments with modulus of 79.95 GPa and tensile strength of 1,357 MPa are shown in Figures 4.41 through 4.44. The parametric study included the number of segmented core layers (up to 100); the level of frictional and adhesional bond, up to 20 and 100 MPa, respectively; and core filament content up to 95% of the bundle.

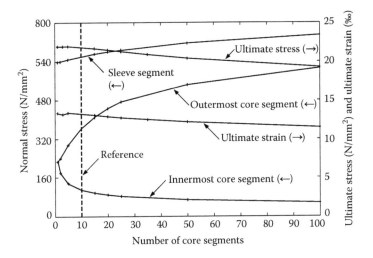

Figure 4.41 Ultimate stresses and strains and normal stress at first crack in the sleeve segment and the inner and outer segments in the core as a function of the segmentation of the core in the model (arrows in parenthesis indicate scale to be used). (After Hartig, J. et al., *Cement Concr. Compos.*, 30, 898, 2008.)

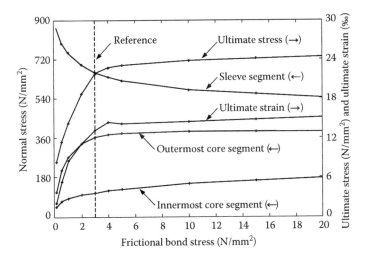

Figure 4.42 Ultimate stresses and strains, and normal stress at first crack in the sleeve segment, the inner and outer segments in the core as a function of the frictional bond stress in the model (arrows in parentheses indicate the scale to be used), in a bundle where 75% of the filaments are core filaments. (After Hartig, J. et al., *Cement Concr. Compos.*, 30, 898, 2008.)

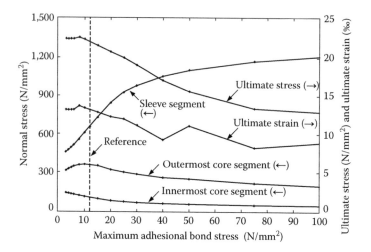

Figure 4.43 Ultimate stresses and strains and normal stress at first crack in the sleeve segment and the inner and outer segments in the core as a function of the adhesional bond stress used in the model (arrows in parentheses indicate scale to be used). (After Hartig, J. et al., *Cement Concr. Compos.*, 30, 898, 2008.)

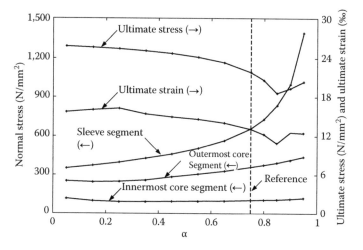

Figure 4.44 Ultimate stresses and strains and normal stress at first crack in the sleeve segment and the inner and outer segments in the core as a function of the portion of core segments in the bundle, α, used in the model (arrows in parentheses indicate scale to be used). (After Hartig, J. et al., *Cement Concr. Compos.*, 30, 898, 2008.)

The modeling of the stress–strain curve suggests that the segmentation of the core does not affect the overall shape. Yet, in the multiple cracking zone, differences could be seen, with greater segmentation resulting in several plateaus of cracking. This difference in behavior was explained in terms of differences in normal stress distributions resulting in greater activation of the length of the reinforcement in the segments between the cracks at larger segmentation. The lower slope in the post multiple cracking zones in the experimental curve has been given several explanations, one of them being gradual softening of the matrix once cracking starts.

Increase in the number of segments results in reduction in the calculated ultimate stress and strain of the composite (Figure 4.41). It has been suggested that this is the result of an earlier failure of sleeve filaments since the tensile stresses in the sleeve are higher when the modeling involves a greater number of core segments.

Increase in frictional bond is expected to increase the ultimate strength and strain (Figure 4.42). The increase is particularly high, in the range of 0–4 MPa, and then it tends to almost level off.

Increase in adhesional bond is expected, according to the model, to result in a reduction of the ultimate stress and strain when the adhesional bond is above 10 MPa, while below that these parameters are not affected by the bond (Figure 4.43). This has been explained by reaching the strength of the sleeve element at a very early stage, and the load that is distributed to the core filaments cannot be supported by them.

Modeling of the effect of the size of the core in the bundle shows a reduction in ultimate stress and strain up to a core size of about 75%, and thereafter, an increase in these parameters (Figure 4.44).

4.4 Bonding in a fabric

The bond of textile fabric to cementitious matrix involves several processes that are generated by the direct interaction between the yarn and the matrix and, in addition to that, a variety of effects associated with the special geometry of the fabric. Several types of mechanical anchoring are invoked by the transverse (fill) yarns that hold together the longitudinal (warp) yarns, as well as effects induced by the geometrical curvature of the reinforcing yarns. The outcome of these effects may be enhanced bonding, which could be much greater than that predicted by the "conventional" yarn–matrix interfacial interaction. This is demonstrated in Figure 4.45 for PP yarns in a woven fabric.

The crimped geometry of the yarn results in a much greater bond compared to the straight yarn, while the bonding of the fabric is even significantly enhanced, being the result of the contribution of the crimping of the reinforcing yarns as well as additional anchoring provided by the fill yarns in the fabric.

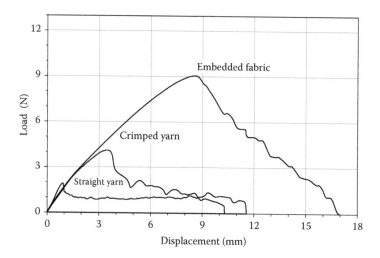

Figure 4.45 Load displacement curves in pullout of a straight yarn, a crimped yarn untied from a fabric, and the fabric, all embedded in a cement matrix. (After Peled, A. et al., *Adv. Cement Based Mater. J.*, 7(1), 1998a.)

These contributions can be materialized if the production process of the composite allows for the formation of intimate contact between all parts of the fabric and matrix. This may not always be the case since the complex geometry of the fabric and the small spaces inherent to it may not enable proper impregnation of the fabric by the particulate matrix. If this happens, the bonding may be compromised and be even lower than that obtained with longitudinal yarns. In consideration to the fabrication process, attention should be given also to the nature of the multifilament yarns, where a production process that involves high shear rates may open up the yarn and force matrix particles in between the filaments. If this happens, additional strengthening of the bond beyond the predictions outlined in the previous section may take place. A benchmark of such influence can be obtained by considering the effect of epoxy impregnation of the fabric prior to embedment in the cementitious matrix, which results in a marked enhancement in bond, as demonstrated in Figure 4.46 for epoxy impregnation of carbon fiber fabric.

4.4.1 Modeling

Two types of models are available. The first is based on the simulation of the fabric pullout response in terms of the behavior of an equivalent bond of a single fiber. The second type is based on attempts to resolve and quantify the actual mechanisms that occur in fabric bonding where the intersection between weft and fill yarns, and their nature, plays a major role in providing mechanical anchorage in addition to fiber–matrix interfacial effects.

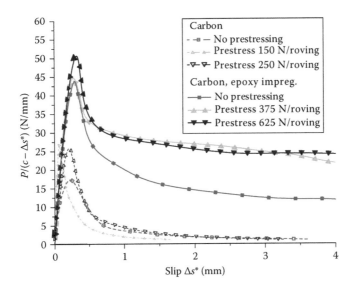

Figure 4.46 Bond stress per unit length versus slip in pullout test of carbon fabric without and with epoxy impregnation. (After Kruger, M. et al., *Otto Graff J.*, 13, 111, 2002.)

4.4.1.1 Equivalent single-fiber modeling

The shapes of the pullout curves of the fabrics are similar in nature to curves of single yarns (or fibers). Therefore, models of single fibers have been used to characterize them in terms of equivalent bond properties, as demonstrated by Sueki et al. (2007) for AR- glass-reinforced TRC produced by three different methods (casting, pultrusion, and vacuum casting), including also a matrix modification with fly ash. The best-fit equivalent pullout–slip relations and the resulting simulations are shown in Figure 4.47.

Internal shear strength parameters, τ_{max} and τ_{frc}, obtained from the best fit of experimental results using back-calculation techniques are compared with nominal shear strength τ_{nom}, which is the average strength at maximum load, obtained from experiments. The nominal shear strength lies in between τ_{max} and τ_{frc}, as shown in Figure 4.48. As the nominal shear strength increases, both the maximum and frictional shear strength values increase as well.

4.4.1.2 Pullout mechanisms in fabric–cement systems

A major contribution to bonding in fabrics is the transverse yarn anchorage mechanism, which can be shown to be made of two stages: intact and failed junction bonds (Figure 4.49). This can be modeled using a periodic arrangement of linear springs providing anchorage at the warp–fill junctions (warp and fill are the yarns parallel and perpendicular to load direction,

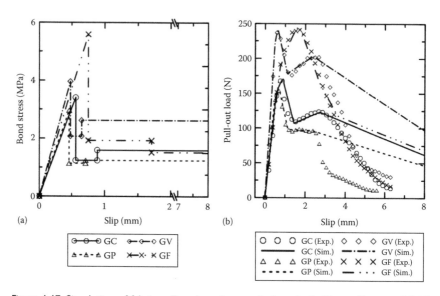

Figure 4.47 Simulation of fabric pullout based on equivalent single-fiber pullout model: (a) best-fit equivalent bond–slip relation and (b) simulation of the whole pullout curve. (After Soranakom, C. and Mobasher, B., *Mater. Struct.*, 42, 765, 2009.)

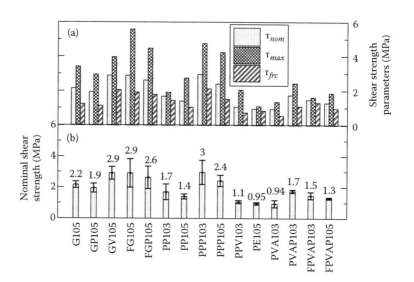

Figure 4.48 Shear strength parameters: (a) experimental nominal shear strength versus maximum and frictional shear strength from simulations and (b) nominal shear strength and associated standard deviation. (After Sueki, S. et al., *J. Mater. Civil Eng.*, 19(9), 718, 2007.)

 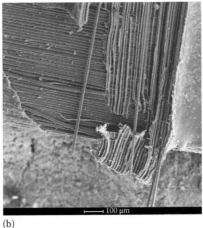

(a)   (b)

Figure 4.49 Two stages of yarn anchorage: intact (a) and a failed junction bond (b). (After Mobasher, B. et al., *Mater. Struct.*, 39, 317, 2006b.)

respectively). Anchorage at the point of intersection of yarns is attributed to the connection of the warp and fill yarns and the restraint offered by the fill yarns in redistributing the load. Anchorage may also be caused by the surface curvature of the yarn in a woven fabric.

The mechanical anchoring at the junction point can be modeled by considering the load transferred to the transverse yarns by means of bending (Mobasher et al., 2006a). The force P_b is obtained by simulating the fill yarn, which rests upon a support as a beam on elastic foundation. The load is calculated using the stiffness of the interface treated as the elastic foundation multiplied by the nodal point displacement:

$$P_b = \sum_{i=1}^{n} K_b u(x_i)$$ (4.24)

where

$$K_b = \frac{2k}{\lambda}\frac{\sinh(\lambda l) + \sin(\lambda l)}{\cosh(\lambda l) + \cos(\lambda l) - 2} \qquad \lambda = \sqrt[4]{\frac{k}{4EI}}$$

The parameter $u(x_i)$ represents the lateral displacement or slip at each of the junction points, while $k = bk_0$. k_0 is the modulus of the foundation, in N/m³, and corresponds to the compressive stiffness of the interface transition zone; b is the width of the beam in contact with the foundation; and EI is the flexural rigidity of the yarn that is treated as a beam, as shown in Figure 4.50a. Using this procedure, the pullout slip is obtained according to the debonded length and the number of active junctions redistributing the load.

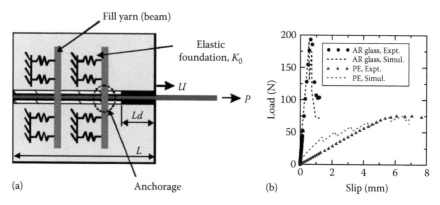

Figure 4.50 Fabric pullout model across a matrix crack (a) and the response of the fabric debonding model for AR glass and PE fabrics (b). (After Mobasher, B. et al., *Mater. Struct.*, 39, 317, 2006b.)

The simulation results were compared with the experimental results of different fabric types and matrix materials and there was good agreement. Figure 4.50b presents a simulation of the fabric pullout force–slip relationship experiments for AR glass and PE fabrics. Note that the initial response up to the peak point is well modeled using the present approach. The incremental drop in the load–response corresponds to the failure of the junction points at each increment of the connection between the warp and fill yarns, or the junctions in the cross yarn attachments.

The mechanical anchorage at the intersection is one process that should be considered in addition to the other interfacial bonding mechanisms. This approach was taken by Soranakom and Mobasher (2009). They developed a quantitative model that takes into account three unique characteristics that differentiate the bonding of a fabric from a straight yarn: (1) the multifilament nature of the yarns (discussed in the previous section); (2) the slack in the longitudinal yarns, which is the result of curvature occurring when no prestressing is applied in the production; and (3) mechanical anchoring induced by the intersection of longitudinal and transverse yarns. The latter may assume different modes of interactions depending on the nature of the junction, whether woven, bonded, or knitted. These three characteristics are shown schematically in Figure 4.51 along with the law used to quantify each of them.

The concepts presented in Figure 4.51 served as the basis for quantification using the finite difference method (Figure 4.52). The treatment quantifies the mechanics of the systems in terms of several mechanisms, each based on a specific equation, and all of these are assembled together to represent the overall pullout behavior, as shown in Figures 4.53 and 4.54 (Soranakom and Mobasher, 2009).

Parametric study was carried out to evaluate the three main features controlling fabric–cement mechanical interaction: bond–slip relation, slack in

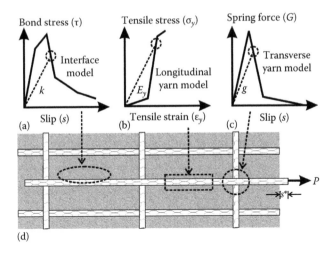

(a)
Bond stress (τ)
Interface model
k
Slip (s)

(b)
Tensile stress (σ_y)
Longitudinal yarn model
E_y
Tensile strain (ε_y)

(c)
Spring force (G)
Transverse yarn model
g
Slip (s)

(d)
P
s^*

Figure 4.51 Fabric pullout mechanism and the law describing each: (a) yarn–matrix bond–slip model, (b) longitudinal yarn model demonstrating initial slack, (c) spring model simulating anchorage at longitudinal and transverse junction, and (d) pullout test scheme of a fabric from the matrix. (After Soranakom, C. and Mobasher, B., *Mater. Struct.*, 42, 765, 2009.)

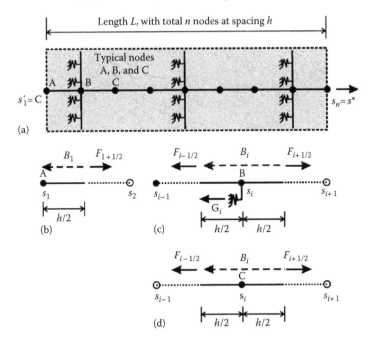

(a)
Length L, with total n nodes at spacing h
Typical nodes A, B, and C
$s_1' = C$
$s_n = s^*$

(b)
B_1 $F_{1+1/2}$
A
s_1 s_2
$h/2$

(c)
$F_{i-1/2}$ B_i $F_{i+1/2}$
B
s_{i-1} s_i s_{i+1}
G_i
$h/2$ $h/2$

(d)
$F_{i-1/2}$ B_i $F_{i+1/2}$
C
s_{i-1} s_i s_{i+1}
$h/2$ $h/2$

Figure 4.52 Finite difference fabric pullout model: (a) discretized fabric pullout model under displacement control; (b–d) free body diagrams of three typical nodes, "A," "B," and "C." (After Soranakom, C. and Mobasher, B., *Mater. Struct.*, 42, 765, 2009.)

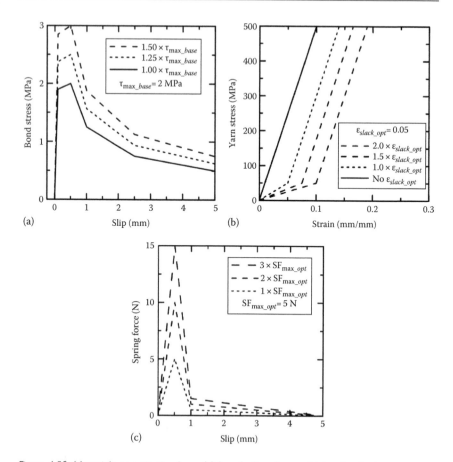

Figure 4.53 Materials constitutive laws: (a) bond–slip of yarn, (b) longitudinal yarn stress–strain showing various degrees of slack, and (c) spring force–slip at junctions. (After Soranakom, C. and Mobasher, B., *Mater. Struct.*, 42, 765, 2009.)

yarn and anchorage at junctions, as shown in Figure 4.51 for the constitutive laws and in Figure 4.54 for the outcome of the parametric study.

The model was applied to study the pullout of AR glass fabric that was previously analyzed by an equivalent single-fiber model (Figure 4.47). The calibrated materials parameters for the finite difference calculation are presented in Figure 4.55, and the simulation results are shown in Figure 4.56. A good match is observed, which is better than the one obtained with the equivalent single-fiber model (Figure 4.47).

The mechanical anchoring component in the textile reinforcement has been treated by Kruger et al. (2003) with an analogy to steel reinforcement in concrete, where anchoring is also an important contributor to bond. They analyzed the bonding of fabrics using models developed for deformed reinforcing bars and

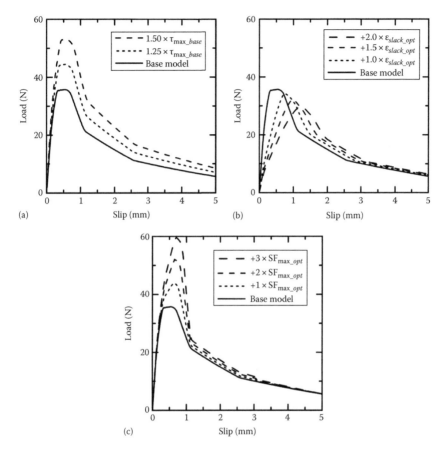

Figure 4.54 Result of parametric studies of materials models for pull-out and slip response: (a) effect of bond strength, (b) effect of slack, and (c) effect of spring strength. (After Soranakom, C. and Mobasher, B., *Mater. Struct.*, 42, 765, 2009.)

considered the effect of mechanical interaction τ_m, the frictional component τ_f, and the influence of the 3D stress field on these values. This effect could be represented by a parameter Ω, by which the characteristic values of τ_m and τ_f should be multiplied. The parameter Ω is a function of three contributions:

$$\Omega = \Omega_s \Omega_c \Omega_{cyc} \tag{4.25}$$

where

    Ω_s is the influence of the yielding of the reinforcement; in the case of fabric $\Omega_s = 1$

    Ω_c is the influence of lateral stresses induced by stresses in the concrete and in the reinforcement

    Ω_{cyc} is the influence of cyclic loading on bond

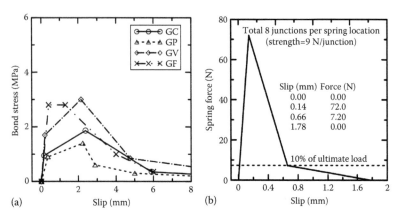

Figure 4.55 Calibrated material parameters for AR glass fabric–reinforced cement: (a) best-fit bond-slip model along the yarns and (b) best-fit spring model at the junctions in the fabric. (After Soranakom, C. and Mobasher, B., *Mater. Struct.*, 42, 765, 2009.)

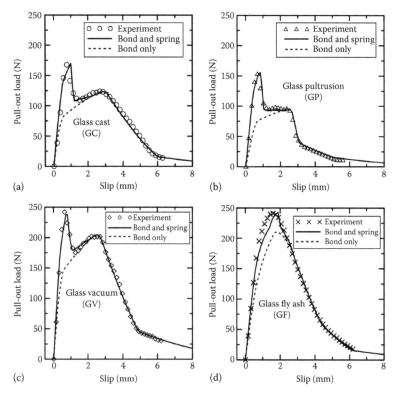

Figure 4.56 Simulation of pullout response of AR glass fabric: (a) cast sample, (b) pultrusion sample, (c) vacuum-processed sample, and (d) mix with fly ash. (After Soranakom, C. and Mobasher, B., *Mater. Struct.*, 42, 765, 2009.)

The main parameter to consider is Ω_c, and this, in turn, is a function of the radial stresses in the matrix and the strains in the reinforcement:

$$\Omega_c = 1.0 + \tanh\left[\alpha_r \cdot \frac{\overline{\sigma_R}}{0.1 \cdot f_c} - \alpha_f \mu_f \left(\varepsilon_f - \varepsilon_{p,0}\right) \frac{1}{1 - \frac{r_f^2}{\left(r_f + h_f\right)^2}}\right] \quad (4.26)$$

where

α_r is the factor controlling the influence of the radial concrete stress (set for 1 in the calculation in Kruger et al., 2003)

$\overline{\sigma_R}$ is the average radial stress in the concrete in the vicinity of the reinforcement

f_c is the uniaxial compressive strength of the concrete

μ_f is Poisson's ratio of the reinforcement

ε_f is the reinforcement strain

$\varepsilon_{p,0}$ is the strain due to the prestressing of the reinforcement

r_f is the radius of reinforcement

h_f is the constant representing the surface roughness of the reinforcement

α_f is a parameter controlling the influence of the roughness of the reinforcement, h_f

Ω_c can theoretically vary between 0 and 2

In an analysis of test results of the bonding of fabrics from different materials (glass, carbon, aramid), they concluded that in a reinforcement with a rough surface, the bonding is mainly influenced by the radial stresses in the surrounding cementitious matrix, while in a reinforcement that is smoother, the influence of the reinforcement becomes greater. They also quoted reports that the impregnation of the fabric with epoxy results in a higher bond and suggested that this may be due to two effects: (1) the ribbed surface formed by the binding yarns in the warp-knitted fabric, which is fixed by the epoxy, and (2) the change in the roving diameter over its length.

4.4.2 Mechanical anchoring induced by fabric geometry

In view of the importance of mechanical anchoring on the bond behavior, studies were conducted to evaluate the influence of fabric structure on the bond behavior and resolve the influence of production process and composition of the matrix. Some of these tests were based on direct evaluation of bond in pullout tests, while others determined the influence of these parameters by studying the efficiency of the reinforcement by flexural and tensile testing of the actual composites.

In fabrics where the reinforcing yarns are straight, anchoring is induced by the interaction at junctions between the weft and fill yarns. In the case

Figure 4.57 Crimped nature of PE yarn in a cement matrix reinforced with woven fabric. (After Peled, A. et al., *Adv. Cement Based Mater. J.*, 7(1), 1998a; Bentur, A. and Peled, A., Cementitious composites reinforced with textile fabrics, in H.W. Reinhardt and A.E. Naaman (eds.), *Workshop on High Performance Fiber Reinforced Cement Composites (RILEM) HPFRCC 3*, Mainz, Germany, May 16–19, 1999, pp. 31–40.)

of woven fabric, additional anchoring might be induced by the shape of the yarn that is forced to be undulating (Peled et al., 1998a,b) as seen in Figure 4.57.

The influence of the crimping on the bond of individual yarns is clearly seen in Figure 4.58, which demonstrates a drastic increase in the strength of the bond and its frictional response after the peak load.

In a study to resolve the geometric parameters that control the enhancement of bond, it was shown that amplitude of the crimp is an important parameter which shows up by a linear relation between the pullout strength and the amplitude (Figure 4.59, Equation 4.27) (Peled and Bentur, 2003a,b):

$$P_{\max} = \left(8Amp + 0.18\right)\frac{1}{\lambda} \tag{4.27}$$

When considering the pullout of the whole woven fabric, where anchoring is induced by the crimping as well as the junctions of the fill and weft yarns, it was resolved that the contribution of the crimping is the more dominant one (Figure 4.60).

The implications of such changes in bond are particularly important when using fabrics made of low-modulus polymeric yarns. The inherent interfacial bond in low-modulus yarns is rather small, and their reinforcing

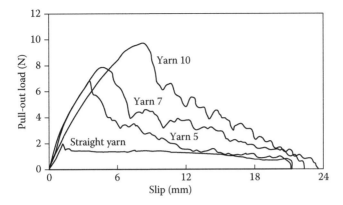

Figure 4.58 Effect of crimping in PE yarn on its pullout–slip curves. The numerical value in the curve denotes the number of crimps per centimeter. (After Peled, A. and Bentur, A., *Composites Part A*, 34, 107, 2003a; Peled, A. and Bentur, A., *J. Mater. Civil Eng.*, 15, 537, 2003b.)

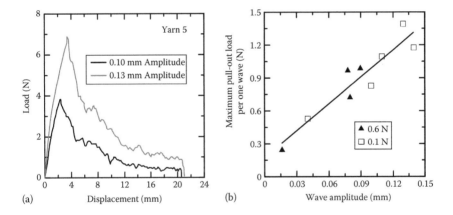

Figure 4.59 Effect of wave amplitude in crimped yarn: (a) pullout–slip curves for yarns of similar wave length but different in amplitude and (b) relation between maximum load per crimp and the crimp (wave) amplitude. (After Peled, A. and Bentur, A., *Composites Part A*, 34, 107, 2003a.)

efficiency is therefore low, resulting usually in strain-softening composites (Figure 4.61). The use of fabrics that due to their geometry provide significant anchoring is the most effective way to obtain a strain-hardening composite with low-modulus yarns. This is demonstrated in Figure 4.61 by load–deflection curves in flexure and in Table 4.4 by calculation of the efficiency of the yarns. Woven and short-weft knit fabrics have geometries that lead to anchoring.

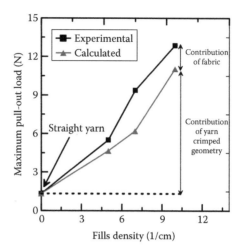

Figure 4.60 Contribution of the crimping and the fabric junction anchoring to the bonding between a cementitious matrix and a woven PE fabric with changing fill density. (After Peled, A. and Bentur, A., *Composites Part A*, 34, 107, 2003a.)

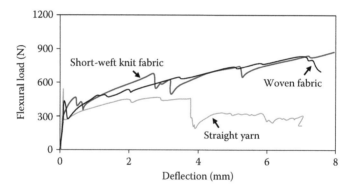

Figure 4.61 Effect of the structure of the fabric on the flexural load–deflection curve of low-modulus PE showing strain-hardening behavior when fabrics that enhance the bond through anchoring (woven and short-weft knit) are being used. (After Peled, A. and Bentur, A., *Composites Part A*, 34, 107, 2003a.)

4.4.3 Coupling of fabric structure and production process

In the previous section, it was demonstrated that the geometry of the fabric can be favorable for enhancing the contribution of the yarns by inducing anchoring effects. However, this is not always the case, and the efficiency of the yarns in the fabric might be reduced when incorporated in the matrix, relative to straight yarns, as seen in the example in Figure 4.62 (Peled and Bentur, 1999, 2000, 2003).

Table 4.4 Relative flexural efficiency factor (efficiency of yarn in a fabric relative to its efficiency when laid as single yarns for reinforcement of the same matrix)

Woven, low-modulus PE		Knitted short weft	Knitted weft insertion	
5 yarns/cm	7 yarns/cm	Low-modulus PE	Low-modulus PP	Low-density PE
1.36	1.64	4.15	0.70	0.74

Source: Bentur, A. and Peled, A., Cementitious composites reinforced with textile fabrics, in H.W. Reinhardt and A.E. Naaman (eds.), *Workshop on High Performance Fiber Reinforced Cement Composites (RILEM) HPFRCC 3*, Mainz, Germany, May 16–19, 1999, pp. 31–40; Peled, A. and Bentur, A., *Cement Concr. Res.*, 30, 781, 2000; Peled, A. and Bentur, A., *Composites Part A*, 34, 107, 2003a; Peled, A. and Bentur, A., *J. Mater. Civil Eng.*, 15, 537, 2003b.

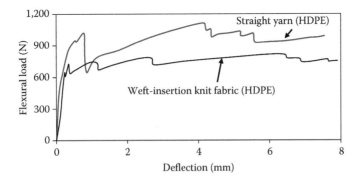

Figure 4.62 Load–deflection curves in flexure of cement matrix reinforced with high density polyethylene (HDPE) yarns in the form of straight yarns and a weft-insertion knit fabric. (After Peled, A. and Bentur, A., *Composites Part A*, 34, 107, 2003a.)

Effects of this kind are the result of the combined influence of the fabric geometry and the production process. The combination between the two can lead to situations where the impregnation of the fabric is compromised, leading to reduced bond, especially at the weft–fill intersections, and thus rather than serving as anchoring points, they become weak links in the bonding. This is the case in Figure 4.62.

The nature of the yarns when in a multifilament form can also have an influence, as seen in Figure 4.63. When the multifilament yarn is in a fabric, the junctions keep the bundle tight and compromise the penetration of the particulate matrix (Figure 4.63a). As a result, the bond in the fabric is smaller than that obtained in the straight yarn itself (Figure 4.63b).

Along this line of reasoning, production processes that involve high shear rates during the processing may be beneficial, as they enable more effective contact with the reinforcement. An example is the pultrusion process, which can more effectively force the particulate matrix to penetrate in between the filaments in the bundle by opening it up. This clearly shows up in the microstructural features in Figure 4.64, where casting and pultrusion techniques

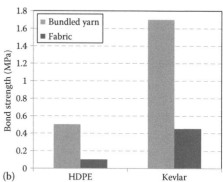

(a)   (b)   HDPE   Kevlar

Figure 4.63 The effect of stitches on clumping together the filaments in a fabric (a), and the resulting bond strength influences between individual multifilament yarns embedded in a cementitious matrix compared with multifilament yarns in a fabric form (b). (Plotted from data of Peled, A. et al., Pultrusion versus casting processes for the production of fabric-cement composites, in M. Di Prisco, R. Felicetti, and G.A. Plizzari, (eds.), *Fibre Reinforced Concrete—BEFIB 2004, Proceedings, RILEM Symposium,* PRO 39, RILEM, 2004, pp. 1495–1504.)

(a)   (b)

Figure 4.64 Microstructure of TRC with multifilament PP knitted fabric, produced by casting (a) and pultrusion (b). (After Peled, A. et al., Pultrusion versus casting processes for the production of fabric-cement composites, in M. Di Prisco, R. Felicetti, and G.A. Plizzari, (eds.), *Fibre Reinforced Concrete—BEFIB 2004, Proceedings, RILEM Symposium,* PRO 39, RILEM, 2004, pp. 1495–1504.)

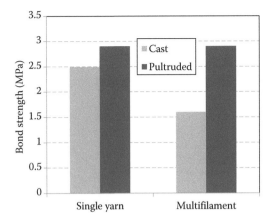

Figure 4.65 The effect of the production process on the bond in single-filament and multi-filament fabric. (Plotted from data of Peled, A. et al., Pultrusion versus casting processes for the production of fabric-cement composites, in M. Di Prisco, R. Felicetti, and G.A. Plizzari, (eds.), *Fibre Reinforced Concrete—BEFIB 2004, Proceedings, RILEM Symposium,* PRO 39, RILEM, 2004, pp. 1495–1504.)

are observed for bundled PP yarns in a fabric. The pultrusion process results in much more effective penetration of matrix particles into the weft intersections, as well as more opening up and penetration in between the filaments (Peled et al., 2004).

These microstructural differences show up in the bond values determined in mono- and multifilament fabric reinforcements produced by casting and pultrusion technologies (Figure 4.65). The production method had only a small influence in the case of the monofilament and a drastic influence in the case of the multifilament fabric, showing much higher values for the pultrusion process.

These characteristics are reflected in the tensile properties of the composite (Figure 4.66). The casting technology produces a composite with lower tensile strength compared to pultrusion, and this can be attributed to the greater penetration of the matrix in the bundle in the pultrusion process, consistent with the bond data in Figure 4.67 and the microstructural observations in Figure 4.64. When monofilament is being used as the reinforcement, the difference in the properties between the two production methods is small (Figure 4.66a). The slightly higher values in the pultrusion process might be attributed to some small enhancement in bond due to the roughening of the surface caused by the high shear rates in the pultrusion process (Peled et al., 2004).

The enhancement in bond and tensile behavior using the pultrusion process was observed in PP and PE but not in glass (Figure 4.67) (Peled et al., 2004; Sueki et al., 2007). The small reduction in the bond in the case

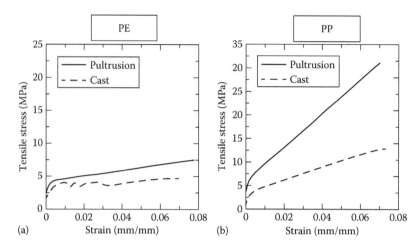

Figure 4.66 Effect of the production process on the tensile stress–strain curve of cementitious composite reinforced with fabrics made of monofilament reinforcement of PE (a) and multifilament reinforcement of PP (b). (After Peled, A. et al., Pultrusion versus casting processes for the production of fabric-cement composites, in M. Di Prisco, R. Felicetti, and G.A. Plizzari (eds.), *Fibre Reinforced Concrete—BEFIB 2004, Proceedings, RILEM Symposium*, PRO 39, RILEM, 2004, pp. 1495–1504.)

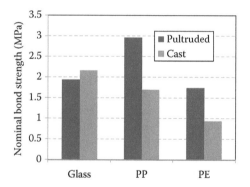

Figure 4.67 Nominal bond values obtained in TRC with bundled reinforcement of glass, PP, and PE under varying production processes. (Plotted from data of Peled, A. et al., Pultrusion versus casting processes for the production of fabric-cement composites, in M. Di Prisco, R. Felicetti, and G.A. Plizzari, (eds.), *Fibre Reinforced Concrete—BEFIB 2004, Proceedings, RILEM Symposium*, PRO 39, RILEM, 2004, pp. 1495–1504 and Sueki, S. et al., *J. Mater. Civil Eng.*, 19(9), 718, 2007.)

of glass might be attributed to damage caused to these filaments during the production, since they are much more brittle than PE and PP and may be more sensitive to abrasion of the surface due to the high shear stresses induced during the pultrusion process. It should be noted that the pullout–slip curve of the pultruded glass and PP fabrics are quite different in nature

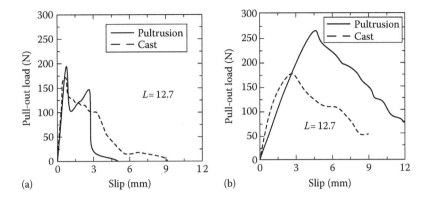

Figure 4.68 Pullout–slip response curves of bundled glass (a) and PP fabrics produced by casting and pultrusion (b). (After Peled, A. et al., Pultrusion versus casting processes for the production of fabric-cement composites, in M. Di Prisco, R. Felicetti, and G.A. Plizzari (eds.), *Fibre Reinforced Concrete—BEFIB 2004, Proceedings, RILEM Symposium*, PRO 39, RILEM, 2004, pp. 1495–1504.)

(Figure 4.68), with the glass showing a double peak (Peled et al., 2004, 2006; Sueki et al., 2007).

Soranakom and Mobasher (2009) modeled pullout responses of fabrics and suggested that the first peak in the glass curve is due to spring failure, while the second one is contributed by the interfacial bond, as seen in the results of the numerical simulation in Figure 4.56.

When considering the structure of the fabric in conjunction with the production process, the density characteristics, such as the density of junctions and the resulting spacings between yarns, should be taken into account. Figure 4.69 shows such spaces that were effectively impregnated with the matrix. In a casting production process, increase in density can have a detrimental effect by compromising the penetration of the matrix into the spaces in the fabric, of the kind shown in Figure 4.64a, resulting in influence such as the one shown in Figure 4.70.

The trend in Figure 4.70 is consistent with microscopical observations showing better penetration of the matrix into the spaces in the bundle in the larger hoop spacing.

There are cases where opposing influences take place, when greater density of the fill yarns in woven fabric are favorable for enhancing anchoring due to crimping. Such influences are shown in Figure 4.71, in which optimal density is observed for woven fabric and continuously reduced efficiency with an increase in density in the case of knitted weft-insertion fabric.

Additional consideration of the influence of fabric density is the structure of the yarns in the fabric with regard to the ease of penetration of the matrix into the bundled reinforcement. Increase in the bundle size, as quantified by the number of filaments per yarn (reflected in the tex value) results in less

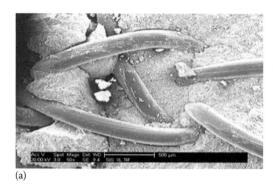
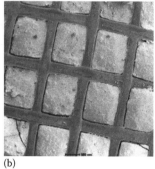

(a)   (b)

Figure 4.69 Spaces that may be of limited accessibility for the impregnation of cementitious matrix in knitted junctions (a) and openings in bonded fabric (b). (After a: From Haim, E. and Peled, A., *ACI Mater. J.*, 103(8), 235, 2011; b: From Peled, A. and Mobasher, B., *ACI Mater. J.*, 102(1), 15, 2005.)

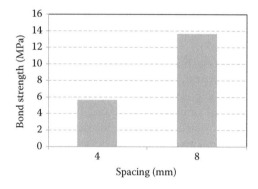

Figure 4.70 Effect of the spacing in the fabric on the bond strength. (Plotted from data of Peled, A. et al., *Cement Concr. Compos.*, 30, 174, 2008.)

efficient penetration of the particulate matrix and a lower bond (Figure 4.72), leading to lower efficiency for the reinforcement of the composite.

When considering the bundle size with respect to its penetrability, consideration should also be given to the pillar stitches around the bundled yarn, as they keep the filaments together and resist penetration of the particulate matrix (Figure 4.73). Dolatabadi et al. (2010, 2011) studied the effect of stitch length (2.1, 4.2, and 6.3 mm) and the stitch tension (tension levels A, B, C representing respectively minimum, normal, and maximum tensions) and determined the packing density of the bundle and the bond to the cementitious matrix.

The effect of the stitch geometrical and tension parameters on the packing density of the bundle is shown in Figure 4.74. As expected, the higher

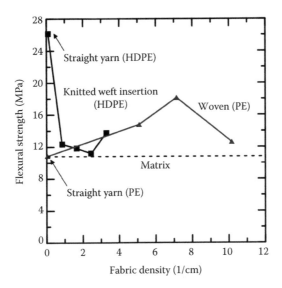

Figure 4.71 The effect of the density of the fill yarns on the flexural performance of TRC with woven and knitted weft-insertion fabric. (After Peled, A. and Bentur, A., *Composites Part A*, 34, 107, 2003a.)

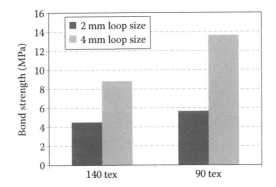

Figure 4.72 The effect of bundle size on the bond strength for HDPE fabric consisting of 90 and 140 tex yarns (200 and 310 filaments per yarn) for 2 and 4 mm loop sizes. (Plotted from data of Peled, A. et al., *Cement Concr. Compos.*, 30, 174, 2008.)

tension and the smaller densities tend to increase the packing. The effect of tension in the 6.3 mm stitch length does not follow the trend of higher packing for higher tension, and this was explained by the higher tension causing wavy shape when the stitch length is big.

If packing density, with regard to penetration of the particulate matrix, was the major consideration, it would have been expected that the bond strength would be higher for the lower packing densities. However, the

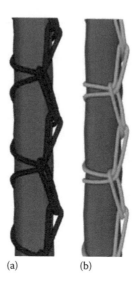

(a)   (b)

Figure 4.73 Simulated pillar stitch around a multifilament bundle: (a) close stitch and (b) open stitch. (After Dolatabadi, M.K. et al., *Mater. Struct.*, 44, 245, 2011.)

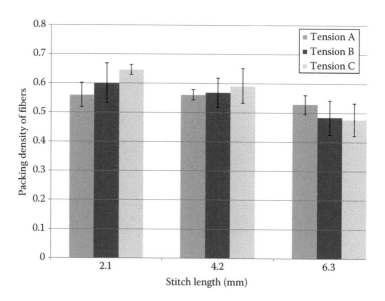

Figure 4.74 The effect of tension and stitch length on packing density. (After Dolatabadi, N. et al., An analytical investigation of cement penetration within bundle of fibers, in *International RILEM Conference on Material Science*, Aachen, Germany, Vol. 1, 2010, pp. 69–78.)

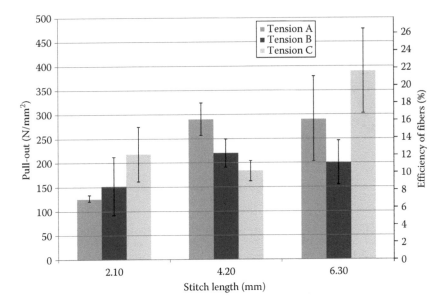

Figure 4.75 The effect of tension and stitch length on bond strength. (After Dolatabadi, N. et al., An analytical investigation of cement penetration within bundle of fibers, in *International RILEM Conference on Material Science,* Aachen, Germany, Vol. 1, 2010, pp. 69–78.)

experimental results do not show this trend (Figure 4.75). Dolatabadi et al. (2010, 2011) suggested that the interpretation of the results is complex, and in addition to the penetration of matrix into the bundle, there is a need to consider the contribution of the friction between the filaments to the bond. This contribution is expected to be higher when the packing density is high.

4.5 Treatments to enhance bond

The limited efficiency of bundled reinforcement has led to attempts to enhance it by inducing more intimate contact with the individual filament in the bundle. The more conventional approaches were based on modification of the rheological properties of the matrix by incorporation of fly ash (Mobasher et al., 2004) or some other means (Mu et al., 2002), showing considerable positive influences such as the one shown in Figure 4.76.

An additional means for treatment to force more intimate contact between the matrix and the reinforcement was based on applying pressure to the composite in its fresh state after the pultrusion production step. Such a treatment gave positive influences of enhancing the strength, which was accompanied by reduction in ultimate strain (Mobasher et al., 2006).

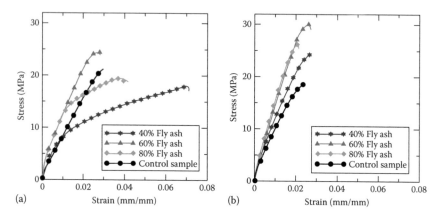

Figure 4.76 Effect of fly ash on the stress–strain curves of glass TRC after 7 and 28 days of curing. (After Mobasher, B. et al., Pultrusion of fabric reinforced high fly ash blended cement composites, in *Sixth RILEM Symposium on Fibre-Reinforced Concrete (FRC), BEFIB 2004*, Varenna, Italy, September 20–22, 2004.)

Different approaches for bond enhancement were taken by modification of the yarns themselves, by impregnation with epoxy, or by infiltration with nanoparticles to modify the internal bonding within the bundle. The rationale for these treatments is the fact that the inner core filaments in the bundle are not impregnated by the particulate matrix, thus reducing the bond efficiency at an early age and leading to durability problems when hydration products gradually deposit in the bundle over time, especially in the case of glass yarns.

Xu et al. (2004) and Kruger et al. (2002) explored the technology of impregnating the fabric bundles with epoxy and prestressing the reinforcing fabric. The bond enhancement due to impregnation can be quite significant (Figure 4.46), enhancing the mechanical anchoring as well as the frictional resistance. This can be the result of the ribbed surface of the impregnated bundled yarns and the change of its diameter over its length, especially at the crossing points of the knitted junctions. Prestressing increased the bond even further due to the Poisson effect and the straightening effect induced by the prestress.

The behavior of such systems was modeled using the bond law shown in Figure 4.77a. τ_{mf} is the bond strength (adhesive), τ_f is the frictional bond, and the difference between the two, τ_m, is called the mechanical bond resistance. This bond law was used for numerical analysis in which the mortar matrix elements are connected to the reinforcement by the FE truss shown in Figure 4.77.

Experimental results were used to determine the parameters of the bond law in Figure 4.77a. Table 4.5 shows the enhancement achieved by

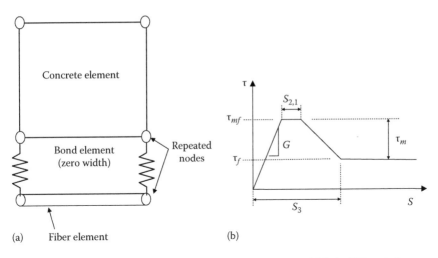

Figure 4.77 (a) Bond law for the impregnated textile bonding and (b) the FE bond element. (After Xu, S. et al., *J. Mater. Civil Eng.*, 16(4), 356, 2004.)

Table 4.5 Bonding properties used for modeling the behavior of plain and epoxy-impregnated AR glass TRC

Material properties	Plain AR glass	Epoxy-impregnated AR glass	
	ARG	ARGE	ARGE prestressed
Prestress (N/roving)	0	0	300
Young's modulus (GPa)	74.0	74.0	74.0
Tensile strength (MPa)	1,600	1,600	1,600
Bond strength, τ_{mf} (MPa)	6.05	16.60	17.5
Mechanical bond strength, τ_m (MPa)	5.70	10.80	12.0
Frictional bond strength, τ_f (MPa)	0.35	5.80	5.50
Shear stiffness, G (MPa/mm)	400	400	400
Slip, $S_{2,1}$ (mm)	0.07	0.12	0.12
Slip, S_3 (mm)	0.65	1.80	1.80
Strain value for applied pre-stress	0.0000	0.0000	0.0038

Source: Xu, S. et al., *J. Mater. Civil Eng.*, 16(4), 356, 2004.

the impregnation, and Figure 4.78 presents the effect of prestressing of impregnated fabric on the bond stress–slip relation. Similar treatments were applied for different types of fabrics made of glass, carbon, and aramid for achieving enhanced mechanical performance. The treatments were especially effective for carbon reinforcement (Reinhardt and Krueger, 2004).

In considering the prestressing of nonimpregnated fabrics, attention should be given to Poisson effects that are induced upon the release of the

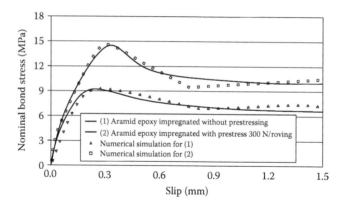

Figure 4.78 Nominal bond stress–slip relations derived in numerical studies of epoxy-impregnated aramid textiles, compared with those measured from pullout tests. (After Xu, S. et al., *J. Mater. Civil Eng.*, 16(4), 356, 2004.)

tension and the expansion of the yarn against the hardened matrix (Peled, 2007; Peled et al., 1998a).

For high-modulus Kevlar yarn, the release after one day resulted in a reduction in bonding, while seven days' tension before release led to increased bond. In this case, there is a significant effect of Poisson expansion, which takes place against a weak matrix at one day, disrupting the interfacial transition zone. However, at seven days, the transition zone is sufficiently strong and normal stresses that enhance frictional bond develop. This explains bond strength values of 0.87 and 3.55 MPa after one and seven days tension before release, respectively (Peled, 2007). In the case of low-modulus yarns such as PE and PP, an effect of this kind will take place when the prestress is relatively high, 4.8 MPa, but not when it is at a low level, for example, of 0.83 MPa (Figure 4.79). Poisson effect and stress relaxation aspect were considered to account for these behaviors.

Impregnation of the strand with nanoparticles was proposed by Bentur et al. (2010) as a means for enhancing the interfilament bonding by frictional effects, to improve the efficiency of the reinforcement. At the same time, such a treatment may prevent the deposition of hydration products, which tend to develop over time a very intimate contact between the filaments, leading to embrittlement of composites made of AR glass (for additional discussion of this topic, see Chapter 9). The impregnation into the yarns was accomplished by taking the multifilament yarns through a slurry in which these nanoparticles were dispersed, with the slurry being absorbed into the bundle. The impregnated bundle exhibited modification in the pullout behavior that shows mainly in much higher residual bond after aging (see Chapter 9), reflecting frictional behavior.

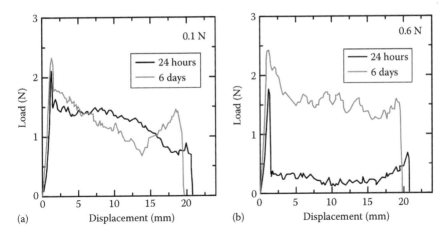

Figure 4.79 Load–slip curves of low-modulus PE, pre-tensioned and released at different sequences: (a) pre-tension at 0.6 MPa and release after 1 and 6 days and (b) pre-tension at 4.8 MPa and release after 1 and 6 days. (After Peled, A. et al., *Cement Concr. Compos.*, 20(4), 319, 1998b.)

The change in the nature in bonding was reflected much more drastically in the overall behavior of the composite, showing a reduction in the stiffness and a marked enhancement in the postpeak load-carrying capacity (Figure 4.80), which might be attributed to induced frictional resistance between the filaments (Cohen and Peled, 2010).

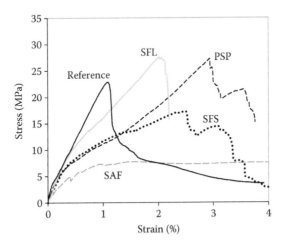

Figure 4.80 Stress–strain curves of AR glass TRC with impregnation treatments of the multifilament bundles by nanoparticles. (After Cohen, Z. and Peled, A., *Cement. Res.*, 40, 1495–1506, 2010.)

References

Ackers, S. A. S. and Garrett, G. G., Fiber-matrix interface effects in asbestos-cement composites, *Journal of Materials Science*, 18, 2200–2208, 1983a.

Ackers, S. A. S. and Garrett, G. G., Observations and predictions of fracture in asbestos cement composites, *Journal of Materials Science*, 18, 2209–2214, 1983b.

Allen, H. G., The purpose and methods of fibre reinforcement, in *Proceedings of the International Building Exhibition Conference on Prospects of Fibre Reinforced Construction Materials*, London, England, Building Research Station, 1971, pp. 3–14.

Aveston, A., Cooper, G. A., and Kelly, A., Single and multiple fracture, in National Physical Laboratories, *The Properties of Fibre Composites, Conference Proceedings*, London, U.K., IPC Science and Technology Press, England, 1971, pp. 15–24.

Banholzer, B., Bond behavior of multi-filament yarn embedded in a cementitious matrix, PhD dissertation, RWTH Aachen University, Aachen, Germany, 2004.

Banholzer, B. and Brameshuber, W., Tailoring of AR glass filament/cement based matrix-bond—Analytical and experimental techniques, in *Sixth RILEM Symposium on Fiber-Reinforced Concretes (FRC)—BRFIB*, Varenna, Italy, September 20–22, 2004.

Banholzer, B., Brameshuber, W., and Jung, W., Analytical simulation of pull-out tests—The direct problem, *Cement and Concrete Composites*, 27, 93–101, 2005.

Banholzer, B., Brockmann, T., and Brameshuber, W., Material and bonding characteristics for dimensioning and modeling of textile reinforced concrete (TRC) elements, *Materials and Structures*, 39, 749–763, 2006.

Bartos, P., Brittle matrix composites reinforced with bundles of fibers, in J. C. Maso (ed.), *From Materials Science to Construction Materials, Proceedings, RILEM Symposium*, Chapman & Hall, the Netherlands, 1987, pp. 539–546.

Bentur, A. and Diamond, S., Fracture of glass fiber reinforced cement, *Cement and Concrete Research*, 14, 31–42, 1984.

Bentur, A. and Mindess, S., *Fibre Reinforced Cementitious Composites*, Taylor & Francis, London and New York, 2007.

Bentur, A. and Peled, A., Cementitious composites reinforced with textile fabrics, in H. W. Reinhardt and A. E. Naaman (eds.), *Workshop on High Performance Fiber Reinforced Cement Composites (RILEM) HPFRCC 3*, Mainz, Germany, May 16–19, 1999, pp. 31–40.

Bentur, A., Tirosh, R., Yardimci, M., Puterman, M., and Peled, A., Bonding and microstructure in textile reinforced concrete, in W. Brameshuber (ed.), *Textile Reinforced Concretes, Proceedings of the International RILEM Conference on Materials Science*, Vol. 1, RILEM Publications, Aachen, Germany, 2010, pp. 23–33.

Brameshuber, W. and Banholzer, B., Bond characteristics of filaments embedded in the fine grained concrete, in *Textile Reinforced Structures, Proceedings of the Second Colloquium on Textile Reinforced Structures (CTRS2)*, Dresden, Germany, 2003, pp. 63–76.

Chudoba, R., Vorechovsky, M., and Konrad, M., Stochastic modeling of multi-filament yarns. I. Random properties within the cross section and size effect, *Solids and Structures*, 43, 413–434, 2006.

Cohen, Z. and Peled, A., Controlled telescopic reinforcement system of fabric-cement composites-durability concerns, *Cement and Concrete Research*, 40, 1495–1506, 2010.

Dolatabadi, M. K., Janetzko, S., Gries, T., Sander, A., and Kang, B. G., Permeability of AR-glass fibers roving embedded in a cementitious matrix, *Materials and Structures*, 44, 245–251, 2011.

Dolatabadi, N., Janetko, S., Gries, T., Kang, B. G., and Sander, A., An analytical investigation of cement penetration within bundle of fibers, in *International RILEM Conference on Material Science*, Aachen, Germany, Vol. 1, 2010, pp. 69–78.

Haim, E. and Peled, A., Impact behavior of textile and hybrid cement-based composites, *ACI Materials Journal*, 103(8), 235–243, 2011.

Hartig, J. and Haussler-Combe, U., A model for Textile Reinforced Concrete under imposed uniaxial deformations, in *CMM-2011, Computer Methods in Mechanics*, Warsaw, Poland, May 9–12, 2011.

Hartig, J., Hausller-Combe, U., and Schicktanz, K., Influence of bond properties on the tensile behavior of textile reinforced concrete, *Cement and Concrete Composites*, 30, 898–906, 2008.

Haussler-Combe, U. and Hartig, J., Bond failure mechanisms of textile reinforced concrete (TRC) under uniaxial tensile loading, *Cement and Concrete Composites*, 29, 279–289, 2007.

Hegger, J., Bruckermann, O., and Chudoba, R., A smeared bond-slip relation for multi-filament yarns embedded in fine concrete, in *Sixth RILEM Symposium of Fiber-Reinforced Concretes (FRC)—BEFIB*, Varenna, Italy, September 20–22, 2004.

Hegger, J., Will, N., Bruckermann, O., and Voss, S., Load-bearing behavior and simulation of textile reinforced concrete, *Materials and Structures*, 39, 765–776, 2006.

Katz, A., Li, V. C., and Kazmer, A., Bond properties of carbon fibers in cementitious matrix, *Journal of Materials in Civil Engineering*, 7, 125–128, 1995.

Kim, D. J., El-Tawil, S., and Naaman, A., Correlation between single fiber pullout and tensile response of FRC composites with high strength steel fiber, in *High Performance Reinforced Cement Composites (HPFRCC5)*, Mainz, Germany, RILEM Publications, France, 2007, pp. 67–76.

Kruger, M., Ozbolt, J., and Reinhardt, H. W., A discrete bond model for 3D analysis of textile reinforced and prestressed concrete elements, *Otto Graff Journal*, 13, 111–128, 2002.

Kruger, M., Ozbolt, J., and Reinhardt, H. W., A new 3D discrete bond model to study the influence of bond on the structural performance of thin reinforced and prestressed concrete plates, in A. E. Naaman and H. W. Reinhardt (eds.), *Fourth International Workshop on High Performance Fiber Reinforced Cement Composites (HPFRCC 4)*, RILEM Publications, France, 2003, pp. 49–63.

Lepenies, I., Meyer, C., Schorn, H., and Zastrau, B., Modeling of load transfer behavior of AR glass rovings in textile reinforced concrete, *ACI SP*, 244-7, 109–124, 2007.

Li, V. C. and Wang, Y., Effect of inclining angel, bundling and surface treatment on synthetic fibre pull-out from cement matrix, *Composites*, 21, 132–140, 1990.

Markovich, I., van Mier, J. G. M., and Walraven, J. C., Single fiber pullout from hybrid fiber reinforced concrete, *Heron*, 46, 191–200, 2001.

Mobasher, B. and Li, C. Y., Modeling of stiffness degradation of the interfacial zone during fiber debonding, *Composites Engineering*, 5, 1349–1365, 1995.

Mobasher, B., Pahilajani, J., and Peled, A., Analytical simulation of tensile response of fabric reinforced cement based composites, *Journal of Cement and Concrete Composites*, 28(1), 77–89, 2006a.

Mobasher, B., Peled, A., and Pahilajani, J., Pultrusion of fabric reinforced high fly ash blended cement composites, in *Sixth Rilem Symposium on Fibre-Reinforced Concrete (FRC), BEFIB 2004*, Varenna, Italy, September 20–22, 2004.

Mobasher, B., Peled, A., and Pahilajanadi, J., Distributed cracking and stiffness degradation in fabric-cement composites, *Materials and Structures*, 39, 317–331, 2006b.

Mobasher, B. and Shah, S. P., Interaction between fibers and the cement matrix in glass fiber reinforced concrete, *American Concrete Institute, ACI*, SP-124, 137–156, 1990.

Mobasher, B. and Soranakom, C., Effect of transverse yarns in the pullout response and toughening mechanisms of textile reinforced cement composites, in *Fracture and Damage of Advanced Fibre-Reinforced Cement-Based Materials, Proceedings of the 18th European Conference on Fracture, ECF18*, Dresden, Germany, Aedificatio Publishers, Freiburg, Germany, 2010, pp. 59–66.

Mu, B., Meyer, C., and Shimanovich, S., Improving the interface bond between fiber mesh and cementitious matrix, *Cement and Concrete Research*, 32, 783–787, 2002.

Naaman, A. E. and Najm, H., Bond-slip mechanisms of steel fibers in concrete, *ACI Materials Journal*, 88, 135–145, 1991.

Naaman, A. E., Namur, G. G., Alwan, J. M., and Najm, H. S., Fiber pullout and bond slip I: Analytical study, *Journal of Structural Engineering*, 117, 2769–2790, 1991a.

Naaman, A. E., Namur, G. G., Alwan, J. M., and Najm, H. S., Fiber pullout and bond slip I: Experimental validation, *Journal of Structural Engineering*, 117, 2791–2800, 1991b.

Oakley, D. R. and Proctor, B. A., Tensile stress-strain behavior of glass fiber reinforced cement composites, in A. Neville (ed.), *Fibre Reinforced Cement and Concrete*, The Construction Press, Lancaster, England, 1975, pp. 347–359.

Ohno, S. and Hannant, D. J., Modeling the stress-strain response of continuous fiber reinforced cement composites, *ACI Materials Journal*, 91(3), 306–312, 1994.

Peled, A., Pre-tensioning of fabrics in cement-based composites, *Cement and Concrete Research*, 37, 805–813, 2007.

Peled, A. and Bentur, A., Geometrical characteristics and efficiency of textile fabrics for reinforcing composites, *Cement and Concrete Research*, 30, 781–790, 2000.

Peled, A. and Bentur, A., Fabric structure and its reinforcing efficiency in textile reinforced cement composites, *Composites Part A*, 34, 107–118, 2003a.

Peled, A. and Bentur, A., Quantitative description of the pull-out behavior of crimped yarns from cement matrix, *Journal of Materials in Civil Engineering*, 15, 537–544, 2003b.

Peled, A., Bentur, A., and Mobasher, B., Pultrusion versus casting processes for the production of fabric-cement composites, in M. Di Prisco, R. Felicetti, and G. A. Plizzari (eds.), *Fibre Reinforced Concrete—BEFIB 2004, Proceedings, RILEM Symposium, PRO 39*, RILEM Publications, France, 2004, pp. 1495–1504.

Peled, A., Bentur, A., and Yankelevsky, D., Effect of woven fabrics geometry on the bonding performance of cementitious composites: Mechanical performance, *Advanced Cement Based Materials Journal*, 7(1), 20–27, 1998a.

Peled, A., Bentur, A., and Yankelevsky, D., The nature of bonding between monofila-ment polyethylene yarns and cement matrices, *Cement and Concrete Composites*, 20(4), 319–328, 1998b.

Peled, A., Cohen, Z., Pasder, Y., Roye, A., and Gries, T., Influences of textile char-acteristics on the tensile properties of warp knitted cement based composites, *Cement and Concrete Composites*, 30, 174–183, 2008.

Peled, A. and Mobasher, B., Pultruded fabric-cement composites, *ACI Materials Journal*, 102(1), 15–23, January–February 2005.

Peled, S., Sueki, S., and Mobasher, B., Bonding in fabric-cement systems: Effects of fabrication methods, *Cement and Concrete Research*, 36, 1661–1671, 2006.

Reinhardt, H. W. and Krueger, M., Prestressed concrete plates with high strength fabrics, in M. Di Prisco, R. Felicetti, and G. A. Plizzari (eds.), *Fibre Reinforced Concrete—BEFIB 2004, Proceedings, RILEM Symposium, PRO 39*, RILEM, France, 2004, pp. 187–196.

Richter, M., Lepenies, I., and Zastrau, B., On the influence of the bond behavior between fiber and matrix on the materials properties of textile reinforced con-crete, in *ABDM 2002, International Symposium on Anisotropic Behaviour of Damaged Material*, Poland, Europe, September 9–11, 2002.

Soranakom, C. and Mobasher, B., Geometrical and mechanical aspects of fab-ric bonding and pullout in cement composites, *Materials and Structures*, 42, 765–777, 2009.

Stang, H., Li, Z., and Shah, S. P., Pullout problem: Stress versus fracture mechanical approach, *Journal of Engineering Mechanics*, 116(10), 2136–2149, 1990.

Stang, H. and Shah, S. P., Failure of fiber-reinforced composites by pull-out fracture, *Journal of Materials Science*, 21, 953–957, 1986.

Stucke, M. S. and Majumdar, A. J., Microstructure of glass fibre-reinforced cement composites, *Cement and Concrete Research*, 11(6), 1019–1030, 1976.

Sueki, S., Soranakum, C., Mobasher, B., and Peled, A., Pullout-slip response of fab-rics embedded in cement paste matrix, *Journal of Materials in Civil Engineering*, 19(9), 718–727, 2007.

Vorechovsky, M. and Chudoba, R., Stochastic modeling of multi-filament yarns. II. Random properties over the length and size effect, *Solids and Structures*, 43, 435–458, 2006.

Xu, S., Krueger, M., Reinhardt, H. W., and Ozbolt, J., Bond characteristics of carbon, alkali resistant glass and aramid textiles in mortar, *Journal of Materials in Civil Engineering*, 16(4), 356–364, 2004.

Yardimci, M. Y., Tirosh, R., Larianovsky, P., Puterman, M., and Bentur, A., Improving the bond characteristics of AR-Glass strands by microstructure modification tech-nique, *Cement and Concrete Composites*, 33(1), 124–130, 2011.

Zastrau, B., Richter, M., and Lepenies, I., On the analytical solution of pullout phenomena in textile reinforced concrete, *Journal of Engineering Materials and Technology*, 125, 38–43, 2003.

Zastrau, Z., Lepenies, I., and Richter, M., The multi-scale modeling of textile rein-forced concrete, *Technische Mechanik*, Band 28 Heft 153–63, 2008.

Zhu, W. and Bartos, P. J. M., Assessment of interfacial microstructure and bond properties in aged GRC using a novel microindentation method, *Cement and Concrete Research*, 27(11), 1701–1711, 1997.

Chapter 5

Mechanical performance under static conditions

5.1 Introduction

Textile fabric reinforcement of a cementitious matrix provides the means for producing a strain-hardening, high-performance cement composite, overcoming the limitations of the brittle cement matrix. The reinforcing mechanisms of fibers in a cement matrix differ from those of polymeric matrices, in which the matrix is usually much more brittle and weaker than the fiber. In polymer composites, the reinforcement content is relatively high, usually above 50% and up to 80% by volume, and as a result, the reinforcing fibers carry most of the load even at early stages of loading. Therefore, polymer-based composite failure is due to fiber rupture, that is, the matrix collapses when the fibers fail. However, in cement composites, the matrix fails long before the full potential of the fibers is realized (i.e., before the fiber reaches its limiting tensile strength). Thus, fiber reinforcement becomes effective mainly after the matrix has cracked and the fibers are bridging the crack. If the volume content of the fibers is large enough (over the critical volume content), they will be able to carry the additional load imposed on them after matrix cracking. Therefore, the addition of fiber reinforcements over their critical volume, will prevent the catastrophic failure of the matrix that happens in the absence of fibers, allowing the composite to sustain higher loads and to maintain its structural integrity under deformation. The micromechanics of these processes and their quantification are discussed in detail in Chapter 4.

The critical volume content of fiber reinforcement in a cement composite is about 1%–3%, much smaller than in polymer matrices. Below this level, strain softening or even catastrophic failure may occur when the composite strain exceeds the ultimate strain of the matrix. Reinforcement level above the critical level will result in a composite that is characterized by multiple cracking and strain hardening. The tensile behaviors of textile-reinforced mortar (TRM) components with various fabric volume fractions of 0.59%, 1.79%, and 2.98% (one, three, and five fabric layers, respectively) of

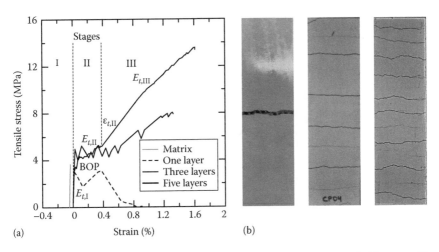

(a)   (b)

Figure 5.1 Cement-based composites reinforced with bidirectional basalt fabrics at different volume contents of 0.59%, 1.79%, and 2.98% (one, three, and five fabric layers, respectively): (a) influence on the tensile behavior and (b) crack patterns of these composites after testing. (From Rambo, D.A.S. et al., *Mater. Des.*, 65, 24, 2015.)

plain cementitous matrix (without textile reinforcement) are presented in Figure 5.1, which also shows the typical cracking patterns of each system (Rambo et al., 2015). In a composite with a single layer of fabric and content below the critical value, only one crack is formed, and the composite exhibits strain-softening behavior with low ductility. In contrast, in composites in which the fabric contents are above the critical volume (three to five fabric layers in the case of Figure 5.1), the fabric can efficiently bridge the cracks that develop under loading and carry the increasing applied loads, leading to strain-hardening behavior with multiple cracking.

The typical tensile behavior of textile-reinforced concrete (TRC) is presented in Figure 5.2, and it can be modeled on the basis of the Aveston–Cooper–Kelly (ACK) model (Aveston and Kelly, 1973; Aveston et al., 1971) (see also Chapter 4). The ACK model was developed to define the theoretical stress–strain behavior of a composite with a brittle matrix and continuous reinforcement (e.g., fabrics in TRC), in which the fiber–matrix bond remains intact even after the matrix has cracked. The typical stress–strain curve is characterized in terms of three stages, briefly described as follows.

Stage I: Linear elastic precracking stage—At this stage, the stresses are below the level of the matrix tensile strength, and it ends when the tensile strength is exceeded and the first crack appears in the matrix. Composite stiffness is dependent on the stiffness of the matrix. After the first crack forms, the tensile load is carried entirely by the textile reinforcement.

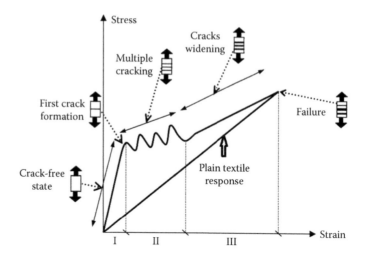

Figure 5.2 Schematic description of the typical response of a cement-based composite reinforced with fabrics and subjected to tensile loading. Crack states and textile responses indicated. (From Barhum, R. and Mechtcherine, V., *Eng. Fract. Mech.*, 92, 56, 2012.)

Stage II: Multiple-cracking stage—After initial matrix cracking, increases in the tensile load result in the formation of new cracks, and their spacing is dependent on the stress transfer between fabric and matrix as quantified by the pullout behavior and bond strength (see Chapter 4). The multiple-cracking stage terminates when all the additional loading is carried by the reinforcement and no further cracks occur in the matrix.

Stage III: Postcracking or crack-widening stage—At this stage, the dominant mechanism is crack widening, which is characterized by a stabilized crack pattern, and the loads are carried mainly by the reinforcing fabric. The fabric filaments can carry the increased load until the stress level exceeds their strength, and complete failure occurs. During this last stage, controlled by the elastic modulus of the fabric itself, the stress–strain curve of the filaments proceeds approximately parallel to that of the plain fabric (see the straight line in Figure 5.2).

The nature of the multiple cracking during stage II—and specifically the spacing between the cracks that develop—has been studied and modeled extensively to quantify it in terms of the pullout behavior (Chapter 4). Greater bond strength leads to more efficient stress transfer between the fabric and matrix and, thus, results in smaller crack spacing and a denser multiple-cracking pattern. The tensile behavior of TRC with alkali-resistant (AR) glass or polypropylene (PP) fabric is presented in Figure 5.3, which also shows how crack spacing changes during loading. Quantification of

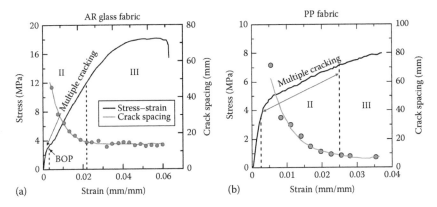

Figure 5.3 Tensile stress–strain curves of cement-based composites reinforced with (a) AR glass fabrics and (b) PP fabrics showing the change in crack spacing during loading. Dashed lines indicate the initiation of multiple cracking and crack widening. (From Peled, A. and Mobasher, B., *J. Mater. Civil Eng.*, 19(4), 340, 2007.)

experimentally obtained crack patterns provides important insights of fabric–matrix bonding, leading to initiation of crack-widening stage. Figure 5.3 shows the point at which the first crack appears, the stress–strain region in which multiple cracking is dominant (stage II), the beginning of crack widening (stage III), marked by the dashed lines, and the strength of the bond between the fabric and the matrix throughout the test, expressed by crack spacing. The two TRC examples illustrated in Figure 5.3 show that the nature of the multiple-cracking, crack-widening, and filament failure processes influence the shape of the stress–strain curve and the resulting strength, ductility, and toughness of a textile–cement composite. These parameters, in turn, are dependent on the fabric properties (including the materials constituting its yarns and yarn tensile properties as well as fabric geometry) and on the matrix structural properties and the strength of the fabric–matrix bond. This chapter will discuss the influences of matrices, fabric materials, fabric geometry, coating and processing on the mechanical performances of textile-reinforced cement-based composites, mainly under tension and flexural loading.

5.2 Influence of the matrix on composite mechanical performance

5.2.1 Matrix compositions

The design of matrices used in TRC must meet the unique requirements of each composite application—including the reinforcement geometry, component shape, production process, and the type of reinforcement material—to promote satisfactory bonding and durability of the component. In addition, resistance to environmental impacts such as frost, and freeze-thaw should be considered for some matrix design applications. Low creep and shrinkage values are also important objectives of matrix design.

A key factor in the design of matrices for TRC applications is the rheology to ensure full matrix penetration in the openings in the fabric to provide good matrix–fabric bonding and efficient composite load-bearing capacity. Due to the relatively dense mesh structure of textile reinforcements, the penetration of the particulate cementitious matrix is not always efficient, unlike in the case of polymeric matrices. To assure proper penetration, the cementitious matrices used are of the mortar type (cement–sand mix with sand particles usually smaller than 1 mm), and they are often referred to as "fine concrete" or "fine-grained concrete" (Brameshuber, 2006; Brameshuber and Brockmann, 2001). An additional parameter of significance in the matrix design is the flowability of the fine-grained mix, which should be adjusted to the textile geometry and the production process. For composites in which an especially dense fabric is used, a fluid matrix is required to ensure that the fabric openings are filled efficiently. However, production techniques such as lamination or pultrusion require matrices that not only fill the fabric openings but also remain fixed within those openings during composite production.

Basic matrix design consists of choice and proportioning of fine aggregates, plasticizers, and a variety of mineral additives, mainly pozzolanic based, such as fly ash, silica fume, and metakaolin (MK). In some cases, in addition to the textile reinforcement, other elements are also incorporated into the matrix, such as short fibers to provide additional reinforcement at the microlevel (discussed later) and polymer additives to improve the component ductility. Table 5.1 lists some examples of matrix compositions of cement-based systems and their design specifications, which are determined by the requirements dictated by the intended application of the TRC. In general, the fine-grained concrete or mortar base of the matrix can be designed for a range of properties required for various applications, from nonstructural to high-performance structural components (Scholzen et al., 2015).

The chemical stability of the fabrics exposed to matrix should be considered for durability. For example, glass or vegetable fibers such as sisal are sensitive to the alkaline environment of the Portland cement–based matrix. Therefore, for TRCs made of alkali-sensitive fibers, special matrix compositions are needed. Critical to matrix design is the selection of an appropriate binder system to reduce its alkalinity when used in conjunction with alkali-sensitive reinforcement. Examples are blast furnace slag and pozzolanic materials such as fly ash, MK, and silica fume that can be incorporated to replace part of the Portland cement (for a more detailed discussion,

Table 5.1 Composition of mixtures developed for textile-reinforced cement composites

	Materials		PZ-0899-01 SFB 532	FA-1200-01 SFB 532	RP-03-2E SFB 532	MF-102-01 SFB 532	PZ-PM 20 SFB 532	Mechtcherine M 030	Mechtcherine M 045	Mechtcherine M45	Reinhardt M7	Hegger
Binder	Cement (c)	kg/m³	490	210	980	441	430	632	554	551	480	490
	Cement type		CEM I 52.5	CEM I 52.5	CEM I 52.5	CEM I 52.5	CEM III B 32.5	CEM III B 32.5	CEM III B 32.5	CEM III/B 32.5 NW-HS-NA	CEM I 42.5 R	CEM I 52.5
	Fly ash (f)		175	455	210	210	154	265	233	248	154	175
	Silica fume (s)		35	35	210	—	31	101	89	—	41	35
	Metakaolin		—	—	—	49	—	—	—	28	—	—
	Total binder		700	700	1,400	700	615	998	876	827	675	700
Additives	Plasticizer	%	1.5	0.9	2.5	2	20	11	2	—	2.5	—
	Polymers		—	—	—	—	—	—	—	—	—	—
	Stabilizer		—	—	—	—	—	—	—	—	—	—
	Admixture											3.8
Aggregates	Siliceous fines 0–0.125 mm	kg/m³	500	470	118	500	438	—	—	—	—	—
	Siliceous sands											
	0–1 mm		—	—	—	—	—	—	—	1,101	—	—
	0.2–0.6 mm		715	670	168	715	626	947	832	—	—	—
	0–0.6 mm		—	—	—	—	—	—	—	—	460	—
	0–0.8 mm		—	—	—	—	—	—	—	—	—	1,249
	0.6–1.2 mm		—	—	—	—	—	—	—	—	920	—

(Continued)

Table 5.1 (Continued) Composition of mixtures developed for textile-reinforced cement composites

Materials		PZ-0899-01 SFB 532	FA-1200-01 SFB 532	RP-03-2E SFB 532	MF-102-01 SFB 532	PZ-PM 20 SFB 532	Mechtcherine M 030	Mechtcherine M 045	Mechtcherine M45	Reinhardt M7	Hegger
Water	Water	kg/m³	280	280	350	280	245	234	330	275	211	280
	w/c	—	0.57	1.33	0.36	0.63	0.57	—	—	0.5	0.44	0.57
	w/B = w/(c + f + s)		0.4	0.4	0.25	0.4	0.4	0.3	0.45	—	0.31	0.44
Hardened concrete properties												
Age: 28 days	Compressive strength	N/mm²	Sealed 74	Sealed 32	Sealed 98	Water, 20°C 66	35	62.3	36.1	—	Water, 20°C 90	89
	Young's modulus		33,000	24,800	26,500	—	—	18,910	16,790	—	—	—

Sources: Barhum, R. and Mechtcherine, V., Mater. Struct., 46, 557, 2013; Brameshuber, W. (ed.), Textile Reinforced Concrete: State of the Art Report, RILEM TC 201-TRC, RILEM Publications, Paris, France, 2006; Brockmann, T., Mechanical and fracture mechanical properties of fine grained concrete for textile reinforced composites, PhD thesis, RWTH Aachen University, Aachen, Germany, 2006; Butler, M. et al., Cement Concr. Compos., 31, 221, 2009; Scholzen, A. et al., Struct. Concr., 16(1), 106, 2015.

see Chapter 9). For example, Silva et al. (2010) designed a fine-grained concrete composition to produce durable, natural fiber composites made of sisal textile and 50% Portland cement CPII F-32 (Brazilian Standard NBR 11578, 1991), 30% MK, and 20% calcined waste crushed clay brick (CWCCB). Butler et al. (2009) reported fine concrete compositions (Table 5.1, Mechtcherine-M45) to produce durable AR glass textile composites. In these composites, the AR glass textile was combined with blast furnace slag cement in which the alkalinity was strongly buffered by added fly ash and silica fume or MK. The use of MK together with fly ash provides good long-term behavior with virtually no changes in composite mechanical performance during aging.

5.2.2 Low-alkalinity cements

Low-alkalinity cements were applied in TRC to produce a durable composite when alkaline-sensitive fibers, such as glass and natural fibers are used (for more details, see Chapter 9). The aging of fibers can be due to direct alkaline attack or mineralization that occurs in vegetable fibers. Low-alkalinity sulfoaluminate cements with pH values of 10.5 were studied by Krüger et al. (2003), consisting of anhydrous calcium sulfoaluminate and dicalcium silicate were studied by Krüger et al. (2003). These cements are characterized by their fast hardening, high strength at late ages, impermeability, frost and corrosion resistance, low pH of their pore solution, and low shrinkage or sometimes expansion. Another example of a cementitious matrix developed for vegetable fibers consists of 50% Portland cement CPII F-32 (Brazilian Standard NBR 11578, 1991), 30% MK, and 20% CWCCB (Silva et al., 2009; Toledo Filho et al., 2009) mixed in a proportion by weight of 1:1:0.4 (cementitious material:sand:water). This specially developed matrix does not show any reduction in strength or toughness in accelerated aging tests.

Inorganic phosphate cements (IPC) were developed by Cuypers et al. (2006) as alternative binders for E-glass textile-reinforced composites. These cements consist of grains with sizes ranging from 10 to 100 μm, and the hardened matrix provides a nonalkaline environment with a pH of about 7, much smaller than the value of ~13, typical of ordinary Portland cement (OPC). Because of its higher viscosity compared to those of other cement pastes, no fillers are introduced in the IPC matrix. E-glass fiber can be incorporated in the matrix up to fiber volume fractions of more than 20% (Orlowsky et al., 2005). One important aspect of these cements is their inherent fast setting, which makes them suitable for repair applications. A specific formulation has been developed with sufficiently retarded setting times (Wastiels, 1999).

5.2.3 Admixtures

Numerous studies have reported the effect of mineral additives such as fly ash, silica fume and polymers on the matrix and composite properties (Mobasher et al., 2004; Peled and Mobasher, 2005; Tsesarsky et al., 2013, 2015). The effects of different cement matrices with fly ash (FA 25 wt.%), silica fume (SF 7 wt.%), polymer (PL 5 wt.%), or short PP fibers (PPF 0.1 wt.%) on the flexural behavior of fabric-based composites were also compared to a reference composite made of plain matrix without additives (PC) (Figure 5.4) (Tsesarsky et al., 2015). The composites were made of

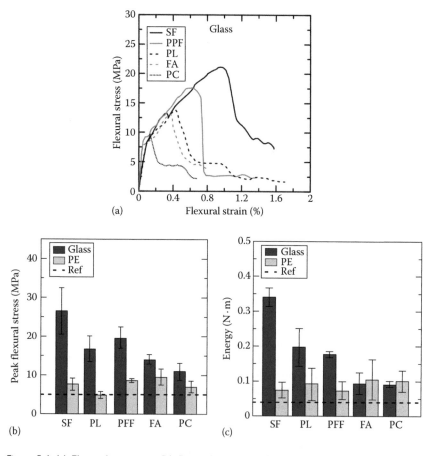

Figure 5.4 (a) Flexural response, (b) flexural strength of AR glass, and (c) energy absorption of AR glass and PE composites with different additives (silica fume, SF; polymer, PL; short fibers, PPF; fly ash, FA; and plain matrix without additives, PC). The dashed horizontal line represents the flexural strength of an equivalently sized cement paste (CEM I 52.5 using ASTM C78). (From Tsesarsky, M. et al., *Mater. Struct.*, 48, 471, 2015.)

laminated AR glass or AR glass and polyethylene (PE) fabrics. Figure 5.4 shows the strong dependence of both flexural strength and toughness on the admixture type when glass fabrics were used as the reinforcement (Figure 5.4). The matrix with silica fume, which achieved the best results, improved load resistance and energy absorption by 242% and 329%, respectively, compared with the PC TRC. However, the positive influence is reduced at silica fume contents of 10% and above, at which the tensile performances of the composite are markedly lower compared to those of composites with a 5% silica fume content (Figure 5.5) (Peled and Mobasher, 2005). The reduced performance of matrices with high silica fume contents was attributed to the properties of the fresh mixture, in which the high percentage of silica fume results in an exceedingly stiff matrix. To prevent the reduction in performance, a high-range water-reducing admixture was added in relatively large amounts, forming air pockets in the matrix. These air pockets, together with the stiff nature of the fresh matrix, can lead to low matrix penetration of the fabric openings that can result in reductions in the matrix properties and in the strength of the bond between fabric and matrix leading to poor mechanical performance of the composite. The reduced bond strength due to the high silica fume content was confirmed by the appearance of fewer cracks with larger crack spacing (Figure 5.5) representing bond strength values of 0.43 and 1.13 MPa for the 10% and 5% silica fume composites, respectively, for AR glass fabric reinforcement. Similar trends were recorded for composites reinforced with PE fabrics (Peled and Mobasher, 2005). The addition of a polymer (PL) admixture to

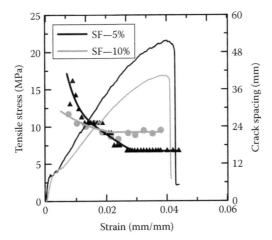

Figure 5.5 Tensile response of composites reinforced with AR glass along with crack spacing development during testing. (From Peled, A. and Mobasher, B., *ACI Mater. J.,* 102(1), 15, January–February 2005.)

AR glass composites conferred twofold improvements in load resistance and energy absorption relative to the PC TRC (Figure 5.4).

A comparative study (Mobasher et al., 2004; Peled and Mobasher, 2005) of AR glass composites with different fly ash contents showed that while the presence of fly ash did not significantly contribute to the early age strength, it markedly improved the 28-day strength. The best results were obtained in composites with 60% fly ash content (Figure 5.6). This improvement was attributed to significantly better mix workability and efficient matrix penetration between the fabric openings, which enhanced bond strength, as verified by mechanical performance and microstructural evaluation (Mobasher et al., 2004).

The trends described indicate that matrix type has a marked influence when AR glass fabrics are the reinforcement in TRC systems. However, in cement matrices reinforced with PE fabric, the additives had only minor influences on load resistance and energy absorption (Figure 5.4). These differences were attributed to the nature of the yarns constituting the fabrics: glass fabric consists of multifilament yarns, whereas the PE has monofilament yarns (for details, see Chapter 4). Thus, the extent of influence is highly dependent on the nature of the yarn—monofilament or multifilament—as well as on the type of fiber material.

The type of matrix, either IPC or OPC, also affects composite mechanical performance due to differences in their chemical makeup, grain size, and fresh mixture viscosity. Inorganic phosphate-based cement (GTR-IPC) composite possesses a higher tensile strength than GTR-OPC

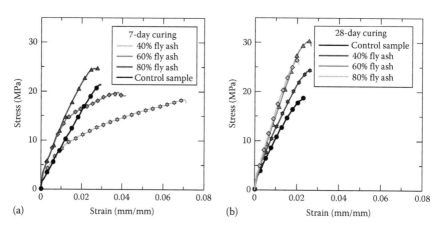

Figure 5.6 Comparison of stress–strain responses of AR glass composites with different fly ash contents and a control sample (a) 7 days and (b) 28 days after casting. (From Mobasher, B. et al., Pultrusion of fabric reinforced high fly ash blended cement composites, in *Sixth RILEM Symposium on Fibre-Reinforced Concretes (FRC)*, BEFIB, Varenna, Italy, September 2004, pp. 1473–1482.)

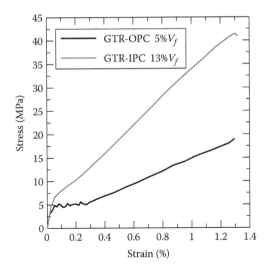

Figure 5.7 Influence of matrix type, OPC or IPC, on the tensile response of glass composites (GTR-OPC with 5 vol.% and GTR-IPC with 13% reinforcement volume). (From Tysmans, T. et al., *Compos. Sci. Technol.*, 69, 1790, 2009.)

due to its larger reinforcement volume fraction, both of which contain glass fabrics (Figure 5.7) (Tysmans et al., 2009).

5.2.4 Short fibers

5.2.4.1 Short fiber incorporation in the matrix

The effect of the addition of short fibers to the matrix on the mechanical behavior of TRC has been investigated by several researchers (Barhum and Mechtcherine, 2012, 2013; Hinzen and Brameshuber, 2006, 2007, 2008, 2009, 2010; Tsesarsky et al., 2015). The fibers enhance the first-crack strength, increase the load-bearing capacity (in addition to the contribution made by the textile), enable a more ductile composite, facilitate optimal crack development and crack patterns, and improve component serviceability and durability. Figure 5.8 is a schematic depiction of the target stress–strain behavior of a TRC composite with short fibers compared to that of a conventional textile-reinforced composite (Hinzen and Brameshuber, 2009), showing an increase in the first-crack stress and enhanced strain-hardening behavior during the multiple cracking stage (stage II in Figure 5.2). To obtain improved composite tensile load-bearing behavior, the bond between the short fibers and the matrix should be strong to achieve efficient crack bridging while minimizing crack widths and crack spacing.

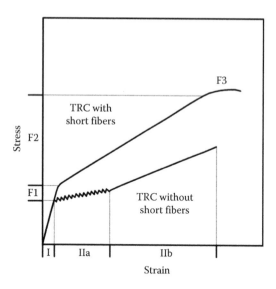

Figure 5.8 Schematic description of the influence of the addition of short fibers to TRC on the stress–strain response. (Adapted from Hinzen, M. and Brameshuber, W., Improvement of serviceability and strength of textile-reinforced concrete elements with short fiber mixes, in C.M. Aldea (ed.), *ACI Convention on Design and Applications of Textile-Reinforced Concrete*, SP-251-1, Farmington Hills, MI, 2008, pp. 7–18; Hinzen, M. and Brameshuber, W., Improvement of serviceability and strength of textile reinforced concrete by using short fibres, in *Fourth Colloquium on Textile Reinforced Structures (CTRS4)*, Dresden, Germany, 2009, pp. 261–272.)

An investigation by Tsesarsky et al. (2015) of AR glass composites with 0.1 wt.% short PP fibers showed improved composite flexural properties, strength, and energy absorption (Figure 5.4). However, no such improvement was recorded for composites reinforced with PE fabric. Positive influence on the flexural behavior (four-point bending test) of the short polyvinyl alcohol (PVA) fibers at larger contents of 1.0%–2.0% (vol.) on TRC with carbon fabrics was reported by Li and Xu (2011). Barhum and Mechtcherine (2012) showed a similar positive trend in AR glass composite when short carbon fibers (SCF) or short AR glass fibers (SGF) were added in amounts of 0.5% or 1.0% (vol.), respectively. The mechanical performance of the TRC with either of the two types of short fibers was clearly improved over the entire course of the stress–strain curves (Figure 5.9), which shows a pronounced enhancement in composite toughness. In both short fiber systems, however, the increase in composite tensile strength is small. The addition of short fibers such as carbon, in particular, also results in the extension of the strain capacity in the multiple-cracking region compared to that of a TRC without short fibers, that is, greater strain capacity between the first-crack initiations and the end of multiple-crack stage.

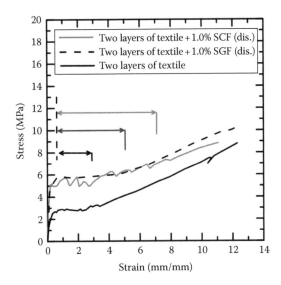

Figure 5.9 Tensile response of AR glass TRC with short fibers of glass (SGF) or carbon (SCF) and without short fibers. (From Barhum, R. and Mechtcherine, V., *Eng. Fract. Mech.*, 92, 56, 2012.)

The nature of the short fibers—that is, individual dispersed versus integral (in fiber bundles within the matrix)—affects the tensile properties of textile-reinforced composites (Figure 5.10). The effects of AR short glass fibers of these two forms were studied (Barhum and Mechtcherine, 2013). Compared to integral short fibers, short dispersed fibers increase the first-crack stress capacity and expand the strain region where multiple cracks

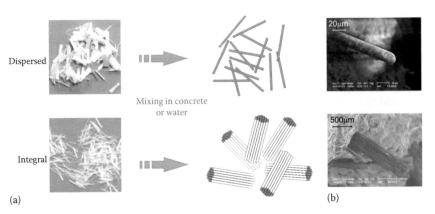

Figure 5.10 (a) Images and graphic representations of dispersed versus integral short fibers and (b) ESEM images of the fibers. (From Barhum, R. and Mechtcherine, V., *Mater. Struct.*, 46, 557, 2013.)

form more efficiently. In contrast, although a less pronounced increase in first-crack stress is obtained with integral fibers, they confer on the composite improved tensile strength and work-to-fracture. These differences in tensile properties are due to the different nature of the bond between each fiber system and the matrix. In the dispersed short fiber system, each individual fiber is well connected and bonded to the matrix, enabling the fibers to bridge microcracks. In contrast, due to the bundled nature of the integral fibers, only the outer fibers of the bundle are in good contact with the matrix, while the inner fibers slide against each other during loading. Therefore, at low deformations or crack openings, mainly the sleeve (outer) fibers are activated, whereas the core (inner) fibers become active only at higher deformations, when the outer fibers start to fail.

The addition to TRC composites of combinations of short fibers of different materials (hybrid), sizes, and shapes with different functions can benefit composite tensile properties. The combination of microfibers and macrofibers may be complementary, such that microfibers may increase the first-crack stress capacity of the matrix by reducing and delaying the microcrack formation and macrofibers may further increase the bridging process of the macrocracks (Banthia and Soleimani, 2005). The idea of reinforcing TRC composites with mixes of different short fiber materials is driven by the potentially different contributions of the materials to the initial stage (until first-crack initiation) and to the multiple-cracking stage of composite stress–strain behavior, as the strength requirements may not be the same for the two stages. Textile-based composites with combinations of different short fiber materials were studied by Hinzen and Brameshuber (2009), using a combination of short carbon (C, 0.5% vol.) and aramid (A, 2% vol.) fibers or with a combination of short glass (G, 1.5% vol., dispersed fibers) and aramid (A, 1% vol.) fibers, which were compared with a composite containing only short glass fibers (G3, 3% vol. integral fibers) and a reference composite lacking short fibers (FC) (Figure 5.11). Although the integral glass fibers (G3) were not as effective at the microscale as the dispersible glass fibers in the FC-1.5G-1A mix, the incorporation of a higher volume of 3% short glass fibers (FC-G3) increased the first-crack load to the same level as that of the composite with dispersible glass fibers. However, the addition of short fibers can also lead to a reduction in the workability of the fine-grained cement-based matrix, a problematic outcome for some practical applications, as the short fibers—which generally feature diameters of only a few microns—can constrain matrix flow significantly. Such trends motivated the development of an optimal short fiber mix (Hinzen and Brameshuber, 2010).

The studies of short fiber addition to the TRC matrix generally showed larger number of crack, finer crack width, and denser crack patterns compared to the results in composites without short fibers, as presented in Figure 5.12.

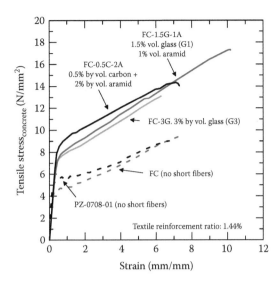

Figure 5.11 Stress–strain response of TRC with different short fiber combinations. (From Hinzen, M. and Brameshuber, W., Improvement of serviceability and strength of textile reinforced concrete by using short fibres, in *Fourth Colloquium on Textile Reinforced Structures (CTRS4)*, Dresden, Germany, 2009, pp. 261–272.)

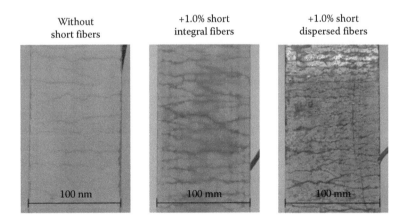

Figure 5.12 Crack pattern of AR glass fabric-reinforced composite with and without added short fibers (dispersed and integral AR glass short fibers). (From Barhum, R. and Mechtcherine, V., *Mater. Struct.*, 46, 557, 2013.)

Figure 5.13 Images of multifilament yarn in the cement matrix with added short carbon fibers and showing a carbon "cross-link" between the multifilament yarn and matrix. (From Barhum, R. and Mechtcherine, V., *Eng. Fract. Mech.*, 92, 56, 2012.)

Several fundamental mechanisms can explain the enhancement of TRC mechanical properties by the addition of short fibers: (1) short fibers can bridge micro- and fine cracks, thus inhibiting their growth and arresting crack propagation; (2) short fibers may restrain matrix shrinkage and, therefore, internal matrix damage; (3) the overall degree of reinforcement is increased; and (4) short fibers may also improve the bond between the multifilament yarns constituting the fabric and the surrounding matrix through formation of "special" adhesive cross-links (Figure 5.13). Generally, the addition of short fibers to TRC can produce a composite with better crack dispersion, thereby improving the durability, service life, and load-bearing capacity of concrete structures.

5.2.4.2 Nonwoven fabrics with short fibers

Short fibers may be used in TRC in the form of nonwoven fabrics, which consist of a web of short fibers (see Chapter 2). In contrast to knitted or woven fabrics, they are not constructed from yarns. Because this method skips yarn production, it allows the use of a high-speed, large-volume production process that can provide a fabric that is of lower cost relative to a knitted or woven fabric. Although nonwoven fabric generally exhibits lower mechanical performance relative to those produced from yarns, it functions well as a low-cost reinforcement alternative for cement-based composites.

Pekmezci et al. (2014) investigated the potential of PVA-based, needle-punched, nonwoven fabrics for used in cement-based composites as a low-cost reinforcement alternative. The needle punch technique results in strongly entangled fibers which produce a relatively dense and highly bonded fabric. In such a dense type of fabric, a self-leveling fine-grained matrix should be used to promote matrix penetration into the spaces between the fibers. Pekmezci et al. (2014) tested two self-leveling matrices,

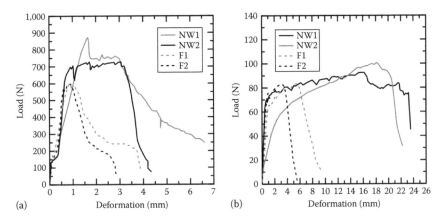

Figure 5.14 (a) Tensile and (b) flexural behaviors of composites reinforced with nonwoven PVA fabrics compared to a composite with short PVA fibers at 1.7 vol.%. (From Pekmezci, B.Y. et al., *J. Compos. Mater.*, 48(25), 3129, 2014.)

with or without quartz-based mineral powder, both of which were found to penetrate the spaces in the nonwoven fabric easily to produce good mechanical interlocking between the fibers and the cement matrix. Composite production comprised the immersion of the nonwoven fabric in a cement matrix bath and embedment of the cement-impregnated fabric in a matrix located in a proper mold to produce a composite containing two layers of fabric with an overall volume fraction of 1.7%. Incorporation of the nonwoven (NW) fabric in the composites improved tensile and flexural properties compared to a composite with randomly dispersed, discrete, short PVA fibers (F) having the same fiber volume fraction (Figure 5.14). The fabric improved both composite strength and toughness due to its stronger mechanical anchoring and enhanced bonding with the matrix. Multiple cracking was observed in the nonwoven fabric composite, while for the discrete fiber-reinforced composite, only a single macrocrack was recorded up to failure, indicating that the nonwoven fabric had a better crack arresting and bridging mechanism.

5.3 Influence of TRC fiber material

5.3.1 Single fiber material

5.3.1.1 Synthetic fibers

The type of fiber materials that constitute the fabric has a significant influence on the mechanical behaviors of TRC. The most important characteristics should be tensile strength, ultimate strain, modulus of elasticity, bonding to the matrix, and durability in the alkaline environment of the cement matrix.

The fiber material should be of relatively low cost and easily processable on standard textile machinery during textile fabric production.

A wide variety of fiber types—including carbon, glass, aramid, PVA, PP, PE of high and low modulus, basalt, and sisal—were evaluated for reinforcement for cement-based matrices (Contamine et al., 2011; Meyer and Vilkner, 2003; Peled and Bentur, 1998; Peled and Mobasher, 2007; Peled et al., 2008a; Reinhardt et al., 2003; Silva et al., 2009; Tsesarsky et al., 2013, 2015). In general, a high-strength, high-modulus fiber, such as glass, carbon, aramid, or high-density polyethylene (HDPE—Dyneema, Spectra), will usually increase the strength and toughness of the cement-based composite, providing strain-hardening behavior. For low-modulus fiber, such as PP and PE, the reinforcement mainly enhances the ductility of the cement composite, but not necessarily its strength, resulting in strain-softening or elastic–plastic behavior.

The two most common and widely studied fiber materials for TRC are AR glass and carbon fibers (Brameshuber, 2006), which have a modulus of elasticity higher than the matrix, thus providing strain-hardening behavior to the TRC. AR glass, although also attractive because of its relatively low cost, is sensitive to the alkaline environment of the matrix. The typical tensile load–strain behaviors of TRC components made of carbon or AR glass, each with two layers of fabric, are shown in Figure 5.15 (Hegger et al., 2007). The stiffness in the elastic zone measured at the uncracked stage, is not influenced by the fiber material, as it depends mainly on the properties of the matrix. The fiber material has a strong influence on the load–strain/deflection response after initiation of the first crack, that is, at

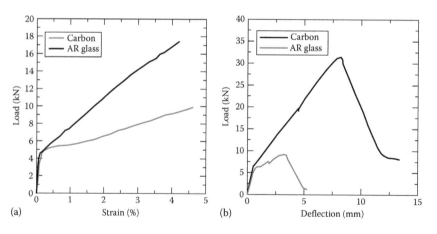

Figure 5.15 (a) Tensile and (b) flexural behaviors of cement-based composites reinforced with AR glass and carbon fabrics (with two fabric layers). (From Hegger, J. et al., *ACI Convention on AR-Glass and Carbon Fibres in Textile Reinforced Concrete—Simulation and Design*, SP-244-4, Farmington Hills, MI, 2007, pp. 57–75.)

the multiple-cracking stage. In that stage, the carbon TRC performed much better than the AR glass TRC. This is due to the greater stiffness of the carbon fiber, about three times higher than that of the AR glass (see Chapter 2). Comparisons of the tensile strengths and reinforcement efficiencies of the two fabric material composites measured as the ratio between the calculated average ultimate tensile strength of the filaments in the component and the tensile strength of the filament showed an effectiveness of about 40% for the AR glass fabrics with an ultimate composite strength of about 470 MPa. The calculated effectiveness of the carbon fabrics was about 69% with an ultimate strength of 770 MPa. These trends correlated well with the overall properties of the carbon fabric (see Section 2.2.1.2.2), which exhibited similar behavior under bending (for I-section beams with a length of 1 m under four-point bending). In that test, a beam reinforced with carbon fabric having the same cross-sectional area as the AR glass beam achieved three times the ultimate load of the latter (Figure 5.15b) (Hegger et al., 2007).

Comparisons of composites reinforced with high-modulus HDPE and AR glass versus low-modulus PP and PE showed marked differences in composite flexural responses (Figure 5.16). Similar differences in behavior were also observed under tension in composites reinforced with fabrics of different modulus of elasticity (AR glass vs. PE) (Figure 5.16a) (Peled and Mobasher, 2005, 2007). In all cases, the values of the initial composite moduli were similar and were not significantly influenced by fiber material because the initial modulus is representative of the matrix modulus.

The influence of fiber type becomes significant in the multiple-cracking zone. Fabric composites with high mechanical and tensile properties

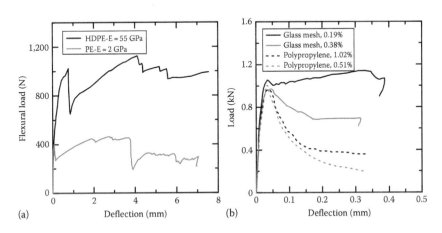

Figure 5.16 Flexural responses of cement composites reinforced with (a) low-modulus versus high-modulus PE fabrics and (b) low-modulus PP versus high-modulus AR glass. (a: From Peled, A. and Bentur, A., *Composites*, 34, 107, 2003; b: From Mu, B. et al., *Cement Concr. Res.*, 32, 783, 2002.)

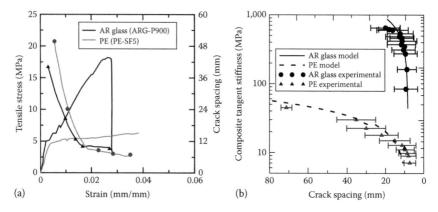

Figure 5.17 (a) Tensile response of composites reinforced with high-modulus AR glass or low-modulus PE fabrics and the change in composite crack spacing during loading and (b) influence of fabric type (PE vs. AR glass) on tangent stiffness of the composite versus crack spacing obtained during loading. (a: From Peled, A. and Mobasher, B., *ACI Mater. J.*, 102(1), 15, January–February 2005; b: From Mobasher, B. et al., *Cement Concr. Compos.*, 28, 77, 2006a.)

demonstrate enhanced strain-hardening behavior compared to the soft hardening or quasiplastic behavior of fabric composites with low mechanical and tensile properties, as demonstrated for composites reinforced with AR glass or PE fabrics (Figure 5.17a). The influence of fiber type on the postcrack region was even clearer when correlating the change in crack spacing during loading with the degradation in stiffness at the postcrack stage (Mobasher et al., 2006a,b) (Figure 5.17b). For the low-modulus PE, the stiffness in the composite postcracking range was an order of magnitude lower compared with that of the high-modulus AR glass. Furthermore, over a small crack distribution range, degradation in the stiffness of the AR glass fabric composite increased by about an order of magnitude. In contrast, the range of cracking in the PE systems was much longer than that of the AR glass fabric systems. The positive effect of the glass fabric on composite performance was even more pronounced when comparing the reinforcing volume fraction of the two composites, which was as high as 9.5% for the PE system but was only 4.4% for the glass composite.

The modulus of elasticity of the fibers in the fabric is also an important parameter controlling the fabric–matrix bond strength (Peled and Bentur, 2000; Stang, 1996). Low-modulus fibers such as PE and PP develop low bond strength, whereas high-modulus fibers such as aramid, carbon, or HDPE bond strongly with a cement matrix (Figure 5.18a) (Peled and Bentur, 2003). This outcome is due in part to the higher clamping stresses that develop around the high-modulus fibers as a result of the autogenous shrinkage of the matrix (Stang, 1996). In addition, the lower bond of low-modulus yarns

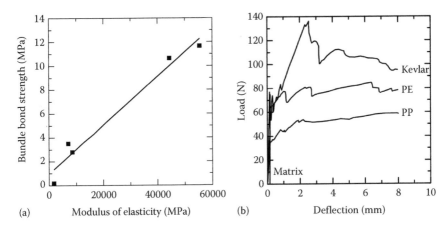

Figure 5.18 (a) Effect of yarn modulus of elasticity on the strength of the bond between the cement matrix and the yarn and (b) the related flexural behavior of the composites reinforced with high-modulus Kevlar, PE (HDPE of Dyneema), or low-modulus PP. (a: From Peled, A. and Bentur, A., *Composites*, 34, 107, 2003; b: From Peled, A. and Bentur, A., *Mater. Struct.*, 31, 543, October 1998.)

is also associated with the Poisson effect (Bentur and Mindess, 2006). The influences on bond strength of both autogenous shrinkage and the Poisson effect are observed when full contact has been achieved between the fibers and the cement matrix.

The effects of the fiber modulus are demonstrated in Figures 5.18b (Peled and Bentur, 1998) and 5.19 (Peled et al., 2008a). The highest average flexural strength and postcracking load-carrying capacity in Figure 5.18b were observed for the high-modulus Kevlar fabric composite, followed by the HDPE and, finally, the low-modulus PP fabric composites. These trends are consistent with the different moduli of elasticity of each fabric's yarns, which are 55, 45, and 8.9 GPa for the Kevlar, HDPE, and PP, respectively.

The tensile behavior of composites with fabrics of aramid, AR glass, or PP (Figure 5.19) could be correlated with the bond strengths they developed with the cement matrix (Table 5.2). The aramid and HDPE fabric TRCs were the strongest, and the PP fabric TRC exhibited the lowest strength. The tensile results also showed good correlation with the bond strength values of the different composite systems: aramid developed the strongest bonding of 5.29 MPa, followed by AR glass (5.13 MPa) and HDPE (4.54 MPa), and the weakest bonding (2.34 MPa) was for the PP composites. Note that the yarn tex and number of filaments used by Peled et al. (2008a) were not the same for all the materials, a scenario that can also affect the tensile behavior and that will be discussed later in this chapter. The brittle behavior observed in the glass fabric composite is due to the inherently brittle behavior of the glass yarn (see Chapter 2).

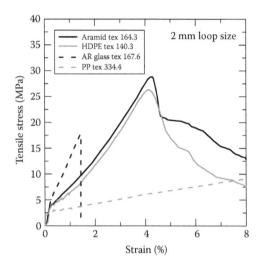

Figure 5.19 Tensile response of composites reinforced with high-modulus Kevlar, AR glass, or HDPE (Dyneema) versus low-modulus PP fabrics. (From Peled, A. et al., *Cement Concr. Compos.*, 30(3), 174, 2008a.)

Table 5.2 Correlation between fabric and composite properties

Yarn type	Yarn modulus of elasticity (GPa)	Fabric tensile strength (MPa)	V_f (%)	Composite tensile strength (MPa)	Composite efficiency factor	Bond strength (MPa)
Aramid	55	2,367	1.9	26	0.58	5.29
AR glass	78	1,591	1.1	18	1.04	5.13
HDPE	45	1,388	2.7	29	0.77	4.54
PP	7	223	6.2	9	0.67	2.34

Source: Peled, A. et al., *Cement Concr. Compos.*, 30(3), 174, 2008a.

Additional influence of fiber composition on the bond is its hydrophilic nature, as demonstrated in the comparison between the performance of TRC with PVA, AR glass, PP, and PE fabrics (Peled and Mobasher, 2007). PVA fabric reinforcement enhanced the initial modulus of the composite, which was about 6,500 MPa, compared to about 2,000 MPa in the other composite systems. This was attributed to the PVA fibers' hydrophilic nature, which confers a stronger bond with the cement matrix (Kanda and Li, 1998; Peled et al., 2008b) and an enhanced initial modulus of the composite. However, after the initiation of the first crack, at the multiple-cracking stage, the PVA fabric conferred no benefit over the other fabrics (Figure 5.20) (Peled and Mobasher, 2007).

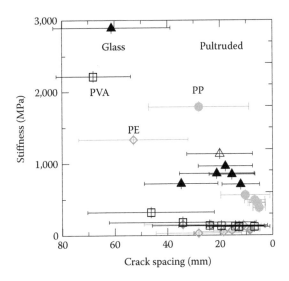

Figure 5.20 Reduction of effective stiffness of the composite as the spacing between cracks decreases due to tensile loading. (From Peled, A. and Mobasher, B., *J. Mater. Civil Eng.*, 19(4), 340, 2007.)

5.3.1.2 Basalt fibers

Basalt fibers are suitable for low-cost fabrics due to their higher modulus than glass and lower cost than carbon (OPERHA, 2006–2008). Furthermore, fabrics made from basalt fibers exhibit high levels of dimensional stability without any environmental restriction and good thermal stability properties, which make them attractive for use in applications in which high temperatures and fires are expected. Several researchers have explored the influence and contribution of basalt fibers as reinforcements in TRM and TRC composites (Larrinaga et al., 2013, 2014), especially for high-temperature applications (Rambo et al., 2015), for the strengthening of structural concrete members (Sim et al., 2005), and for the rehabilitation of masonry elements (Garmendia et al., 2011).

The strain-hardening behavior of basalt-based composites exhibits the three typical stages of TRC components, that is, elastic, multiple cracking, and crack widening (see Figure 5.2). When short, polymeric fibers are added to the matrix (3–5 wt.%), one layer of basalt fabric is sufficient to obtain a composite with quasiplastic behavior, whereas reinforcement with two or more fabric layers induced strain-hardening behavior (Larrinaga et al., 2013, 2014), even without the addition of short fibers. The basalt fabric was also found to control the composite modulus at stage III (crack-widening regime), in which modulus values of 60 and 67 GPa for crack widening and of the fabric itself, respectively, were obtained.

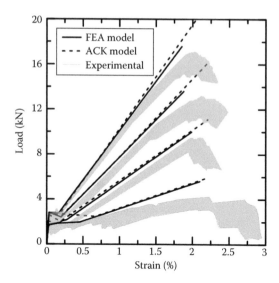

Figure 5.21 Experimental (shown by the gray regions) and predicted tensile behaviors of basalt fabric TRM calculated using the ACK and FEA models. (From Larrinaga, P. et al., *Mater. Des.*, 55, 66, 2014.)

The tensile behavior of basalt fabric composites was predicted by the ACK model and finite element analysis (FEA) using the FE code of Cervenka et al. (2009) while assuming a rigid interface (Larrinaga et al., 2014), which correlated well with experimental behavior (Figure 5.21). Some discrepancy is apparent between the predicted and experimental values mainly at stage III, at the end of the multiple-cracking stage, while the stiffness during the third stage (crack widening) appears to be greater in the models. The difference was explained by the progressive rupture of filaments inside the rovings during testing and debonding at the textile–matrix interface, both of which were not accounted for in the assumptions of either models.

The superior performance of TRC with basalt fabric at high temperature is demonstrated in Figure 5.22 (Rambo et al., 2015) for fabric coated with styrene–acrylic latex. The tensile performance of composites produced under heated conditions were better than those at room temperature (Figure 5.22a). Exposure to heat during the production process improves the interlocking mechanism between the matrix and the polymer coating of the basalt fibers, which in turn, strengthens the filament–matrix bond. This was evident by the denser crack pattern, that is, lower crack spacing, obtained for the elevated temperature systems (Figure 5.22b). Additionally, heating the coated yarns (within the composite) improved penetration of the coating between the bundle filaments to promote a better distribution of the load between them. At temperatures higher than 150°C, however,

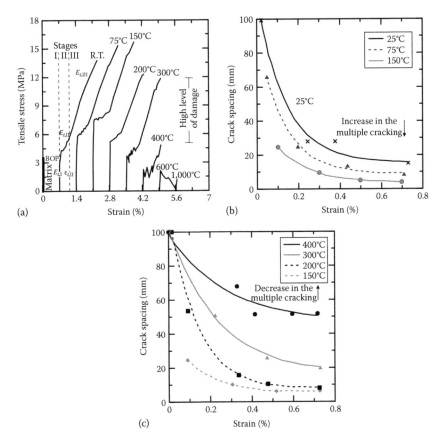

Figure 5.22 Influence of temperatures (preheating) on basalt fabric composites based on (a) the tensile behavior, (b) the crack spacing of TRCs tested at temperatures up to 150°C and (c) above 150°C. (From Rambo, D.A.S. et al., *Mater. Des.*, 65, 24, 2015.)

the heating process reduced the tensile performance (Figure 5.22a) and increased crack spacing (Figure 5.22c), which was a result of the coating's thermal decomposition and of the dehydration process of the matrix. At very high temperatures above 600°C, the basalt-based TRC exhibited brittle behavior, and its tensile strength was even lower than that of the plain matrix (Rambo et al., 2015).

Sim et al. (2005) reported that basalt TRM is a good alternative to a fiber-reinforced polymer as a strengthening layer for concrete when moderate structural strengthening and high resistance to fire are required simultaneously (such as for building structures). At high temperatures (over 600°C), the basalt maintained its volumetric integrity and 90% of its strength, which was not the case for carbon or S-glass fibers.

Garmendia et al. (2011) reported that basalt TRM is effective in the reinforcement of stone masonry arch-shaped structures due to the limited applicability of traditional strengthening systems.

5.3.1.3 Vegetable fibers

The potential applicability of textile fabrics made of vegetable fibers for semistructural and structural applications was demonstrated in several studies (Almeida et al., 2010; Silva et al., 2006, 2009, 2010). Vegetable fibers are attractive reinforcement material because they are relatively cheap, have high strength and low density, and originate from renewable resources. Their manufacturing process requires low energy consumption, and their production and use are associated with much lower CO_2 emission than glass, carbon, or other man-made fibers. On the other hand, the alkaline environment of Portland cement matrices can result in mineralization of vegetable fibers due to the migration of calcium hydroxide (CH) into lumen and the fiber walls. Their alkali sensitivity dictates that a special matrix composition may help reduce their potential aging. To that end, OPC was modified by adding MK and CWCCB to reduce the relative amounts of CH generated during Portland cement hydration and provide a matrix free of CH (Silva et al., 2009).

Most of the research on the use of vegetable fibers as the reinforcing material in cement-based laminates has been dedicated to composites reinforced with sisal fabrics, but jute fabrics have also been investigated (Almeida et al., 2010). TRC with these fabrics exhibited multiple cracking with strain-hardening behaviors.

Sisal fabric composites were evaluated using Portland cement matrix and CH-free matrix (Silva et al., 2010). The composites exhibited multiple-cracking and strain-hardening behavior under tension and bending loadings. The properties of the TRC with the CH-free matrix were superior with respect to tensile strength, toughness, and durability evaluated by the accelerated test of hot water immersion (Figure 5.23). Microstructural analysis showed mineralization in the Portland cement matrix composite, while in the CH-free composite, no sign of fiber deterioration was observed. A similar trend was reported by Toledo Filho et al. (2009) after wet–dry accelerated aging using the same materials.

5.3.2 Hybrid fiber materials

5.3.2.1 Introduction

The combination of two or more fiber materials in a hybrid formation in a single composite can exploit the advantages of each type of fiber. Most hybrid composites are formed using a mixture of different short fibers that

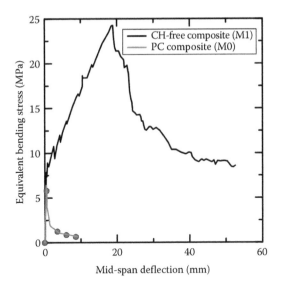

Figure 5.23 Influence of matrix type (Portland cement [PC] vs. CH-free) on the bending response of sisal-based fabric composites immersed in hot water of 60°C. (From Silva, F.A. et al., *Construct. Build. Mater.*, 24, 777, 2010.)

are randomly dispersed in the cement-based composite (Banthia, 2003; Cyr et al., 2001; Kakemi et al., 1998; Kobayashi and Cho, 1982; Lawler et al., 2003; Mobasher and Li, 1996; Peled et al., 2004; Perez-Pena and Mobasher, 1994; Xu et al., 1998). The optimization of strength and toughness in hybrid composites has been studied extensively using combinations of different fiber types with low or high modulus of elasticity. A high-strength, high-modulus fiber primarily tends to increase the composite strength with nominal improvements in toughness, and a low-modulus fiber can only be expected to improve toughness and ductility while offering only a limited improvement in composite strength. The combination of two or more types of fiber can produce a composite that is both strong and tough compared to a monofiber-type composite. However, because the short fibers in hybrid systems are randomly dispersed in the cement-based composite, control over their exact location and orientation within the composite is restricted, a limitation that may rule out the use of hybrid composites for many applications. The benefit of hybridization, therefore, can be combined with the mechanical and tensile contributions of the fabric structure to provide a stronger, tougher, and more durable and economical composite than a conventional fabric-based composite made of a single fiber material.

5.3.2.2 Hybrid fabrics

The production of hybrid composites that use fibers in their fabric form can be achieved using two main approaches: (1) combination of several fabric layers in a single composite, where each layer is made from a different material (Figure 5.24 shows a schematic representation of a hybrid composite made of two different fabrics in two main zones, one at the surface and the other at the composite core; other fabric layer arrangements are possible), and (2) utilization of various yarn types of different materials arranged in different directions within a single fabric, which can allow full control over the exact location of each yarn and its orientation in the composite during production. An example of such fabric is schematically presented in Figure 5.25, which shows how two yarn types can be incorporated into the fabric in an alternating manner in both fabric directions. These approaches provide full control of the exact location of each fabric or yarn, the orientation of which in the composite during production facilitates the design of hybrid fabric-based composites that meet the loading direction and magnitude requirements for specific application. Hybrid reinforcement combinations are also suitable for the manufacture of economically viable

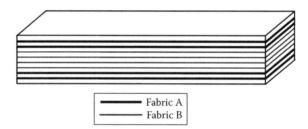

Figure 5.24 Schematic description of a hybrid laminated composite of two different fabric materials located in different composite zones.

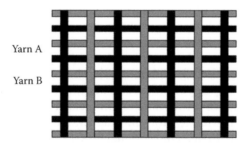

Figure 5.25 Schematic description of hybrid fabric made of two alternating yarn materials in both directions.

products in which less costly fibers, such as PE or PP fibers, substitute for more expensive carbon, aramid, or even AR glass fibers. The benefits of hybrid combinations can also be viewed from the perspectives of durability and sustainability: fibers with low durability, such as glass, can be partially replaced with more durable PE or PP fibers. Also, in terms of the energy consumed in their production, low energy consumption fibers such as PP or PE can be partially substituted for higher energy consumption fibers such as carbon, aramid, or glass.

The potential improvement in reinforcement offered by hybrid fabrics compared to monofabric composites was demonstrated for fabric-based cement composites by Peled et al. (2009), who studied laminated hybrid composites in sandwich construction made with a combination of different layers of low-modulus PE or PP and high-modulus AR glass (G) fabrics. The hybrid composite, which comprised a sandwich hybridization of brittle glass fabrics and ductile PE fabrics, sustained strains better than the single glass fabric composites and had a greater strength than the mono-PE fabric composites. Although this method greatly facilitates composite production, it results in different zones within the composite depending on the relative location of each material layer (Figure 5.26a). For example, a hybrid composite in which brittle and strong fabrics were located at the center of the composite and ductile fabrics on the surfaces of the composite (PE–G–PE) performed better in tension than a composite with the opposite arrangement (G–PE–G), that is, ductile fabrics at the center and strong/stiff fabrics on the surfaces. One of the main drawbacks of using fabrics with significant differences in their properties and bonding in a sandwich arrangement is delamination, because during tensile loading, the ductile fabric can sustain greater

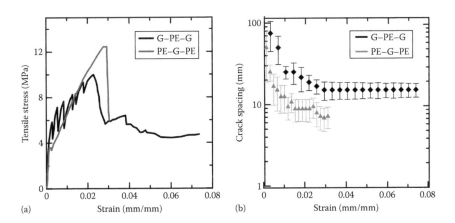

Figure 5.26 Tensile behavior of the PE–glass hybrid sandwich composites: (a) stress–strain response and (b) crack spacing behavior during loading. (From Peled, A. et al., *Cement Concr. Compos.*, 31(9), 647, 2009.)

 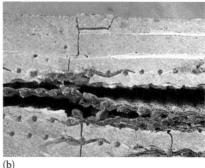

(a)   (b)

Figure 5.27 Crack pattern of glass–PE hybrid composites, side view of the longitudinal section: (a) PE–G–PE and (b) G–PE–G. (From Peled, A. et al., *Cement Concr. Compos.*, 31(9), 647, 2009.)

strain than the brittle fabric, leading to separation and sliding between the two fabric regions (Figure 5.27). The interlayer delamination was more pronounced when the ductile fabric was located at the center and the strong/stiff fabric was on the surfaces of the composite, leading to larger crack spacing and lower tensile behavior of the G–PE–G hybrid composite compared with the PE–G–PE hybrid composite (Figure 5.26b). When using a sandwich hybrid combination of PP and glass fabrics, although the hybrid composites had better ductility than a monoglass composite and no delamination occurred, no benefit in performance was observed compared with the mono-PP composite (Peled et al., 2009).

The potential for delamination in hybrid material combinations constructed using the sandwich approach can be addressed by combining two or more yarn materials within a single fabric. This setup can eliminate composite delamination because there is no significant difference in properties between the layers in the composite. Hybrid fabrics combining ductile, low-cost yarns of PP with more expensive, stiff aramid yarns in various combinations were developed for cement-based composite reinforcement (Mobasher et al., 2014; Peled et al., 2009). Using this approach, different yarn types are mechanically connected in a single fabric unit, thereby reducing the potential for delamination. The benefit of the aramid-PP hybrid fabric as a cement reinforcement is demonstrated in Figure 5.28, which shows similar tensile behaviors of the hybrid fabric and monoaramid composites, that is, the hybrid fabric enables partial replacement of expensive fibers with more economical fibers while maintaining composite performance. Furthermore, the hybrid fabric composite exhibits a much greater reinforcing efficiency than those of the monofabric composites, as quantified by calculated efficiency factors (ratio between the tensile strength of the composite and the tensile strength of the yarn making up the fabric and composite volume fraction)

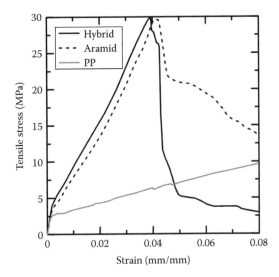

Figure 5.28 Comparison of the tensile responses of composites reinforced with hybrid fabric comprising equal numbers of aramid and PP yarns and mono-material fabrics made of PP or aramid yarns. (From Peled, A. et al., *Cement Concr. Compos.*, 31(9), 647, 2009.)

of 0.94, 0.67, and 0.58 for the hybrid, mono-PP, and monoaramid fabric composites, respectively (Peled et al., 2009).

An analytical parametric model of aramid-PP hybrid fabric composites with different hybrid yarn ratios was developed to simulate composite flexural behavior using a closed-form solution based on the composite tensile stress–strain constitutive relation and compared with experimental data (Figure 5.29) (Mobasher et al., 2014). A gradual change in postcrack stiffness, noted for higher PP content composites, is due to the replacement of a stiff fiber (aramid) with a more compliant fiber system (PP). The combination of different yarns in a hybrid format allows one to design hybrid fabric composites for a desired stiffness by properly aligning the yarns and optimizing the response in accordance with the stiffness required. Mobasher et al. (2014) showed that altering the relative proportions of the aramid and PP yarns in the hybrid fabrics affected the back-calculated tension responses of those fabrics in terms of elastic modulus, postcrack stiffness, first-cracking strain, and ultimate strain (Figure 5.30). Decreasing the proportion of aramid to PP yarns reduced the elastic modulus by almost 50% (Figure 5.30a) and increased the first-cracking strain (Figure 5.30b). In the post initial cracking range, the stiffness increased (Figure 5.30c) and the ultimate strain capacity decreased (Figure 5.30d). These relationships can be exploited to develop effective design tools for the customization of material

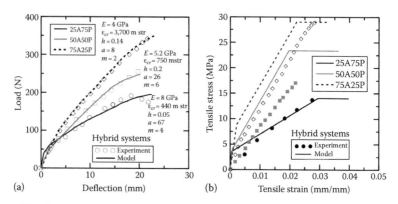

(a)

(b)

Figure 5.29 Comparison between hybrid systems of aramid (A) and polypropylene (P): experimental and analytical (a) flexural and (b) tension responses of different yarn material ratios. (From Mobasher, B. et al., *Cement Concr. Compos.*, 53, 148, 2014.)

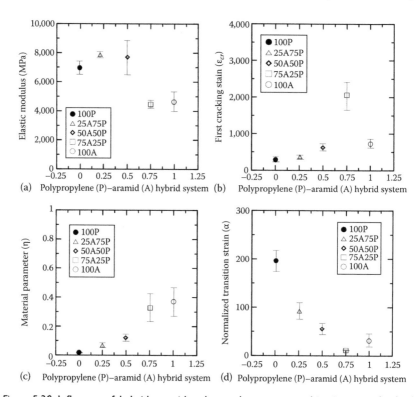

(a) Polypropylene (P)–aramid (A) hybrid system

(b) Polypropylene (P)–aramid (A) hybrid system

(c) Polypropylene (P)–aramid (A) hybrid system

(d) Polypropylene (P)–aramid (A) hybrid system

Figure 5.30 Influence of hybrid aramid–polypropylene yarn combinations on the back-calculated tension responses in terms of (a) elastic modulus, (b) first-cracking strain, (c) parameter showing the ratio of post-crack stiffness normalized with respect to the initial elastic stiffness, and (d) ultimate strain. (From Mobasher, B. et al., *Cement Concr. Compos.*, 53, 148, 2014.)

properties for a given application. The ability to produce hybrid textiles thus confers the possibility to use a specific fiber for a given loading criteria.

Hybrid fabric composites consisting of varying contents of brittle glass and ductile PP yarns were studied by Cohen et al. (2012), Mobasher et al. (2014), and Peled et al. (2011a). The combination of ductile PP and brittle glass yarns in a hybrid fabric was found to have a strong influence on the tensile behavior of cement-based composites, such that adjusting the PP:glass yarn ratio facilitates the production of composites with a range of properties, from brittle to highly ductile, capable of withstanding very high strains. Interestingly, a correlation was obtained between the tensile behavior modes of the fabrics and those of the composites. For both the fabric and composite systems, two stress peaks were observed (Figure 5.31). The first peak is controlled by the brittle glass yarns, while the second is governed by the ductile PP yarns. Increasing the PP content, therefore, enhanced the second peak, while the inclusion of larger glass content increased the first peak. This observation indicates that the glass yarns influence composite properties mainly at low strains, whereas the PP yarns primarily govern the final composite properties, especially its ductility and toughness, at high strains. Thus, composites with high PP:glass yarn ratios are ductile and those with high glass:PP ratios are brittle. Optimum composite performances were obtained when PP yarns constituted 50%–75% of the fabric hybrid content to produce a ductile composite with relatively high strength. It was suggested that in the hybrid system with 64% PP yarn content, equal

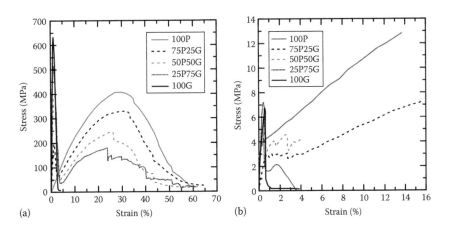

Figure 5.31 Tensile behaviors of (a) hybrid fabrics (without cement) made of different combinations of polypropylene (PP) and AR glass (G) yarns and of (b) cement composites reinforced with these hybrid fabrics. (From Cohen, Z. et al., Hybrid cement-based composites: Dynamic and static tensile behaviors, in J. Barros, I. Valente, M. Azenha, and S. Diasl (eds.), *Eighth RILEM Symposium on Fibre-Reinforced Concretes (FRC)*, BEFIB, Guimarães, Portugal, September 19–21, 2012, pp. 139–141.)

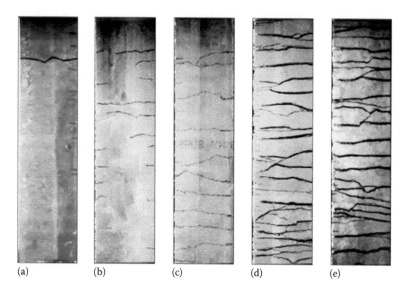

(a)   (b)   (c)   (d)   (e)

Figure 5.32 Cracking patterns of hybrid fabric composite systems with different combinations of PP (P) and glass (G) yarns at maximum stress. (a) 100G. (b) 25P75G. (c) 50P50G. (d) 75P25G. (e) 100P. (From Cohen, Z. et al., Hybrid cement-based composites: Dynamic and static tensile behaviors, in J. Barros, I.Valente, M. Azenha, and S. Diasl (eds.), *Eighth RILEM Symposium on Fibre-Reinforced Concretes (FRC), BEFIB*, Guimarães, Portugal, September 19–21, 2012, pp. 139–141.)

stresses would be carried by the glass and the PP yarns (Cohen et al., 2012; Peled et al., 2011a).

All hybrid and monofabric composites exhibited multiple-cracking behaviors, but the type of fabric material used affected the crack width (Figure 5.32). For example, a very fine, dense crack pattern—hardly visible to the naked eye until a major single crack widened at failure (Figure 5.32a)—was reported for monoglass fabric (100G) composites. Increasing the amount of PP yarns in the hybrid composites led to wider and more visible cracks (Figure 5.32b through e). When the dominant yarns are AR glass, the relatively high modulus and good bonding of AR glass yarns with the cement matrix typically causes them to break rather than pull out, resulting in very fine cracks and low strain. In contrast, the weaker bond between PP yarns and the cement matrix causes them to pull out and straighten during loading, a process that enlarges crack width. Such hybrid combinations can thus be exploited to tailor composites for any given application based on the mechanical and tensile properties of the yarn material.

5.4 Influences of fabric geometry and yarn direction on composite mechanical performance

5.4.1 Introduction

Fabrics for cement-based composites must be constructed in a net structure to allow penetration of the particulate cement matrix between the fabric openings and provide mechanical anchoring (Figure 5.33). For such a system, a variety of geometries that differ mainly in the way the yarns are connected together at the junction points are possible. The most common net structures for cement-based composites are leno (see Figure 2.13) and weft insertion warp-knitted (see Figure 2.28) fabrics due to the complete control conferred by these fabrics on yarn spacing and fabric openings in both fabric directions. However, other types of fabrics such as plain weaves and short-weft warp-knitted and nonwoven fabrics have also been investigated. The influences of fabric type, geometry, and fabric direction on the tensile and flexural behaviors of cement-based composites are discussed as follows.

5.4.2 Fabric structure and yarn shape

5.4.2.1 General concept

Fabric types mainly differ in how the yarns are connected at the junction points. The interlacing of the yarns to form the fabrics affects not only the geometry of the fabric itself but also the geometry of the individual yarns

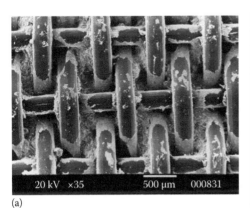

(a)   (b)

Figure 5.33 Mechanical anchoring of fabrics embedded in a cement-based matrix: (a) PE plain weave, (b) AR glass bonded. (a: From Peled, A. et al., *J. Mater. Civil Eng.*, 11(4), 325, 1999; b: From Peled, A. and Mobasher, B., *ACI Mater. J.*, 102(1), 15, January–February 2005.)

that make up the fabric. Yarn geometrical characteristics play an important role in composite properties and performance, and therefore, they must be considered when designing fabrics. In general, the yarn constituting the fabric in composite materials should be as straight as possible, without crimping or any other irregular shape, to achieve high reinforcing efficiency. If the reinforcing yarns are not straight, the transfer of the applied load along the yarn length is not efficient because the yarn is not oriented in parallel with the direction of the main stresses (see also Chapter 2), a scenario that prevents the full reinforcing potential of the yarn from being utilized. A study by Ko (1987) clearly showed that yarn reinforcing effectiveness was reduced in polymer-based composites reinforced with fabrics whose yarns did not maintain a straight form, for example, woven fabrics. In woven fabrics, the yarns are crimped due to their interlacing manner, where two perpendicular sets of yarn (0°/90°) pass over and under each other. Increasing the crimping of the reinforcing yarns in the fabrics—that is, increasing the density of the yarns perpendicular to load direction—reduced the flexural strength of polymer-based matrix composites (Figure 5.34).

The geometry of the woven fabric also affects the flexural strength of polymer-based composites. For example, comparing a fabric whose reinforcing warp yarn passes over three fill yarns and then under one fill yarn (crowfoot) to one in which the reinforcing warp yarn passes over eight yarns and then under one fill yarn (eight-harness) shows that the former has

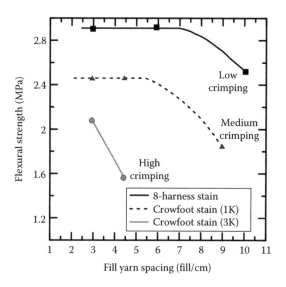

Figure 5.34 Influence of woven fabric density on polymer matrix composite flexural strength (the density is varied perpendicular to the load, or fill, direction). (From Shibata, N.A. et al., *SAMPE Quart.*, 7(4), 25, July 1976.)

greater crimping, which results in a composite with lower flexural strength. A similar trend is also obtained for woven fabrics with increasing fill yarn diameters such that the larger the fill diameter (3K vs. 1K), the higher the crimping of the reinforcing warp yarn and the lower the flexural strength of the composite. Therefore, in the case of polymer-based composites, predictions of fabric reinforcement efficiency usually consider only the fabric's longitudinal yarns, which are in the loading direction. The sole objective of the perpendicular yarns is to provide a mechanism to hold the longitudinal yarns in place during the production of the composite, and as such, they are treated as "nonstructural" elements.

The relation between fabric geometry and composite flexural strength in cement composites is markedly different from that for polymer composites. The nature of the interaction between the cement matrix and the fabric and its individual yarns is more complex than for polymers because the interfacial shear bond that develops between the cement matrix and the fabric is small in many cases (Bentur and Mindess, 2006; Peled and Bentur, 2003; Peled et al., 1998a). This can be attributed to the particulate nature of the matrix, which is markedly different from the more continuous microstructure of the polymer matrix, leading to low interfacial bond, and thus inducing mechanical anchoring becomes important for control of stress transfer (see Chapter 4). The influence of fabric structure should be considered from several perspectives: (1) the fabric openings, that is, the distance between the yarns constituting the fabric in all directions; (2) the density of the connection points in the loading direction, that is, transverse yarn density; (3) the nature of the connection between the yarns, that is, bonding, either by glue, friction, or stitches; and (4) the shape of the yarn in the fabric, that is, either straight, curved, or complex. Some of these issues are discussed as follows.

5.4.2.2 Mesh openings

The size of the fabric openings—fabric density—determines the extent to which the cement matrix can penetrate between the yarn spacing. Higher fabric penetrability leads to its stronger anchoring and, subsequently, better composite performance (Peled and Bentur, 2003; Peled et al., 1999). The influence of woven fabric density on cement penetration is shown in Figure 5.35 (Peled, 1995). Greater transverse yarn density (10 fill yarns/cm with fabric openings measuring less than 1 mm vs. 5 fill yarns/cm with larger fabric openings) results in a denser fabric structure that leads to lower matrix penetration, mainly at the yarn junctions. The use of highly dense fabric, as shown in Figure 5.36 (PVA plain weave), in which a large diameter (high bundle width) PVA yarn is used, results in only limited penetration of the fabric openings by the cement matrix and correspondingly low fabric

(a)   (b)

Figure 5.35 SEM micrographs of plain-weave fabrics in cement matrix with different fill yarn densities: (a) 5 fill yarns/cm and (b) 10 fill yarns/cm. (From Peled, A., Reinforcement with textile fabrics of cement materials, DSc thesis, Technion, Israel Institute of Technology, Haifa, Israel, 1995.)

Figure 5.36 PVA plain-weave fabric embedded in a cement matrix. (From Peled, A. and Mobasher, B., *J. Mater. Civil Eng.*, 19(4), 340, 2007.)

reinforcing efficiency (Peled and Mobasher, 2007). Poor matrix penetration reduces the efficiency of the mechanical connection between the dense fabric and the matrix, and in addition to reducing matrix quality by creating cavities, it can even lead to composite delamination. This scenario accounts for the relatively low tensile strength of the PVA composite compared to the PP composite (9.5 vs. 7.9 MPa), in spite of the much higher modulus of the PVA relative to PP (36 vs. 7 GPa), which should provide much greater reinforcing efficiency for the PVA (Peled and Mobasher, 2007).

5.4.2.3 Density of the transverse yarns

The number of yarn junctions in the loading direction (weft–warp connection), that is, the density of the yarns perpendicular to the load direction, can have a significant influence on the composite mechanical properties and behavior. The influence of the yarns perpendicular to the load direction on composite performance is complex, involving several processes. On the one hand, the transverse yarns directly anchor the reinforcing yarns (Figure 5.37), limiting the sliding of the latter. The number of fabric connections affects anchorage quality: larger number of yarn junctions in the loading direction lead to more anchoring points. The perpendicular anchoring mechanism is explained and discussed in detail in Chapter 4. On the other hand, the transverse yarns may act as defects in the matrix that can cause delamination and/or crack propagation in the direction transverse to the applied loads. The zones in the vicinity of the transverse yarns can thus be considered weak points in the matrix, with high concentrations of stress, and therefore, cracks tend to initiate at these sites and to propagate from one weak spot to the next, that is, from one yarn junction to the next, as clearly demonstrated in Figure 5.38a and b (Mobasher et al., 2006b). The inclusion of transverse yarns in a woven fabric can cause local stresses to develop in the matrix around the perpendicular yarns, leading to severe damage and cracking near the yarn junctions of the fabric as their wavy geometry is straightened during tensile loading (Figure 5.38c).

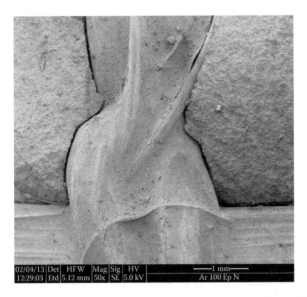

Figure 5.37 Yarn junction point anchoring within the cement matrix (knitted fabric). (From Adiel Sasi, E. and Peled, A., *Composites*, 74, 153, 2015.)

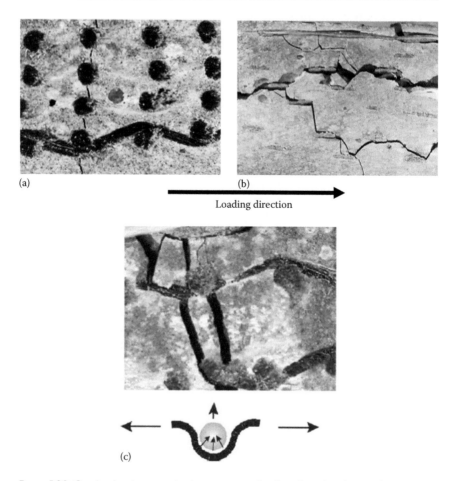

Figure 5.38 Cracks development in the transverse loading direction due to the presence of the yarn perpendicular to loading: (a) low-modulus PE woven fabric, (b) AR glass warp-knitted fabric, and (c) crack propagation across the loading direction of woven PE fabric in a cement composite with schematic representation of the stresses developed at the yarn junction. (From Mobasher, B. et al., *Mater. Struct.*, 39(3), 317, 2006b.)

To achieve the best composite performance, transverse yarn density must therefore be optimized so that the anchorage mechanism will be the dominant contributor to composite mechanical behavior. Hence, fabric design should take into account three contradictory parameters that affect overall composite performance: (1) the structure must be sufficiently open to ensure proper matrix penetration, on the one hand, but on the other hand, (2) the number of fabric joints should be large, that is, a high density of transverse yarns, to improve fabric anchorage within the matrix (but a yarn density

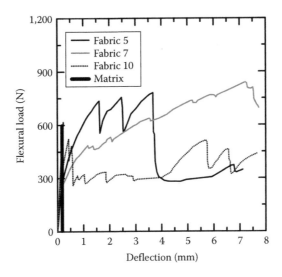

Figure 5.39 Flexural response of composites reinforced with fabrics made with different fill yarn densities perpendicular to the load direction (5, 7, and 10 fill yarns/cm). (From Peled, A. et al., *J. Mater. Civil Eng.*, 11(4), 325, 1999.)

that is too high can reduce cement penetrability and, consequently, matrix properties), and (3) the weakening of the matrix at the yarn junctions where the concentration of stress is high due to the presence of the transverse yarns.

Following these design parameters, Peled et al. (1994, 1999) evaluated the influence of the fill yarn density (fill yarn per cm) on the efficiency of TRC composite with PE plain-weave fabric (Figure 5.39). Optimal flexural strength of 18 MPa was recorded for fabric with 7 fill yarns/cm, compared to similar fabrics with densities of 5 and 10 fill yarns/cm that exhibited flexural strengths of 15 and 13 MPa, respectively. The reduced flexural strength of the high-density fabric composite was related to the reduced matrix compaction (Figure 5.35b) and matrix weakening by the fill yarns that may function as defects. The improved flexural performance observed in the fabric with a larger number of fill yarns, that is, 7 versus 5 fill yarns/cm, was related to the anchoring of the warps by the fills and the resultant strengthening of the bond between fabric and matrix. The reinforcing warp yarn content was the same in all composites, and these fabrics were of a plain weave whose warp and fill yarns were in a crimped shape, which may further increase mechanical anchoring. This influence will be discussed later in more detail.

Optimal tensile performance was reported for leno woven fabric composites with 30 mm spacing between the transverse yarns by Colombo et al. (2013), who compared fabrics with a transverse weft yarn spacing of 10, 20, 30, or 50 mm (Figure 5.40). The warp contents in this study were also similar in all the composites, but the weft yarns were in straight form and the

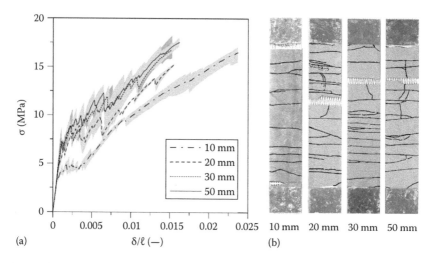

Figure 5.40 Influence of transverse yarn density (spacing of 10–50 mm between transverse yarns) on (a) tensile behavior (stress vs. normalized displacement, calculated by applied stroke displacement divided by the initial distance between clamping edges) and (b) crack pattern of composites reinforced with AR glass fabrics. (From Colombo, I.G. et al., *Mater. Struct.*, 46, 1933, 2013.)

fabric opening was relatively large even for the fabric with smaller spacing. The best performances were observed for the 30 and 50 mm spacing fabric composites, which exhibited similar tensile behaviors, including a relatively stiff behavior at the multiple-cracking branch, but the 30 mm spacing fabric had a denser crack pattern (Figure 5.40b). In addition, cracks also appeared between the weft yarns, an observation that implies improved bonding, consistent with the preceding discussion. Also, the first crack was greater for the fabrics with the wider spacing, indicating improved bonding by a more efficient mechanical anchoring of the fabric within the matrix.

5.4.2.4 Type of junction connection

The binding mode and strength of yarn junctions, that is, the connection between weft and warp yarns, is a design factor that can have a significant influence on composite performance. Particularly for virgin fabrics without coating, junction connections markedly affect the stability and anchoring ability of the fabric within the matrix. Junction binding can be the result of friction between warp and weft yarns in woven fabrics or, in knitted types of fabric, anchoring due to stitches that hold the reinforcing yarns in place. Since an open structure is required for fabrics to be used in TRC, woven fabrics are mainly of the leno type and knitted fabrics are primarily of the weft insertion warp-knitted type. The stitches present in the latter usually

Figure 5.41 Anchorage patterns by the loops of a knitted matrix: (a) loops made of multi-filament bundles, (b) bundle loop imprint in the matrix, and (c) loops made of monofilament yarn. (a: From Peled, A. and Mobasher, B., *J. Mater. Civil Eng.*, 19(4), 340, 2007; b: From Adiel Sasi, E., Flexural behavior of cement based element reinforced with 3D fabric, MSc thesis, Ben Gurion University, Beer-Sheva, Israel, 2014; c: From Haim, E. and Peled, A., *ACI Mater. J.*, 103(8), 235, 2011.)

promote stronger fabric connections than are possible in a woven fabric, rendering the knitted fabric the preferable of the two from an anchoring point of view, as is clearly observed in Figure 5.41. The loop imprints (Figure 5.41b) indicate that the cementitious products penetrated the open spaces formed by the loops of the knitted fabric. Regardless of whether the loops are made of multifilament yarn (Figure 5.41a and b) or monofilament yarn (Figure 5.41c), strong mechanical anchoring of the fabric in the matrix, which results in a high reinforcement efficiency of the knitted fabric, can be obtained. In fabrics with loops made of multifilament yarn, the matrix can also penetrate between the filaments of the bundle, which can enhance fabric anchorage. The higher the level of matrix penetration, the better the bonding and mechanical anchoring of the bundles and loops within the matrix.

Note that the wavy shape of the stitches in the warp direction may further increase knitted fabric anchoring in the cement matrix. When the bundles that constitute the fabric are coated with polymer, however, the cement cannot penetrate the loops or the filaments that make up the loops, and therefore, the effect of the loop structure per se is smaller. Instead, when using coated bundles, the loops around the warp yarns confer on the reinforcing unit in the warp direction a complex shape, that is, with alternating narrow and wide sections along its length (Figure 5.37). Compared to a smooth and straight shape, it can promote much stronger anchoring of the warp reinforcing unit in the matrix, which can further improve composite mechanical performance.

When virgin, noncoated knitted fabrics are employed as composite reinforcement, the anchorage by the loops and junctions is dependent on the stitch density in the reinforcing direction and stitch tightening. In general, the tightening effect of the stitches in weft insertion warp-knitted fabrics is controlled by the number of stitches per given length: increasing their number reduces their size and usually produces a more intense tightening effect (Dolatabadi et al., 2011). The size of the loop can also influence composite performance. On the one hand, each loop effectively functions as an anchorage point once it is filled with the cement matrix, as discussed earlier, and therefore, increasing loop density can enhance composite performance by virtue of the correspondingly greater number of anchoring points in the fabric. On the other hand, increasing the number of loops tightens the filaments of the reinforcing bundle (Figure 5.42), which reduces the spaces between them, thereby lowering fabric penetrability by the cement and, consequently reducing bond strength and composite performance. It was shown experimentally (Peled et al., 2008a) that a loop size of 4 mm was clearly preferable compared with a smaller loop size of 2 mm as

Figure 5.42 Knitted fabric showing loops tightening the filaments of the reinforcing warp yarns. (From Peled, A. and Bentur, A., *Mater. Struct.*, 31, 543, October 1998.)

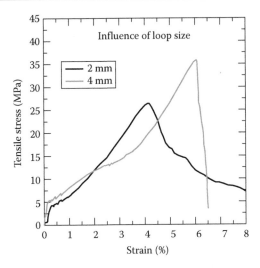

Figure 5.43 Influence of the loop size, 2 versus 4 mm, on composite tensile behavior in a knitted fabric of 138 tex bundles. (From Peled, A. et al., *Cement Concr. Compos.*, 30(3), 174, 2008a.)

reinforcement for a cement composite. The fabric with the 4 mm loop size exhibited higher (20%) composite tensile strength than that with the 2 mm loop size (Figure 5.43). The difference between the tensile strengths of the fabrics with the two loop sizes (without cement) was much less, 1,460 versus 1,388 MPa for the 4 and 2 mm loop sizes, respectively, indicating that the differences between the two composites based on the different loop sizes were not the result of the fabric properties alone.

The same trends are even greater when comparing the strength efficiency factors (calculated as the ratio of composite tensile strength to fabric strength and composite volume fraction; Table 5.3), which shows that the efficiency factor of the 4 mm loop size systems is markedly greater than that of the 2 mm loop size systems for reinforcing yarns with 140 and 90 tex yarns.

Table 5.3 Effect of fabric loop size using 140 or 90 tex yarns on fabric and composite properties

Yarn tex	Fabric loop length (mm)	Fabric tensile strength (MPa)	V_f (%)	Bond strength (MPa)	Composite efficiency factor	Composite tensile strength (MPa)
140	2	1,388	2.7	4.5	0.77	28.8
140	4	1,460	1.5	8.8	1.60	35.1
90	2	1,349	1.6	5.7	1.14	24.6
90	4	1,631	0.9	13.7	1.42	20.8

This observation was explained on the basis of the greater bond strengths between the 4 mm loop size fabric and the cement matrix (see also Chapter 4). The 4 mm loop size fabric composite exhibited greater bond strength of about twice that of the 2 mm loop size fabric due to better cement penetration between the filaments, resulting in looser bundles (i.e., less tightening) generated by the relatively large loop size (see Figure 4.72). It should be noted that the number of stitches, that is, the size of the loop, also affects the shape of the reinforcing yarn (bundle). Dense stitches result in a crimped yarn shape (see Chapter 2) that, in turn, can lead to lower efficiency, as the stresses are not necessarily developed in the loading direction, leading to a further reduction in composite performance.

For knitted fabrics, the type of fabric binding, either tricot or chain (i.e., the warps are interleaved between the overlap of the tricot stitches or loops surround the warp yarn), can also influence composite performance (see Chapter 2), in addition to the presence and size of the loops. Tricot binding produces a relatively flat bundle structure with low filament tightening (relatively open bundle structure), whereas chain binding results in a bundle possessing a rounder diameter with compacted filaments (see Figure 2.29). These differences in bundle geometry can influence the penetration depth of the matrix between the filaments of the bundle. Hegger and Voss (2008) reported that the tensile strengths of composites made with the two textile binding types, tricot or chain, showed significant differences. Both warp-knitted fabrics were made of AR glass yarns with openings of 8/8 mm in the 0°/90° directions. The calculated average ultimate strengths were 813 and 550 MPa for the tricot and chain binding, respectively. The reinforcement efficiencies (i.e., the ratio between the calculated average ultimate strength of the filaments in the component and the tensile strength of the filament), were 0.40 and 0.27 for tricot and chain, respectively.

The binding of tricot-type fabric achieves significantly higher ultimate strengths due to (1) preferable bundle openings, which promote greater matrix penetration between the bundle filaments, and (2) oval roving cross sections that provide larger numbers of filaments located at the bundle surface, and as such, they are in direct contact with the matrix. This is in contrast to the more compact and rounder filament structure of the bundle in the fabric with chain-type binding. The impact of the tricot and chain binding types in AR glass fabrics was also explored under bending conditions (four-point bending on I-section beams 1 m in length) (Hegger and Voss, 2008). A comparison of the ultimate composite strengths of bending relative to those obtained in tension with respect to the reinforcement ratio is presented in Figure 5.44. The bearing capacity of the tricot fabric under bending ranged above the values measured in tensile tests. However, for the chain binding fabric, the bearing capacity tended to be equal in tension and bending. In chain-type binding, the filaments in the bundle are strongly tightened and compressed by the stitches, and therefore, friction between

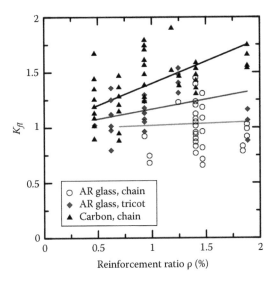

Figure 5.44 Influence of knitted fabric binding on the ultimate composite strengths, bending relative to tension K_{fl}, with respect to reinforcement ratio (ratio between textile and matrix cross section, A_t/A_m). (From Hegger, J. and Voss, S., *Eng. Struct.*, 30, 2050, 2008.)

the inner filaments is intensively activated whether tested under tension or bending. In contrast, the bonding of the inner filaments in the tricot-type binding is greater under bending than under tension due to deflection pressures caused by the bending of the beam at the crack edges, which leads to the better performance of the tricot-type fabric in bending.

The conventional chain-stitch-bonding process can produce spalling on the concrete surface due to deflections of the warp yarns in the knitted fabric. Spalling occurs because the knitting thread (stitches) only fixes the warps at the cross-points with the weft yarns, while between the wefts, the warp is widened and deflected (see Figures 2.30a and 2.31a). This limitation was solved with the development of extended stitch-bonding, in which the warp yarns are fixed on both sides of the fabric (Hausding et al., 2007), and the knitting thread also runs diagonally on the backside of the fabric (see Figures 2.30c and 2.31b), which prevents warp yarn deflection and waviness. The symmetry of this arrangement and the straight warp yarn formation provide stiffer composites when reinforced with the extended stitch-bonding fabric (5-2 and 5-3 in Figure 5.45) than those produced using the conventional process (5-1 in Figure 5.45). In addition, no spalling was observed on the concrete surface with the modified binding fabric (Hausding et al., 2011), an important characteristic for the practical use of such fabrics.

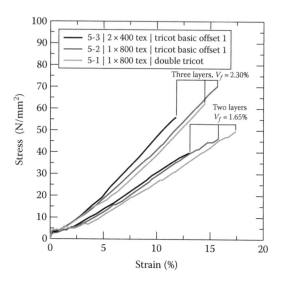

Figure 5.45 Tensile response of composites reinforced with knitted fabric with conventional tricot stitch (double tricot 5-1) or extended stitch-bonding fabric (5-2 and 5-3). (From Hausding, J. et al., Concrete reinforced with stitch bonded multi-plies—A review, in C. Aldea (ed.), *Thin Fibre and Textile Reinforced Cementitious Systems*, ACI Symposium Publication 244, Farmington Hills, MI, 2007, pp. 1–16.)

5.4.2.5 Bundle diameter: number of filaments

The influence of the cross-sectional area of the reinforcing bundles within the fabric (altered by the inclusion of yarns with different tex values) was studied by several researchers (Hegger and Voss, 2008; Peled and Bentur, 1998; Peled et al., 2008a) for noncoated fabrics. In general, ultimate composite strength rises as the bundle tex value decreases due to the increased ratio of the surface area of the bundle to its cross-sectional area. Therefore, compared to high-tex yarn bundles, in the smaller diameter, low-tex bundle, a greater number of filaments are in direct contact with the cement matrix, leading to a stronger bond between the bundle and the matrix. Furthermore, in bundles made of low-tex value yarns with fine diameter, which exhibit matrix penetration depths similar to those for bundles made with high diameter yarns, more filaments of the low-tex yarn bundle are in intimate contact with the cement matrix, leading to improved composite performance.

To investigate the effects of the cross-sectional area, chain-type knitted fabrics made of AR glass comprising 1,200 or 2,400 tex yarns (Hegger and Voss, 2008) and HDPE made of 140 or 90 tex yarns (Peled et al., 2008a) were studied. Greater composite efficiency factors were recorded for the fabric with the low-tex values (Tables 5.3 and 5.4). Also, the bond strength was greater for the lower tex fabric than for the fabric made of high-tex

Table 5.4 Influence of yarn tex on composite tensile properties

Fabric material	Yarn tex	Filament tensile strength (MPa)	Tensile strength of reinforcement in the composite (MPa)	Composite coefficient of efficiency $(k_1)^a$
AR glass	2,400	1,808	460	0.25
AR glass	1,200	2,018	550	0.27
Carbon	1,600	3,912	753	0.19

Source: Hegger, J. and Voss, S., *Eng. Struct.*, 30, 2050, 2008.

[a] k_1 = the ratio between the calculated average ultimate strength of the filaments in the component and the tensile strength of the filament.

yarns (see also Figure 4.72). Note that in terms of strength, the larger (140 tex) bundle size composite outperformed that with the smaller size bundle due to the higher reinforcement content in the 140 tex composite. The improved tensile properties of composites reinforced with chain-type fabric made with low cross-sectional area yarn (low tex) can also be related to the higher structural properties of the fabric itself when loaded in the warp direction (see Figure 2.42) (Peled et al., 2008a). Bundle tightening by the stitches can produce some misalignment and crimping along the warp yarn (Figure 2.41). More severe for thicker bundles, that is, with higher tex value, crimping causes a greater reduction in the tensile properties of high-tex yarns compared to those of the low-tex yarns (2,400 vs. 1,200 tex in the warp direction in line with the stitches) (Table 5.4).

The effect of the yarn geometry (tex) and material (AR glass vs. carbon) on the tensile behavior of TRC composites possessing the same mesh size was studied by Hegger and Voss (2008). The carbon bundles contained 1,600 tex yarns consisting of 21,000 single filaments each (filament diameter 7 μm), and the AR glass comprised 2,400 tex bundles consisting of 1,600 single filaments each (filament diameter 29 μm). The better tensile properties of the carbon fabric discussed earlier yields a 60% higher calculated ultimate strength than the comparable fabric of AR glass (Table 5.4). However, the reinforcement efficiency of the carbon fabric (k_1) is only 19% compared to 25% for the AR glass. These differences are related mainly to low matrix penetration into the carbon bundle due to the large number of filaments in the bundle. A similar trend of lower mechanical performance of composites made with chain-type warp-knitted fabrics consisting of multifilament yarns with a large number of filaments was also reported by Peled and Bentur (1998), comparing the bending behavior of TRC reinforced with Kevlar yarns containing 325 filaments with HDPE composites made of 900-filament yarns. The Kevlar bundle, which had the lower number of filaments, exhibited greater matrix penetration compared to the HDPE (Figure 5.46), leading to greater flexural strength, 37 versus 19 MPa, for the Kevlar versus

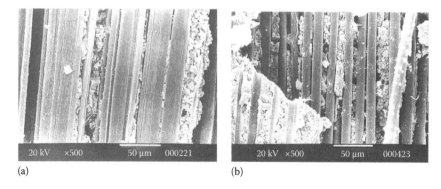

(a)   (b)

Figure 5.46 Bundle in a cement matrix: (a) HDPE 900 tex and (b) Kevlar 325 tex. (From Peled, A. and Bentur, A., *Mater. Struct.*, 31, 543, October 1998.)

the HDPE, respectively. It should be noted that both fabrics had similar high moduli of 59 and 45 GPa, respectively. The reduction in composite tensile response due to low matrix penetration into bundles with a large number of filaments was also observed when the tensile responses of a carbon composite was compared with its fabric alone (Figure 5.47) (Zhu et al., 2011). The low matrix penetration (Figure 5.47a) reduced the reinforcing efficiency of the fabric when it was part of the composite, leading to the significantly lower tensile behavior of the composite compared to that of the fabric not embedded in the cement matrix (Figure 5.47b). The composite exhibited greater strain compared with the fabric alone due to pullout of the filaments from the bundle of the former; lacking direct contact with the matrix, the composite filaments were thus free to slide.

5.4.2.6 Yarn shape

The shape of individual yarns in the fabric can have a considerable influence on the composite performance. Yarn shape depends on the mode of their incorporation in the fabric production method. For example, in woven fabrics the yarns in both directions are in a crimped shape, whereas in knitted fabrics they are in a curved shape induced by the loops. The effect of the crimped shape of the individual yarns of plain-weave fabric on composite flexural behavior is demonstrated in Figure 5.48 (Peled and Bentur, 2003), in which low-modulus PE in plain-weave fabric is compared with the same PE (but not in a fabric form) in individual crimped yarns untied from the fabric and in individual straight yarns. The individual yarns in the plain-weave fabric have assumed an intensively crimped shape relative to other woven fabric structures (see Section 2.5.3.1 and Figure 2.25). The results clearly demonstrate the benefit of the woven fabric structure compared with the single straight yarn composite, where the improved performance of the

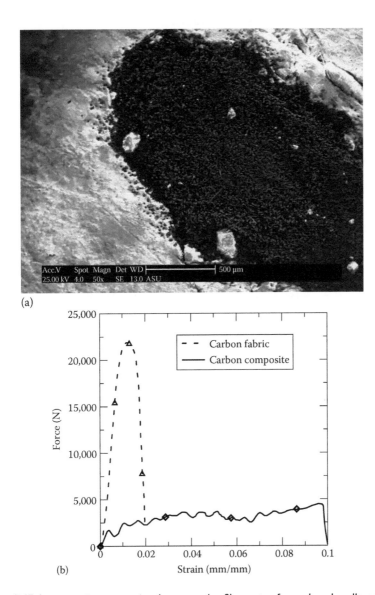

Figure 5.47 Low matrix penetration between the filaments of a carbon bundle embedded in the cement matrix (a), and (b) comparison of the tensile responses of the carbon composite and of the carbon fabric not embedded in the cement matrix. (From Zhu, D. et al., *Construct. Build. Mater.*, 25, 385, 2011.)

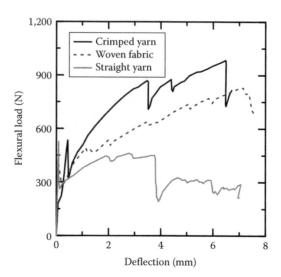

Figure 5.48 Flexural behavior of composites reinforced with woven fabric, untied crimped yarns of this fabric, and straight yarns used to produce the fabric, all from PE. (From Peled, A. and Bentur, A., *Composites*, 34, 107, 2003.)

former is due to the crimped geometry of the individual yarns within the woven fabric. The crimped geometry confers improved bonding by promoting strong mechanical anchoring, as can be clearly seen in Figure 5.49. The influences of crimping and anchoring were discussed in Chapter 4 (see also Section 4.4.2). It should be noted that straight low-modulus PE yarn reinforcement provides elastic–plastic response with some strain softening, while the crimping generates strain-hardening behavior (Figure 5.48).

The contribution by the crimped geometry of the individual yarns within the woven fabric to composite flexural performance is highlighted when comparing the influence of crimping intensity on the flexural strength of a composite reinforced with crimped yarns (untied from the woven fabrics) to that reinforced with woven fabrics, straight yarn (defined as zero in the figure), or plain matrix (Figure 5.50) (Peled et al., 1999). The improved composite flexural strength induced by the crimp geometry of the individual yarns is obvious in Figure 5.50, which shows optimal crimping density for seven crimps or fills/cm for both the fabrics and the untied crimped yarns. The better tensile properties of the crimped yarn composites and the reduction in composite strength at high fill/crimp density were explained based on matrix weakening due to less efficient compaction at the yarn junction points (Figure 5.51) and high yarn curvature. The adoption of a more suitable production method may improve composite strength.

The positive effect induced by the fabric and yarn structures is even more pronounced for composites reinforced with short-weft-knitted fabric

Figure 5.49 Mechanical anchoring of the reinforcing crimped yarn promoted embedment of the woven fabric in the cement matrix. (From Peled, A., Reinforcement with textile fabrics of cement materials, DSc thesis, Technion, Israel Institute of Technology, Haifa, Israel, 1995.)

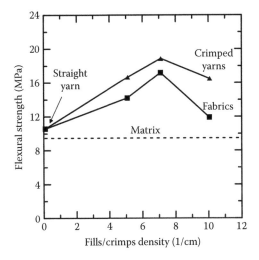

Figure 5.50 Influence of fill yarn density (transverse to load direction) on the flexural strength of a composite reinforced with woven fabric and untied crimped yarns (not in a fabric structure). (From Peled, A. et al., *J. Mater. Civil Eng.*, 11(4), 325, 1999.)

(Figure 2.26), in which the low-modulus PE yarns assume a more complex, "zigzag" geometry than in the woven fabric (Figure 5.52). Strain-hardening behavior is observed for both woven and short-weft-knitted fabric, despite the low-modulus PE yarn content, while the composite with straight yarns of the same PE material, not in a fabric form, exhibits strain-softening behavior.

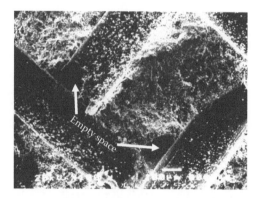

Figure 5.51 Empty spaces near woven yarn junctions embedded in cement matrix. (From Peled, A. et al., *Adv. Cement Based Mater. J.*, 1, 216, 1994.)

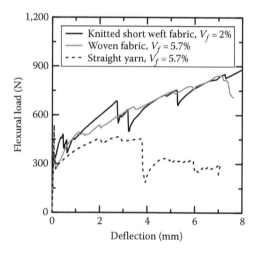

Figure 5.52 Comparison of the flexural behaviors of woven fabric, short-weft-knitted fabric, and straight yarn composites. (From Peled, A. and Bentur, A., *Composites*, 34, 107, 2003.)

The contribution of fabric structure to composite mechanical properties was also evident when the flexural behavior of composites reinforced with low- and high-modulus fabrics was compared (Peled and Bentur, 2003). In general, high-modulus fabric composites are expected to exhibit better mechanical performance than low-modulus fabric composites (see Section 5.3.1), which holds true when composites reinforced with individual yarns not in fabric forms are compared. However, when comparing the flexural performances of different fabric systems with different geometries, the trend is not necessarily the same: the composite with high-modulus

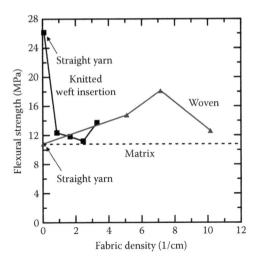

Figure 5.53 Influence of yarn density transverse to the load direction on composite flexural strength made of high modulus HDPE knitted fabric vs. low modulus PE woven fabric. (From Peled, A. and Bentur, A., *Composites*, 34, 107, 2003.)

fabric does not always exhibit better performance than one with low-modulus fabric as demonstrated in Figure 5.53. The composite with the high-modulus fabric (HDPE) did not perform much better than the one with the woven fabrics (PE), in spite of the fact that the latter fabric was made of low-modulus yarns, that is, the HDPE knitted fabric performed poorly despite the superior properties of its yarn; moreover, it did not perform better, or its performance was even lower, than that of the low-modulus PE fabric. In addition, in spite of the relatively low reinforcing yarn content of the low-modulus, PE-knitted, short-weft fabric composite (2% by vol.), the flexural properties of this composite were similar to those of the high-modulus HDPE fabric composite (3.5% by vol.) (Tables 5.2 vs. 5.5), showing a trend that is completely different from that observed for composites with the straight yarns in similar systems.

The counterintuitive trends found in comparisons of low-modulus with high-modulus PE fabrics are even more striking when considering strength efficiencies (calculated as the ratio between the postcracking flexural strength and the product of the yarn volume and its tensile strength) (Figure 5.54). Such trends highlight the fact that the geometries of the fabric and its yarn can have a significant impact on the mechanical performance, as much as that of the yarn's modulus of elasticity and strength, or even bigger. Some of these influences are outlined as follows.

Fabric/yarn geometry improves the performance of woven and short-weft-knitted fabrics (compared to the straight yarn), while it drastically reduces the performance of weft insertion knitted fabric (compared to the straight yarn).

Table 5.5 Effect of fabric structure on composite flexural properties

Fabric type	Yarn type	Nature of yarn	Reinforcing yarn content in the composite (vol.%)	Flexural strength (MPa)	Yarn efficiency coefficient in the fabric in the composite
Woven					
7 yarns/cm	PE (low modulus)	Monofilament	5.7	18	1.21
5 yarns/cm				15	1.01
Knitted short weft					
3 yarns/cm	PE (low modulus)	Monofilament	2.0	16	3.08
		Monofilament		10	1.92
		Film		12	2.31
Knitted weft insertion					
3 yarns/cm	PP (low modulus)	Bundle 100 filaments	3.5	13	0.74
	Kevlar (high modulus)	Bundle 325 filaments		36	0.45
	HDPE (high modulus)	Bundle 900 filaments		19	0.28

Source: Peled, A. and Bentur, A., *Composites*, 34, 107, 2003.

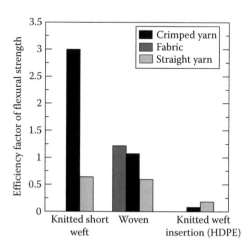

Figure 5.54 Strength efficiency factors of composites reinforced with woven (PE), weft insertion (HDPE), and short-weft-knitted (PE) fabrics. (From Peled, A. and Bentur, A., *Composites*, 34, 107, 2003.)

Yarns in the weft insertion knitted fabrics are in a bundled form and a straight geometry, whereas in the woven and short-weft-knitted fabrics, they are monofilament and do not maintain a straight geometry (i.e., they adopt a crimped geometry in woven fabric and a relatively complex, "zigzag" geometry in the short-weft-knitted fabric). Whereas the crimped and complex shape of the yarns in the woven and short-weft-knitted fabrics tends to increase the matrix–fabric bond, the straight and multifilament nature of the yarns in the weft insertion knitted fabric tends to decrease the bond. Further comparison of the woven and short-weft-knitted fabrics shows that the shape of the reinforcing yarns in the latter is more complex than the crimping geometry of the yarn in the woven fabric. Moreover, the reinforcing yarns of the short-weft-knitted fabrics are held in place tightly by the fabric structure, which apparently induces extremely strong anchoring effects. Such enhanced bonding can confer better performance to the composite with the short-weft-knitted fabric as observed by the high flexural strength efficiency factor (Figure 5.54) and the strain-hardening behavior that occurs even in composites with low reinforcement content volumes (Figure 5.52). However, since the yarns in the weft insertion knitted fabric are bundled and straight and are connected at yarn junctions by stitches, the efficiency of the bundles in a weft insertion knitted fabric used as reinforcement in cement composites is low due to poor matrix penetration between the filaments (see Sections 5.4.2.5 and 4.3.3). When these bundles are kept in a straight form as part of a knitted fabric structure (chain-stitch-bonding), matrix penetration is even lower due to the presence of the bulky stitches and the tightening effect of the stitches, which firmly hold the filaments in the bundle and prevent matrix penetration into the spaces between them.

5.4.2.7 Fabric orientation

The effect of the orientation of the yarns comprising the fabric relative to the load direction was reported in several studies (Hegger and Voss, 2004; Hegger et al., 2006; Mobasher et al., 2007; Mu and Meyer, 2002; Peled et al., 2009). For individual single yarns (not in a fabric form), increasing the angle between the yarns and the load direction is expected to reduce the composite performance (Hull and Clyne, 1996). However, fabric structure consisting of at least two sets of yarns that can be perpendicular (or at angles less than 90°) to each other may generate in the case of cementitious matrices mechanical anchoring effects that may more than compensate for the reduced efficiency due to the orientation effect (see Chapter 4).

Mobasher et al. (2007) showed that for the very small orientation of 6°, the composite exhibited higher tensile load-carrying capacity than a composite tested in the fiber orientation, that is, 0° (Figure 5.55). The minor orientation of the yarns increases the anchorage of its filaments in the matrix leading to higher bond and reinforcing efficiency. However, at a larger yarn

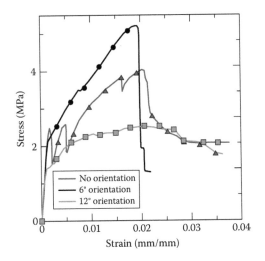

Figure 5.55 Effect of fabric orientation on composite tensile stress response. (From Mobasher, B. et al., *Am. Concr. Inst.*, SP-244-8, 124, 2007.)

angle of 12°, composite mechanical performance was reduced. Mu and Meyer (2002) reported the same trend under bending conditions for several fabric types: AR glass, PP, and PVA. At small angles up to 30°, the composite first-crack strength under bending increased, but for angles above 30°, composite first-crack strength declined with the increase in the warp yarn angle. Yet the highest toughness indices (following ASTM C 1018) have consistently been obtained for the 0° orientation fabric for all three tested fabrics: AR glass, PVA, and PP (Figure 5.56).

Hegger et al. (2006) (Figure 5.57) reported reductions in composite load-bearing capacity under tensile loading with increases in textile inclination. They showed that the effectiveness factor $k_{0,\alpha}$ was also reduced for very small yarn inclinations (defined as the ratio between the load-carrying capacity of the reinforcement with and without inclination). In addition, they also found a reduction in the effectiveness factor of about 50% for an angle of 45°. The effect of textile inclination on the load-bearing capacity was nearly equal for AR glass and carbon fabrics.

The three studies discussed—Mu and Meyer (2002), Hegger et al. (2006), and Mobasher et al. (2007)—each employed different fabrics with different properties (e.g., yarn tex, fabric mesh opening, matrix composition), which can, in some cases, cause better performance at small angles and reductions in composite properties in other cases. For example, mesh size openings of 8 × 8, 5 × 5 to 10 × 10, and 2.5 × 2.5 mm² were studied by Hegger et al., Mu and Meyer, and Mobasher et al., respectively. Yet in general, it can still be concluded that the use of a large reinforcing yarn angle relative to the load direction reduces composite performance, although

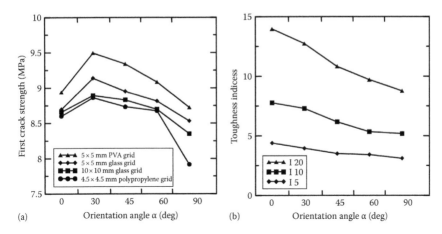

Figure 5.56 First-crack strength (a) and toughness (b) versus fabric orientation of composites reinforced with AR glass 5 × 5 mm² grid when tested under bending. (From Mu, B. and Meyer, C., *ACI Mater. J.*, 425, September–October 2002.)

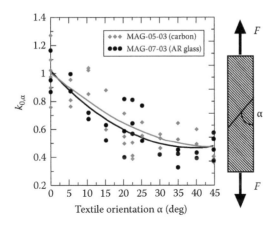

Figure 5.57 Effectiveness factor $k_{0,\alpha}$ of AR glass and carbon fabric composites relative to reinforcing yarn orientation to load direction. (From Hegger, J. et al., *Mater. Struct.*, 39, 765, 2006.)

the exact angle at which such reductions are observed varies depending on yarn and fabric type and geometry.

It should be noted that trade-off between positive and negative effects of yarn orientation has also been reported for individual fiber reinforcement, in terms of competing mechanisms of enhanced anchoring due to inclination and reduced efficiency due to orientation away from the loading direction. Mechanisms of this kind were labeled as snubbing and kink effects

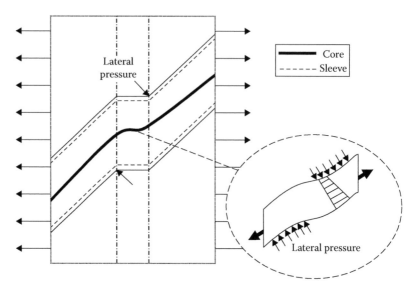

Figure 5.58 Inclined reinforcing bundle at the crack during the multiple-cracking stage. (From Hegger, J. et al., *ACI Convention on AR-Glass and Carbon Fibres in Textile Reinforced Concrete—Simulation and Design*, SP-244-4, Farmington Hills, MI, 2007, pp. 57–75.)

(Leung and Li, 1992; Leung and Ybanez, 1997). Figure 5.58 by Hegger et al. (2006) provides an illustration of the local stresses developed in an inclined fiber bridging a crack, which can either provide anchoring or cause premature fracture of the fiber, depending on its ductility. Such mechanisms were considered in the analysis of orientation effect in fabrics (Figures 5.55 and 5.56) (Hegger et al., 2006; Mobasher et al., 2007; Mu and Meyer, 2002). For small mesh openings, yarn anchoring may increase in magnitude. However, at larger angles, the dominant factors affecting composite properties are those related to a decrease in the number of effective yarns that bridge the cracks and a reduction in the embedded warp yarn length, and as a result, composite mechanical properties are degraded as the warp yarn angle increases.

Mu and Meyer (2002) developed a model to predict the resistance moment during the postcracking stage of TRC as a function of the warp yarn orientation angle α when the corresponding quantities for the 0° and 90° orientation angles are known (Equation 5.1). The model is insensitive to fabric structure, wavy yarn structure, or grid size.

$$M(\alpha)=\left[k_1 \cdot jd \cdot \alpha+M(0)\right]\left(-\frac{2}{\pi}\alpha+1\right)\cdot\frac{2l'}{b}+\left[k_2 \cdot jd \cdot \left(\frac{\pi}{2}-\alpha\right)+M\left(\frac{\pi}{2}\right)\right]\cdot\frac{2}{\pi}\alpha\cdot\frac{2l''}{b}$$

$$(5.1)$$

where k_1 and k_2 are material constants reflecting the frictional resistance of the yarn surface, which can be assumed to relate to yarn diameters $d1$ and $d2$ as follows: jd is the lever arm, which is readily determined, $M(\alpha)$ is the average postcrack moment capacity, and l' and l'' are the warp and fill embedded lengths, respectively.

Composite properties in the two fabric directions, 0° and 90°, are expected to be identical if the same yarns with equal densities are used to produce the fabric in those two directions. However, as discussed in Chapter 2, fabric performance is more complex than that of its single yarns, and it is not always the same in the two perpendicular directions, even though the yarns are identical. Composite performance in the two fabric directions, therefore, may also vary, and several researchers have investigated this phenomenon (Hegger et al., 2006; Mobasher et al., 2007; Peled et al., 2009).

Greater composite tensile strengths were reported along the weft (90°) direction compared with the warp (0°) direction (Figure 5.59). This observation is in stark contrast to the tensile strengths of AR glass fabric alone (not in cement), which were greater in the warp (974 MPa) than in the weft (533 MPa) direction (Hegger et al., 2006). A similar trend for AR glass fabric was reported by Adiel Sasi (2014), who ran tests on chain-type knitted fabrics without any coating on the fabric yarns (Figure 5.60). In such fabrics, the warp direction is stronger than the weft direction (Figure 5.60b) due to the tightening of the warp filaments by the stitches surrounding them, which produces a rounder diameter bundle in cross section due to bundle compaction (Figure 5.61a) and results in greater friction between the

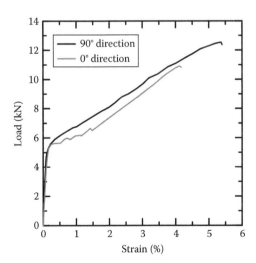

Figure 5.59 Tensile response of composites reinforced with AR glass fabrics tested in two fabric directions, 0° and 90°. (From Hegger, J. et al., *Mater. Struct.*, 39, 765, 2006.)

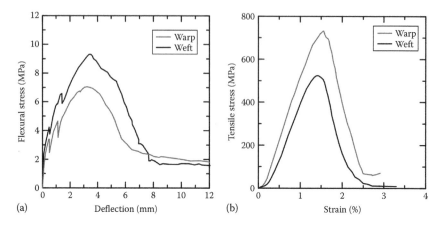

Figure 5.60 Tensile responses of (a) a composite and (b) fabric not embedded in a cement matrix of AR glass fabrics tested in the 0° (warp) and 90° (weft) directions. (From Adiel Sasi, E., Flexural behavior of cement based element reinforced with 3D fabric, MSc thesis, Ben Gurion University, Beer-Sheva, Israel, 2014.)

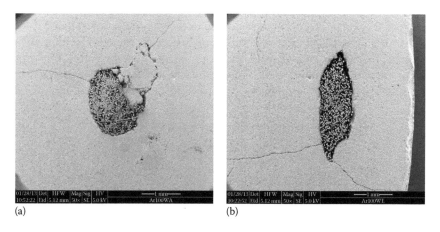

Figure 5.61 Images of AR bundles embedded in a cement matrix: (a) warp with a round cross section and (b) weft with an oval cross section. (From Adiel Sasi, E., Flexural behavior of cement based element reinforced with 3D fabric, MSc thesis, Ben Gurion University, Beer-Sheva, Israel, 2014.)

filaments. In the weft direction, in contrast, the yarns are looser, exhibiting a flattened, ovular shape in cross section (Figure 5.61b), and free to slide, conferring ultimately weaker tensile properties on the whole fabric.

When fabrics are embedded in a cement matrix in TRC composite, an additional factor, the extent to which the matrix penetrates the bundle cross section (and related parameters such as the number of filaments in direct contact with the cement matrix and the resultant matrix–fabric bond strength), must also be considered. The flat cross section of the weft bundle places more filaments at the bundle perimeter, that is, in direct contact with the matrix, and their loose binding promotes more efficient matrix penetration into the bundle interior, thereby improving the bond between the outer and inner filaments. The compaction and tightening of the filaments along the warp direction, in contrast, results in a round cross section that limits cement penetration and reduces bonding with the matrix. This scenario leads to a more homogeneous activation of the total bundle cross section and higher effectiveness of the weft yarns compared to the warp yarns. Such differences in the bonding of the bundles in the two fabric directions lead to higher composite tensile strength along the weft 90° direction than along the warp 0° direction of a chain-type fabric (Figures 5.59 and 5.60a).

Likewise, a composite reinforced with polymer-coated fabric was found to be stronger in the weft direction than in the warp direction (Mobasher et al., 2007). Flexural strengths of 14.1 and 12.6 MPa and greater deflection of 16.7 versus 12.2 mm in the weft versus the warp direction, respectively, were recorded for a coated AR glass composite. These results correlate strongly with those for the behavior of a coated fabric not in a cement matrix (see also Figure 2.38). Polymer penetration is greater in the weft direction, leading to improved fabric performance in this direction. Furthermore, when used as a reinforcing unit in a cement-based composite, the curvature of the warp yarns due to the presence of the stitches reduces their reinforcing efficiency in the warp direction, and consequently, composite mechanical performance is also reduced. The straightness of the weft yarns provide better alignment and, as a result, greater reinforcing efficiency when located in the load direction of a composite.

The trends of the effects of fabric orientation discussed earlier are valid mainly for noncoated fabrics made by the conventional casting process. Composites produced by the pultrusion process exhibit markedly different trends with regard to tensile behavior (Peled et al., 2009). In the pultrusion process, the fabric is passed through a cement bath to produce a laminated composite. They had better mechanical properties in the pultrusion direction compared to the transverse direction (Figure 5.62). This improvement was explained to be due to more efficient matrix penetration between the bundle filaments located in the processing direction, that is, the pultrusion direction (Figure 5.63), resulting in stronger mechanical anchorage and, subsequently, better bonding. In the work by Peled et al. (2009), the warps were oriented in the pultrusion direction during the impregnation process, which led to enhanced tensile performance in the warp (0°) direction (Figure 5.62). Cement matrix penetration between the yarns in the 90° direction during the impregnation bath stage was incomplete (Figure 5.63b), which resulted in poor bonding and reduced tensile performance in that direction.

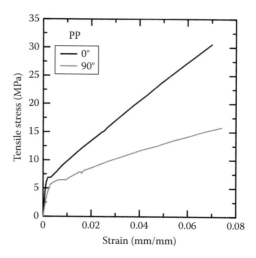

Figure 5.62 Tensile behavior of uncoated multifilament PP fabric composites produced by the pultrusion process under testing in the warp (0°) and weft (90°) directions. The warp yarns were located in the pultrusion direction and the weft yarns were transverse to the pultrusion direction. (From Peled, A. et al., *Cement Concr. Compos.*, 31(9), 647, 2009.)

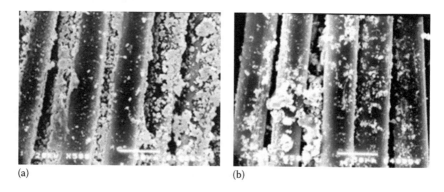

(a)   (b)

Figure 5.63 SEM images of PP fabric composites produced by the pultrusion process: (a) improved matrix penetrability in the pultrusion (warp) direction and (b) poor matrix penetrability in the transverse (weft) direction. (From Peled, A. et al., *Cement Concr. Compos.*, 31(9), 647, 2009.)

5.4.3 Three-dimensional (3D) fabrics

Fabrics can be produced in two-dimensional (2D) forms, in which the yarns are located only at the fabric plain, or in three-dimensional (3D) forms, in which a third set of yarns is inserted into the fabric in the direction transverse to the plan, or z-direction (see Chapter 2). A large variety of 3D fabrics

Figure 5.64 Warp-knitted 3D fabric made of AR glass yarns at the fabric surfaces and aramid yarn along the fabric thickness. (From Adiel Sasi, E. and Peled, A., *Composites*, 74, 153, 2015.)

of various shapes can be produced by exploiting different textile production methods. Warp-knitted spacer fabrics are ideal for concrete applications because the net-like grid produced on the fabric surface when using this knitting method (Figure 5.64) facilitates high cement penetration between the fabric openings (Gries and Roye, 2003; Roye, 2007; Roye and Gries, 2004, 2005; Roye et al., 2004, 2008). In addition, fabric stability and strength of the 3D textile can be improved by the inclusion of straight, multifilament yarns on the fabric surface (Roye and Gries, 2007a).

The production of 3D warp-knitted spacer fabric entails the simultaneous fabrication of two chain-type warp-knitted fabric units (having the same or different structures) that are connected to each other during fabric production by another set of yarns situated in the fabric's z-direction, to obtain a 3D structure (see Section 2.5.2.2). The lengths of the connecting (spacer) yarns can be varied to obtain different fabric thicknesses, which are dictated by the composite geometry. In addition, this approach allows for the incorporation of several independent spacer distances in a single textile structure (Figure 5.65), that is, one fabric can have two or more thicknesses based on the geometrical requirements of the composite element.

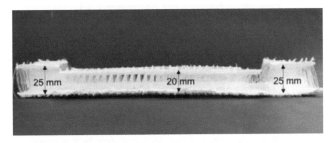

Figure 5.65 Warp-knitted 3D fabric with various thicknesses at different locations. (From Roye, A. and Gries T., *Am. Concr. Inst.*, SP-244-5, 77, 2007b.)

3D spacer knitted fabrics can be used to produce TRC composites that have one layer of fabric on each of the two surfaces of the component by using a single textile structure comprising reinforcement yarns along its thickness and with the desired cross-sectional shape. Using this method, the fabric structures are produced with the desired shape and thickness dependent on the mechanical performance required by the TRC component, and the cement matrix is poured in between the openings of the 3D fabric, without the need to apply any tension to the fabric. From a production point of view, this approach can be beneficial as the desired concrete structure can be directly produced in a single step. In addition, a large range of yarn materials—and different yarn types in the three directions—can be integrated in this type of knitted fabric (Figure 5.64). Practical applications range from sandwich panels, claddings, and simple facade elements to complexly shaped components. For such applications, the distances between each set of spacer yarns can be varied along the fabric thickness to obtain different structures. For example, a fabric with an open shape can be obtained by using spacer yarns in only a few locations (Figure 5.66a) to fabricate TRC elements with a horseshoe profile (Figure 5.66b and c) (Brameshuber et al., 2008). The good

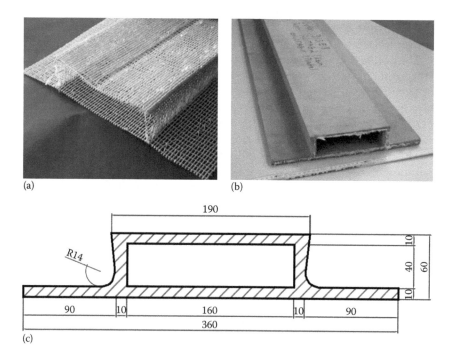

(a)

(b)

(c)

Figure 5.66 Spacer fabric connected at two locations only by the spacer yarns (a), concrete element made with this 3D fabric (b), and (c) scheme of concrete element design. (From Brameshuber, W. et al., *Am. Concr. Inst.*, SP-250-3, 35, 2008.)

stability of 3D spacer fabrics permits the design of extremely thin TRC structural elements with exceptional mechanical performance (Mecit and Roye, 2009; Naaman, 2012; Zhu et al., 2013).

The flexural behavior of TRC reinforced with 3D warp-knitted spacer fabric, particularly the influence of the yarns situated in the through-thickness (z-direction) of the fabric, was explored by several investigators (Adiel Sasi and Peled, 2015; Amzaleg et al., 2012, 2013). The effects of the through-thickness yarns were examined in terms of four parameters, including (1) yarn properties, that is, high-performance aramid yarn (Ar) versus low-performance polyester yarn (PES) (3D Ar vs. 3D REF); (2) variation of the z yarn content, that is, 100% versus 50%, of the high-performance aramid yarn (3D Ar 100 or 3D Ar 50); (3) treatment of the fabric with epoxy; and (4) inclusion of the yarns in 2D versus 3D fabric composites. Overall, the 3D fabric composites performed better than the 2D fabric composites (Figure 5.67), as the latter tended to delaminate under bending tests (Figure 5.68). The superior mechanical properties observed for the 3D fabric composites are due to the complex geometries of the 3D fabrics, which facilitated stronger anchoring mechanisms with the cement matrix and good connections between the yarns along the fabric planes, ultimately producing a more efficient reinforcing unit. Adiel Sasi and Peled (2015) reported that the yarns in the fabric z-direction, that is, composite thickness, markedly influence composite performance (Figure 5.67). They found that (1) high-performance z yarns such as aramid greatly improved the strength and toughness of the cement-based composite compared to a composite with low-performance yarns such as PES and that (2) the greater the relative

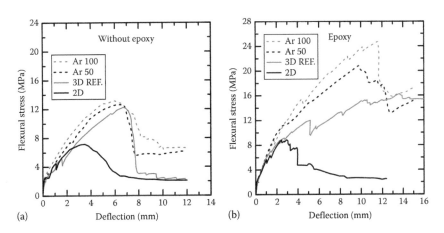

Figure 5.67 Flexural behavior of composites reinforced with 3D fabrics having different z yarn contents relative to a 2D fabric composite of the same material: (a) uncoated versus and (b) epoxy-coated fabric. (From Adiel Sasi, E. and Peled, A., *Composites*, 74, 153, 2015.)

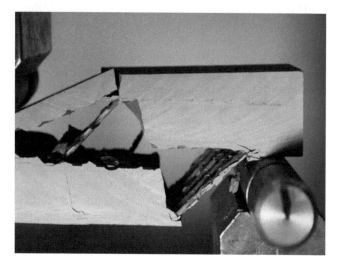

Figure 5.68 Delamination of 2D fabric composite with two fabric layers tested in bending. (From Adiel Sasi, E. and Peled, A., *Composites*, 74, 153, 2015.)

content of high-performance yarns in the *z*-direction, the better the composite performance (Ar 50, Ar 100 in Figure 5.67).

The influence of *z*-direction yarns was more pronounced in fabric treated in epoxy before its inclusion in the cement composite (Figure 5.67b). The benefit of the epoxy treatment was induced by its ability to bind the different elements of the fabric together, enabling them to behave as a single unit in all three directions and thus to carry the applied loads together, as observed by photoelastic measurements during bending tests (Figure 5.69).

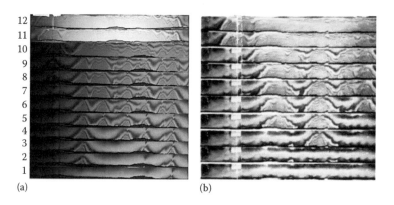

Figure 5.69 Crack propagation shown by photoelastic responses of 3D fabric composites during bending tests: (a) without and (b) with epoxy (only PES yarns in the *z*-direction). (From Adiel Sasi, E. and Peled, A., *Composites*, 74, 153, 2015.)

For the epoxy systems, small cracks were initiated immediately when loading began along the entire length of the specimen, that is, reflecting behavior of a whole fabric unit. Starting from the center of the specimen and progressing in both directions gradually, the initially tiny cracks then grew during loading. In the system without epoxy coating, cracks were created individually and separately during loading, starting from the weak point of the specimen, that is, reflecting the different behaviors of individual yarns. The reinforcing efficiency of the epoxy-treated composite, however, was limited, in that the bond between epoxy and matrix was not perfect, such that cracking at the yarn junction was more likely (Figure 5.70).

Within the cement-based composite, the arrangement of the 3D fabric relative to the applied force was shown to influence composite flexural behavior (Amzaleg et al., 2012). In the vertical arrangement, the 2D fabric layers (connected by the spacer yarns) are located at the sides of the composite during loading, and the spacer yarns are situated along the width of the composite. For the horizontal system, the fabrics are at the bottom and top of the specimen, and the spacer yarns run from top to bottom, that is, through thickness (Figure 5.71). Tests with the two arrangements, with low-modulus PES yarns in the z-direction, showed virtually no difference between the two

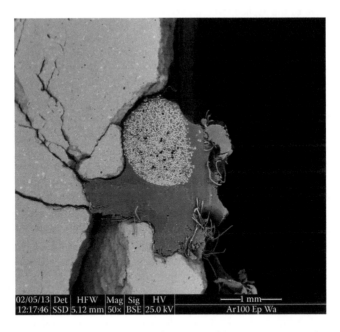

Figure 5.70 Back-scattered SEM image of 3D aramid fabric composite with epoxy; view of the fabric cross section showing the warp glass yarn bundle (represent by the white dots) with the aramid loop around it. (From Adiel Sasi, E., Flexural behavior of cement based element reinforced with 3D fabric, MSc thesis, Ben Gurion University, Beer-Sheva, Israel, 2014.)

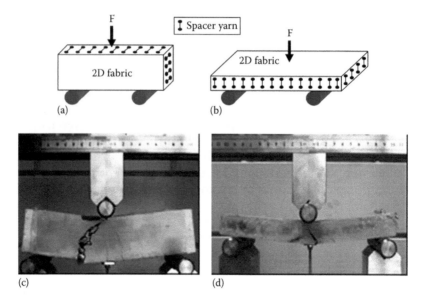

Figure 5.71 Testing arrangement of the 3D fabric within the cement-based composite relative to the applied force: (a) vertical and (b) horizontal arrangements. Images of 3D Ar 100 fabric composite taken at the end of testing (8 mm deflection): (c) vertical and (d) horizontal arrangements. (From Amzaleg, E. et al., Bending behavior of 3d fabric reinforced cementitious composites, in J. Barros, I. Valente, M. Azenha, and S. Diasl (eds.), *Eighth RILEM Symposium on Fibre-Reinforced Concretes (FRC)*, BEFIB, Guimarães, Portugal, 2012, pp. 71–73.)

(Figure 5.72a), yet significantly better mechanical properties were obtained for the horizontal arrangement systems when high-performance aramid yarns constituting 50% of the fabric were located in the z-direction (Figure 5.72b). The differences in the behaviors of the two systems were attributed to the different locations of the fabrics and spacer yarns relative to the applied loads and to the differences in composite heights, all of which led to divergent crack bridging and crack propagation behaviors between the two systems (Figure 5.71c and d) and to variations in the reinforcing efficiency of each system. Impact tests (e.g., drop-weight impact testing with a hammer) of similar 3D REF fabric composites with low-modulus PES yarns in the z-direction showed trends different than those observed under static bending (Figure 5.73) (Peled et al., 2011b). Under impact, the vertical arrangement specimens clearly exhibited rebound behavior, that is, after reaching their maximum deflection, the specimens shifted back to some extent, and no failure occurred. On the other hand, no such rebound behavior was observed in the horizontally arranged composites. Under impact, toughness was improved by about 50% in the vertically arranged versus the horizontally arranged 3D fabric, which is different than the trend in static loading.

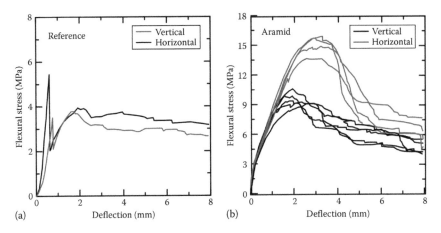

(a) Deflection (mm)

(b) Deflection (mm)

Figure 5.72 Flexural responses of the different test arrangement systems for (a) 3D reference (low-modulus PES yarns along the fabric thickness) and (b) 3D Ar (high-modulus aramid yarns along the fabric thickness). (From Amzaleg, E. et al., Bending behavior of 3d fabric reinforced cementitious composites, in J. Barros, I. Valente, M. Azenha, and S. Diasl (eds.), *Eighth RILEM Symposium on Fibre-Reinforced Concretes (FRC)*, BEFIB, Guimarães, Portugal, 2012, pp. 71–73.)

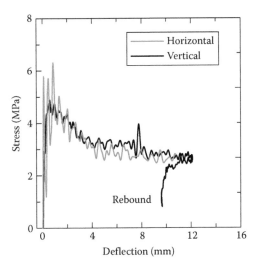

Deflection (mm)

Figure 5.73 Impact behavior of AR glass 3D reference fabric composites. (From Peled, A. et al., Impact behavior of 3D fabric reinforced cementitious composites, in H.W. Reinhardt and G. Parra-Montesinos (eds.), *Workshop on High Performance Fibre Reinforced Cement Composites (RILEM) HPFRCC-6*, Ann Arbor, MI, June 19–22, 2011b, pp. 543–551.)

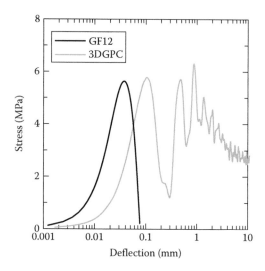

Figure 5.74 Impact response of an AR glass reference 3D fabric composite (3DGPC) compared to a composite with short 12 mm glass fibers (GF12). (From Zhu, D. et al., *J. Sust. Cement Based Mater.*, 2(1), 1, 2013.)

3D reference AR glass fabric ($V_f = 1.8\%$) was found to markedly improve composite toughness and energy absorption by as much as 200-fold compared to short 12 mm AR glass fiber composites ($V_f = 3.5\%$) (Figure 5.74), but this improvement was much smaller for the composite maximum stress. The short fiber composite clearly exhibited brittle behavior (i.e., test duration was <0.1 ms) in contrast to the strongly ductile behavior of the 3D fabric composite (i.e., test duration was >10 ms). Maximum deflections of 0.08 and 10 mm were recorded for the GF12 composite and for the 3DGPC, respectively (Zhu et al., 2013).

In the conventional applications of the 3D fabric composites, the fabric is placed in a mold and thereafter impregnated with the matrix that is cast into it. 3D TRC can be produced in a drastically different way such as concrete canvas (CC), first proposed by Brewin and Crawford in 2005 (Concrete Canvas Ltd., 2012). CC is produced by impregnating first the 3D spacer fabric with dry cement powder (without water). At this stage, the CC remains flexible and soft like any cloth, and as such, it can be used to cover virtually any structural surface or element, regardless of surface geometry. Once the CC has been fitted to the structure, it is sprayed with water or water is added to the fabric's top surface, and hardening takes place. The result is a thin composite layer conforming to the shape of the outer profile of the structure or element. This relatively quick and efficient production method of CC is ideal for civil

Table 5.6 Tensile properties of concrete canvas made with 3D spacer fabrics

CCs	Yarn direction	Volume fraction of spacer yarns (%)	Orientation angle (warp/weft)	Tensile strength (MPa)
T20	Warp	3.89	$0/0.35\pi$	1.16
	Weft	1.99	$0.5\pi/0.17\pi$	1.08
N15-I	Warp	2.91	$0.17\pi/0.17\pi$	1.14
	Weft	2.91	$0.33\pi/0.33\pi$	1.04
N15-II	Warp	1.46	$0.17\pi/0.17\pi$	1.07
	Weft	1.46	$0.33\pi/0.33\pi$	0.96
N15-III	Warp	1.46	$0/0.5\pi$	0.97
	Weft	1.46	$0.5\pi/0$	1.10
N15-IV	Warp	2.91	$0/0.5\pi$	1.12
	Weft	2.91	$0.5\pi/0$	0.57

Source: Han, F. et al., *Construct. Build. Mater.*, 65, 620, 2014.

engineering applications such as prefabricated shelter coverings, stable paths for vehicles and pedestrians, and protective layers for pipes.

The tensile behaviors of CC composites were studied by Han et al. (2014), using five polyethylene terephthalate (PET)-based 3D spacer fabrics with varying amounts of spacer yarns (Table 5.6) and in which the orientation angles of the warp/weft yarns and the tightness of the solid fabric architecture were also varied. The T20 3D fabric had a thickness of 20 mm, and the fabrics on its two faces were identical mesh structures that contained 0.28 yarn volume fractions in both directions. All the N15 3D fabrics had thicknesses of 15 mm, and one face of the fabric had a mesh structure while the other comprised solid (dense) fabric with 0.35 volume fraction in both directions. The inclusion of the solid (dense) fabric on only one face of the 3D fabric aids CC production, as it prevents the fresh cementitious mixture from "leaking" or flowing out of the fabric structure while simultaneously allowing the water to easily penetrate all the cement-impregnated fabric when poured from the loose face of the fabric. From a mechanical point of view, the 3D fabric with the solid (dense) fabric (N15) improved significantly the tensile performance—tensile strength, ductility, reinforcing efficiency, and crack pattern—in a manner comparable to the 3D fabric with net structures on both composite faces, T20 (Figure 5.75). Increasing the number of spacer yarns improved the ductility and tensile strength (Table 5.6). This trend was attributed to the higher bridging stress induced by the spacer yarns. In addition, the better the warp/weft yarn alignment in the loading direction, the greater the improvement in 3D composite tensile performance due to their higher reinforcing efficiency. In conclusion, the 3D spacer fabric with one solid outer textile substrate was found to be preferable to that with open faces as it conferred more efficient reinforcement on CCs.

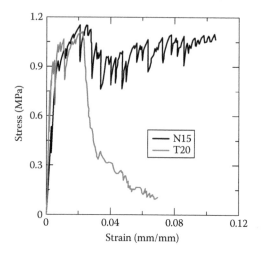

Figure 5.75 Tensile responses of 3D fabrics with different thicknesses, one of 20 mm (T20, open structure on both fabric faces) and the second of 15 mm (N15, solid dense structure on one fabric face) used for concrete canvas applications, tested in the warp direction. (From Han, F. et al., *Construct. Build. Mater.*, 65, 620, 2014.)

5.5 Influence of coating on composite mechanical performance

The limitation of multifilament yarns in the fabric with regard to impregnation by the particulate matrix has been extensively discussed (Banholzer, 2004; Bartos, 1987; Bentur, 1989; Bentur and Mindess, 2006; Zhu and Bartos, 1997) and highlighted in Chapter 4. It leads to stress distributions over the total cross section of the bundle that is inhomogeneous. The outer filaments are exposed to larger strains and are therefore subjected to higher stresses in the cracks than the inner filaments (Figure 5.76).

To improve reinforcing efficiency of the bundle, treatments with polymers to coat and bond the individual filament in the bundle or impregnate the spaces between them have been evaluated. A wide range of polymers with different viscosities were studied to evaluate their efficiency for enhancing the properties of TRC: PVA, ethyl acetate (PVAc) (Weichold, 2010), poly(ethylene oxide) monomethyl ether (PEO-MME), acrylate dispersions, nonionic polyurethane (Glowania and Gries, 2010; Glowania et al., 2011), styrene–butadiene copolymers, and a variety of epoxy resins (Gao et al., 2004; Hartig et al., 2012; Hegger et al., 2006; Schleser, 2008; Schorn et al., 2003). In AR glass TRC composites, epoxy resin impregnation provided greater reinforcing efficiency than acrylate dispersion (Figure 5.77) (Hegger et al., 2006). A denser crack spacing pattern and finer cracks were observed

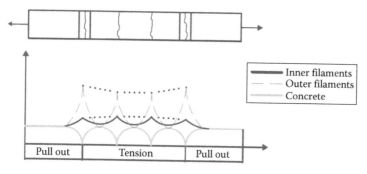

Figure 5.76 Strain distributions on the inner and outer filaments within the bundle relative to crack development during loading. (From Hegger, J. et al., *Mater. Struct.*, 39, 765, 2006.)

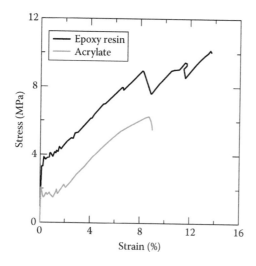

Figure 5.77 Tensile responses of composites reinforced with AR glass fabric coated with epoxy resin and with acrylate. (From Hegger, J. et al., *Mater. Struct.*, 39, 765, 2006.)

in the epoxy composite. Also, the two coating materials exhibited different failure modes: splitting of the concrete along the reinforcement for the epoxy coating and fracture of the reinforcement in the case of the acrylate coating.

A greater improvement in load-bearing capacity was reported for carbon fabric with the polymer coating compared to that found for the noncoated fabric by Koeckritz et al. (2010). The treatment was achieved by integration of the polymer coating in the fabric production process (i.e., in the warp knitting machine) using an in situ polymer coating and an infrared drying device. One limitation of the polymer coating, however, is its relatively low bonding with the cement matrix, which results in composite delamination. Modification of the composite with sand that is spread on the polymer coating promotes stronger mechanical anchoring of the carbon textile with the cement matrix to provide greater composite load-bearing capacity and bending performance (Li and Xu, 2011).

Composite mechanical performance and crack width can also be affected by the interaction between the sizing applied on the individual filaments during spinning with the coating applied on the entire bundle (Scheffler et al., 2009). The addition of nanoclay to the coating led to a reduction in composite crack widths and improved maximum load-bearing capacity (Figure 5.78) due to the development of hydration products at the yarn–matrix interface and the corresponding enhancement of interface stiffness in the composites.

Brameshuber et al. (2008) introduced a coating process for AR glass friction spun hybrid (FSH) yarn to surround the AR glass yarn with PP fibers (Figure 5.79a). The entire yarn is then exposed to high temperatures of 180°C–200°C to melt the PP. Each core-reinforcing yarn (e.g., AR glass) is thus surrounded with a molten PP coating. Moreover, the melted PP also penetrates to the interior of the reinforcing yarn core, binding the individual AR glass filaments together once it cools down. Composites in which the fabric reinforcements, made of AR glass yarns, have been coated with PP exhibit markedly greater bending performance, as shown by fabrics 2, 3, and 4

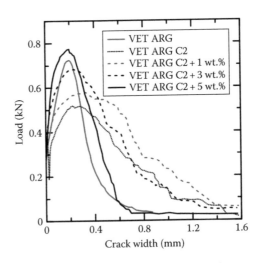

Figure 5.78 Load crack width response of composites reinforced with AR glass yarns coated with a range of coating materials, including the addition of nanoclay at varied levels (1%, 3%, 5%) to the coating. (Adapted from Scheffler, C. et al., *Compos. Sci. Technol.*, 69, 905, 2009.)

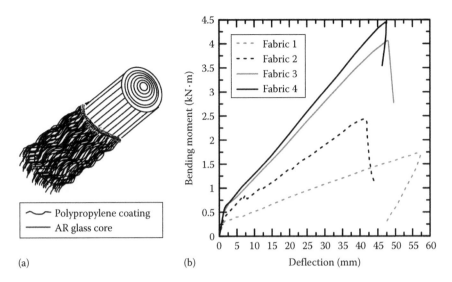

Figure 5.79 FSH yarn made of AR glass reinforcement yarn at the core with PP coating (a) and the bending behavior of composites reinforced with different fabrics made of FSH-coated AR glass yarns and uncoated yarns (b). (From Brameshuber, W. et al., *Am. Concr. Inst.*, SP-250-3, 35, 2008.)

in Figure 5.79b, compared to fabric 1, which was made with regular AR glass yarns (without PP coating). Furthermore, the composites reinforced with fabrics 3 and 4—whose yarns comprise three thin FSH yarns knitted together into a single strand to achieve a greater proportion of PP coating relative to yarn amounts and good connection of the individual filaments to each other—exhibited the best flexural performances of the fabrics tested. The composite reinforced with fabric 4 was able to sustain an ultimate load that was 250% greater than that recorded for fabric 1.

The transfer of stress between the inner filaments of the bundle and the cement matrix can be improved without using a polymer coating by filling the spaces between filaments with nanoparticles, which can increase the stress transfer between the inner filaments and the cement matrix while preserving the composite's telescopic mechanism to maintain its ductility (see also Chapters 4 and 9). Mineral nanoparticles (fillers) were studied and compared to polymer-based coating (Bentur et al., 2010, 2013; Cohen and Peled, 2010, 2012; Zamir et al., 2014) for AR glass and carbon fabrics made of multifilament yarns. The nanoparticles were inserted into the bundle spaces using the impregnation process, in which the impregnated fabric was dried (two steps, dry) or kept wet (one step, wet) in preparation for the composite production. In general, the use of particulate fillers led to

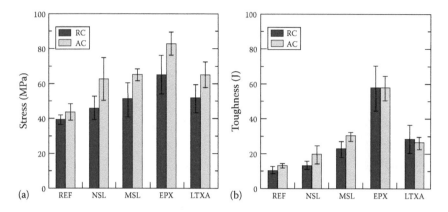

Figure 5.80 Average tensile properties of carbon fabric composites filled with silica fume of different particle sizes (MSL, NSL) or with polymer (EPX and LTXA) relative to nonimpregnated fabric (REF) under regular curing (RC) and accelerated curing (AC): (a) tensile strength, (b) toughness. (From Zamir, M. et al., Fabric cement-based composites with nanoparticles filler, interfacial characteristics, in E. Schlangen, M.G. Sierra Beltram, M. Lukovic, and G. Ye (eds.), *Third International RILEM Conference on Strain Hardening Cementitious Composites (SHCC3)*, Dordrecht, the Netherlands, November 3–5, 2014, pp. 171–178.)

improved composite tensile properties in terms of strength and toughness (Figure 5.80), particularly for carbon fabric composites (Zamir et al., 2014) due to effective filling of the bundle spaces (Figure 5.81) and improved bonding. Although the polymer filler composites obtained the most impressive composite mechanical performance, they suffered from low bonding between fabric and matrix as evidenced by large crack spacing, inconsistent crack patterns, and critical delamination (Figure 5.82). In contrast, composites that contained mineral filler exhibited denser and more even cracking patterns due to the pozzolanic reaction of the filler with the cement matrix, which promotes greater bonding and improved tensile behavior. Another important advantage of the inorganic fillers, such as silica fume, as opposed to a polymer-based coating is the ability of the former to withstand high temperatures and fire.

Composite tensile behavior is strongly dependent on the production method used, impregnation of the fabric in the matrix either under dry condition (24 hours after the filler is applied) or under wet condition (immersion in the matrix immediately after filler impregnation) (Cohen and Peled, 2012). In general, the dry process resulted in the deposition of fillers within the bundle that limited cement penetration, while the wet process washed some of those fillers out, leaving empty spaces within the bundle, thereby leading to a reduction in composite tensile performance (Figure 5.83).

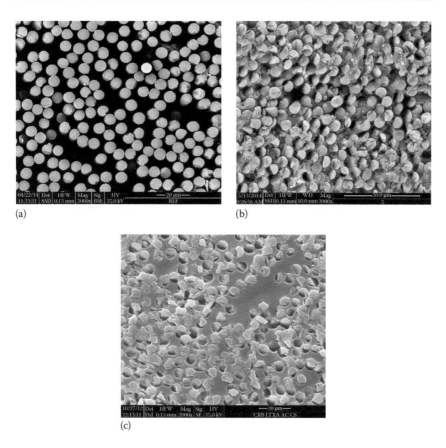

Figure 5.81 SEM micrographs of the cross sections of composites reinforced with carbon bundles (a) with no filling, (b) filled with silica fume, and (c) filled with epoxy. (From Zamir, M. et al., Fabric cement-based composites with nanoparticles filler, interfacial characteristics, in E. Schlangen, M.G. Sierra Beltram, M. Lukovic, and G. Ye (eds.), *Third International RILEM Conference on Strain Hardening Cementitious Composites (SHCC3)*, Dordrecht, the Netherlands, November 3–5, 2014, pp. 171–178.)

5.6 Influence of processing on composite mechanical performance

The manufacturing process can significantly affect the mechanical properties of fabric cement-based composites, and therefore, it should be adapted to the structure of the fabric in order to optimize its reinforcing efficiency.

Laminated fabric cement-based composite can be produced by the pultrusion process, in which the fabrics are passed through a slurry infiltration chamber and then pulled through a set of rollers to squeeze the cement paste

(a)

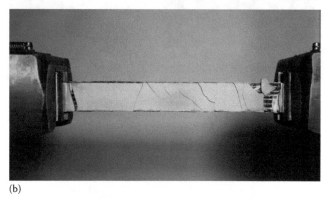

(b)

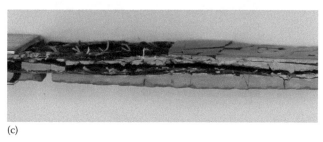

(c)

Figure 5.82 Crack pattern of composites reinforced with carbon fabrics (a) filled with silica fume (dense crack pattern), (b) filled with epoxy (few cracks only), and (c) filled with epoxy (intensive delamination). (From Zamir, M. et al., Fabric cement-based composites with nanoparticles filler, interfacial characteristics, in E. Schlangen, M.G. Sierra Beltram, M. Lukovic, and G. Ye (eds.), *Third International RILEM Conference on Strain Hardening Cementitious Composites (SHCC3)*, Dordrecht, the Netherlands, November 3–5, 2014, pp. 171–178.)

between the fabric openings while removing excess paste (see Chapter 3). A controlling factor in the success of the pultrusion process is the rheology of the fresh cement mixture, which should be stiff enough to remain in the fabric and bundle spaces but not too stiff that it limits cement matrix penetration. After the pultrusion stage, the laminated sheets are formed on a plate-shaped mandrel.

The advantages of the pultrusion technique compared to the conventional hand lay-up process have been clearly demonstrated, especially for

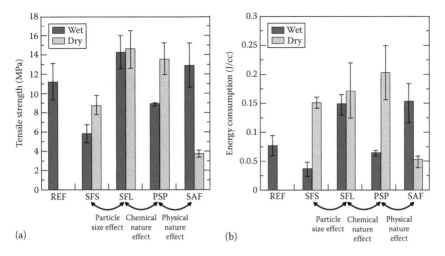

Figure 5.83 Tensile properties of composites reinforced with AR glass fabrics impregnated with various filling materials (silica fume, SFS or SFL; polymers, PSP or SAF): (a) tensile strength and (b) toughness, calculated as the area under the stress–strain curve. (From Cohen, Z. and Peled, A., *Composites*, 43, 962, 2012.)

composites made of knitted fabrics with multifilament yarns and open junction points (Mobasher et al., 2006; Peled and Mobasher, 2005, 2006, 2007; Peled et al., 2006). In such knitted fabrics, intensive shear forces are needed during processing to open up the spaces between the filaments and ensure they are filled with the matrix. The pultrusion process does that effectively (Figure 5.84), resulting in better utilization of the filaments to maximize their efficiency, ultimately conferring improved tensile performance to the composite (Figure 5.85). Mobilization of the filaments in the pultrusion process results in strain-hardening composite even when the yarn modulus is relatively low, as demonstrated by tests of PP fabric composites (Figure 5.86). Here the positive effect on composite performance was due to the improvement in bonding achieved by the impregnation of cement paste during the pultrusion process, which was evidenced by the dense crack pattern and fine crack width over the entire loading process (Figure 5.86) (see also Chapter 4).

Although the pultrusion process works well with relatively open fabrics, it is a less effective method for very densely woven fabrics that contain large numbers of filaments, between which the matrix cannot penetrate even when an intensive method such as pultrusion is used, as shown for dense PVA fabric (Figure 5.87). In fabrics composed of monofilament yarns (such as PE) or of bundles of filaments coated with sizing (such as glass), pultrusion offers no advantage from a mechanical point of view, since both fabric types lack fine spaces into which the matrix can be forced to penetrate, and therefore,

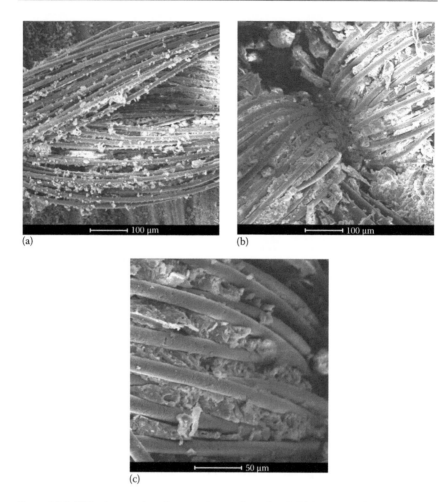

Figure 5.84 SEM micrographs of composite reinforced with (a) knitted PP fabric produced by the cast process and (b and c) knitted PP fabric produced by the pultrusion process. (From Peled, A. and Mobasher, B., *J. Mater. Civil Eng.*, 19(4), 340, 2007.)

the casting and pultrusion methods produce similar results (Mobasher et al., 2006; Peled and Mobasher, 2005, 2006, 2007; Peled et al., 2006).

The effectiveness of the pultrusion process could be enhanced by applying pressure to the top of the laminated composite (immediately after the pultrusion stage, while the composite is in the fresh stage) to push more effectively the matrix into the spaces between the bundled filaments and the fabric openings (Peled and Mobasher, 2005, 2006). Increasing the pressure from 100 N (1.7 kPa) to 900 N (15 kPa) improved the tensile strength of the composite by approximately 40% and reduced its ductility (Figure 5.88).

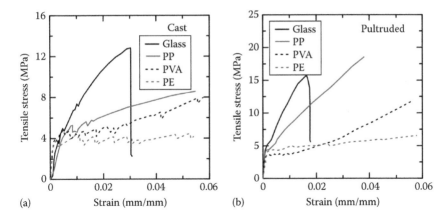

Figure 5.85 Tensile responses of composites reinforced with different fabric materials produced by (a) cast process and (b) pultrusion process. (From Peled, A. and Mobasher, B., *J. Mater. Civil Eng.*, 19(4), 340, 2007.)

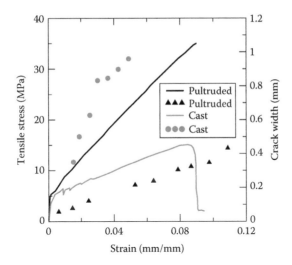

Figure 5.86 Comparison of the tensile responses of composites reinforced with PP fabric produced by the cast and pultrusion processes, with the crack width developed during loading. (From Peled, A. and Mobasher, B., *J. Mater. Civil Eng.*, 19(4), 340, 2007.)

This observation was explained by the better bonding achieved between the fabric and the matrix and was supported by the observed dense crack spacing over the duration of the applied load (Figure 5.88) and the relatively small gap obtained between the matrix and the fabric under the higher pressure (Figure 5.89). In addition, extending the duration of the applied load

Figure 5.87 Composite reinforced with woven PVA fabric produced by the pultrusion process. (From Peled, A. and Mobasher, B., *J. Mater. Civil Eng.*, 19(4), 340, 2007.)

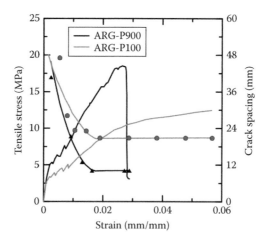

Figure 5.88 Influence of pressure applied on the laminated composite immediately after the pultrusion process. (From Peled, A. and Mobasher, B., *ACI Mater. J.*, 102(1), 15, January–February 2005.)

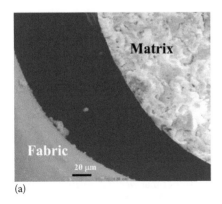
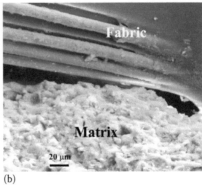

(a)   (b)

Figure 5.89 Effect on the fabric–cement interface of pressure applied on an AR glass fabric–cement composite after the pultrusion process: (a) 100 N and (b) 900 N. (From Peled, A. and Mobasher, B., *ACI Mater. J.*, 102(1), 15, January–February 2005.)

was found to slightly improve composite mechanical performance (Peled and Mobasher, 2005).

During composite production under the conventional casting process, fabric pretensioning can be exploited to enhance the composite mechanical performance (Krüger, 2004; Mobasher et al., 2007; Peled, 2007; Peled et al., 1998b; Reinhardt and Kruger, 2004; Vilkner, 2003; Xu et al., 2004). Pretensioning can promote yarn alignment in the loading direction (Mobasher et al., 2007), and when applied at a high level, it can also improve bonding (see also Chapter 4) and composite mechanical performance (Peled, 2007; Reinhardt and Kruger, 2004; Xu et al., 2004). The improvement in bonding can be caused by the development of compression and friction between the fabric and the surrounding matrix due to the lateral expansion of the yarns after the pretension is released and the fabric shifts back. Increasing the prestress level subsequently increases yarn lateral expansion and compression forces that, in turn, lead to improved fabric–cement bonding, such that composite flexural first-crack stress and strength are also enhanced (Reinhardt et al., 2003). The prestressed composite, therefore, typically exhibits a denser crack pattern and fine cracks (Figure 5.90) (Reinhardt and Kruger, 2004; Xu et al., 2004).

The dynamics of pretensioning, however, are more complex than the preceding discussion indicates. For example, yarn properties, fabric geometry, and the time at which the pretension is removed significantly affect the mechanical performance of prestressed fabric–cement systems (Peled, 2007). Whereas the flexural properties of a composite containing high-modulus fabric made of pretensioned straight yarns are reduced when the pretension is released one day after casting compared to that produced in the

(a)   (b)

Figure 5.90 Composites reinforced with AR glass fabrics that were (a) prestressed and (b) not prestressed. (From Reinhardt, H.W. and Krueger, M., Prestressed concrete plates with high strength fabrics, in M. Di Prisco, R. Felicetti, and G.A. Plizzari (eds.), *Proceedings of the RILEM Symposium on Fibre Reinforced-Concrete, BEFIB*, Varenna, Italy, RILEM PRO 39, 2004, pp. 187–196.)

Table 5.7 Flexural properties of composites reinforced with different fabrics and times to pretension release

Material type	Fabric geometry	V_f (%)	Flexural strength (MPa)			Toughness (N·mm)		
Initial load release (days)			0	1	7	0	1	7
PE	Woven	5.7	10.2	4.2	—	1,046	994	—
PE	Knitted short weft	2.0	6.0	6.8	—	547	684	—
PP	Knitted chain	3.5	17.7	16.7	—	1,856	1,937	—
Kevlar	Knitted chain	3.5	19.5	10.8	27.2	2,286	2,286	3,200

Source: Peled, A., *Cement Concr. Res.*, 37(5), 805, May 2007.

immediately released system, they are significantly improved when the pretension is released at a later stage seven days after casting (Table 5.7).

Similarly, improved flexural properties were also reported for a composite with low-modulus PE in a woven structure produced using the immediately released system. However, when low-modulus PE or PP is part of a knitted fabric structure, the time at which the pretension load is released does not significantly affect the flexural properties of the composite. These divergent trends between composites with different yarn properties and fabric geometries are related to the bond strength developed under each condition. Damage to the matrix–fabric interface can occur at a very early stage—one day after casting when the matrix is weak—due to lateral expansion of the yarn when the pretension is released. At later pretension release times, for example, seven days after casting, the matrix is stronger and is not easily damaged, and therefore, such lateral expansion of the yarn can lead to improved bonding and provide good flexural behavior (Table 5.7), an outcome that is more pronounced for a high-modulus, low-creep fabric such as aramid. In low-modulus, high-creep yarns, the stresses that develop within the yarns can be relaxed during tensioning, and therefore, significant lateral expansion of the yarn is not expected, that is, no damage to the matrix. Thus, a similar bond is obtained whether pretension is released one or seven days after casting, and no significant differences are observed in composite flexural performances. However, when low-modulus, high-creep PE yarns are part of a woven fabric, the tensioning effectively "straightens" the crimped shapes of the yarns. Release of the pretension immediately after the casting causes the yarns to "recoil" back to their original crimped shape, and since at this stage the matrix is still fresh, it can overcome these changes in yarn geometry. This is not the case when the tension is released one day after casting, when the matrix has hardened. For woven structures, therefore, the immediate-release system leads to greater mechanical anchoring of the reinforcing PE yarns with the cement matrix and better overall composite flexural performance than that of the one-day-to-release system.

References

Adiel Sasi, E., Flexural behavior of cement based element reinforced with 3D fabric, MSc thesis, Ben Gurion University, Beer-Sheva, Israel, 2014.

Adiel Sasi, E. and Peled, A., 3D fabrics as reinforcement for cement-based composites, *Composites Part A*, 74, 153–165, 2015.

Almeida, A. L. F. S. D., Toledo Filho, R. D., and Melo Filho, J. A., Bending behavior of cement-based composites reinforced with jute textile, in *International RILEM Conference on Material Science (MATSCI)*, Aachen, Germany, ICTRC, Vol. 1, 2010, pp. 177–184.

Amzaleg, E., Peled, A., Janetzko, S., and Gries, T., Bending behavior of 3d fabric reinforced cementitious composites, in J. Barros, I. Valente, M. Azenha, and S. Diasl (eds.), *Eighth RILEM Symposium on Fibre-Reinforced Concretes (FRC), BEFIB*, Guimarães, Portugal, 2012, pp. 71–73.

Amzaleg, E., Peled, A., Janetzko, S., and Gries, T., Flexural behavior of cement based element reinforced with 3d fabric, in J. G. M. Van Mier, G. Ruiz, C. Andrade, R. C. Yu, and X. X. Zhang (eds.), *Eighth International Conference on Fracture Mechanics of Concrete and Concrete Structures, FraMCoS-8*, Toledo, Spain, 2013.

Aveston, A., Cooper, G. A., and Kelly, A., Single and multiple fracture, in *Proceedings of Conference on the Properties of Fibre Composites*, National Physical Laboratories, IPC, Guildford, England, Science and Technology Press, Teddington, England, 1971, pp. 15–24.

Aveston, J. and Kelly, A., Theory of multiple fracture of fibrous composites, *Journal of Material Science*, 8, 411–461, 1973.

Banholzer, B., Bond behaviour of a multi-filament yarn embedded in a cementitious matrix, Dissertation, RWTH Aachen, Aachen, Germany, 2004.

Banthia, N., Crack growth resistance of hybrid fibre reinforced cement composites, *Cement and Concrete Composites*, 25(1), 3–9, 2003.

Banthia, N. and Soleimani, S. M., Flexural response of hybrid fibre-reinforced cementitious composites, *ACI Materials Journal*, 102(6), 382–389, 2005.

Barhum, R. and Mechtcherine, V., Effect of short, dispersed glass and carbon fibres on the behavior of textile-reinforced concrete under tensile loading, *Engineering Fracture Mechanics*, 92, 56–71, 2012.

Barhum, R. and Mechtcherine, V., Influence of short dispersed and short integral glass fibres on the mechanical behaviour of textile-reinforced concrete, *Materials and Structures*, 46, 557–572, 2013.

Bartos, P., Brittle matrix composites reinforced with bundles of fibres, in J. C. Maso (ed.), *Proceedings RILEM Symposium: From Material Science to Construction Materials*, Versailles, France, Chapman & Hall, London, U.K., 1987, pp. 539–546.

Bentur, A., Silica fume treatments as means for improving durability of glass fibre reinforced cements, *Journal of Materials in Civil Engineering*, 1(3), 167–183, 1989.

Bentur, A. and Mindess, S., *Fibre Reinforced Cementitious Composites*, 2nd edn., Taylor & Francis Group, Routledge, U.K., 2006.

Bentur, A., Tirosh, R., Yardimci, M., Puterman, M., and Peled, A., Bonding and microstructure in textile reinforced concrete, in W. Brameshuber, (ed.), *Textile Reinforced Concretes: Proceedings of the International RILEM Conference on Materials Science*, Aachen, Germany, RILEM Publications, Vol. 1, 2010, pp. 23–33.

Bentur, A., Yardımcı, M. Y., and Tirosh, R., Preservation of telescopic bonding upon aging of bundled glass filaments by treatments with nano-particles, *Cement and Concrete Research*, 47, 69–77, May 2013.

Brameshuber, W., *Textile Reinforced Concrete: State of the Art Report, RILEM TC 201-TRC*, RILEM Publications, Paris, France, 2006.

Brameshuber, W. and Brockmann, T., Development and optimization of cementitious matrices for textile reinforced elements, in *Proceedings of the 12th International Congress of the International Glass Fibre Reinforced Concrete Association*, Dublin, Ireland, Concrete Society, London, U.K., May 14–16, 2001, pp. 237–249.

Brameshuber, W., Koster, M., Hegger, J., and Voss, S., Integrated formwork elements made of textile-reinforced concrete, *American Concrete Institute*, SP-250-3, 35–47, 2008.

Brazilian Standard NBR 11578, Cimento Portland Composto. Associac, ao Brasileira de Normas Tecnicas (ABNT), 1991 (In Portuguese).

Brockmann, T., Mechanical and fracture mechanical properties of fine grained concrete for textile reinforced composites, PhD thesis, RWTH Aachen University, Aachen, Germany, 2006.

Butler, M., Mechtcherine, V., and Hempel, S., Experimental investigations on the durability of fibre–matrix interfaces in textile-reinforced concrete, *Cement and Concrete Composites*, 31, 221–231, 2009.

Cervenka, V., Jendele, L., and Cervenka, J., ATENA program documentation. Part 1: Theory, Cervenka Consulting, Prague, Czech Republic, 2009.

Cohen, Z. and Peled, A., Controlled telescopic reinforcement system of fabric-cement composites—Durability concerns, *Cement and Concrete Research*, 40, 1495–1506, 2010.

Cohen, Z. and Peled, A., Effect of nanofillers and production methods to control the interfacial characteristics of glass bundles in textile fabric cement-based composites, *Composites: Part A*, 43, 962–972, 2012.

Cohen, Z., Peled, A., Mobasher, B., Janetzko, S., and Gries, T., Hybrid cement-based composites: Dynamic and static tensile behaviors, in J. Barros, I. Valente, M. Azenha, and S. Diasl (eds.), *Eighth RILEM Symposium on Fibre-Reinforced Concretes (FRC), BEFIB*, Guimarães, Portugal, September 19–21, 2012, pp. 139–141.

Colombo, I. G., Magri, A., Zani, G., Colombo, M., and di Prisco, M., Textile reinforced concrete: Experimental investigation on design parameters, *Materials and Structures*, 46, 1933–1951, 2013.

Concrete Canvas Ltd., Concrete canvas. http://concretecanvas.de/index.html, Accessed July 22, 2012.

Contamine, R., Si Larbi, A., and Hamelin, P., Contribution to direct tensile testing of textile reinforced concrete (TRC) composites, *Materials Science and Engineering A*, 528, 8589–8598, 2011.

Cuypers, H., Wastiels, J., Van Itterbeeck, P., De Bolster, E., Orlowsky, J., and Raupach, M., Durability of glass fibre reinforced composites experimental methods and results, *Composites: Part A*, 37, 207–215, 2006.

Cyr, M. F., Peled, A., and Shah, S. P., Extruded hybrid fibre reinforced cementitious composites, in M.A. Mansur and K. C. G. Ong (eds.), *Proceedings FERRO-7: Ferrocement and Thin Fiber Reinforced Cement Composites*, Singapore, 2001, pp. 199–207.

Dolatabadi, M. K., Janetzko, S., and Gries, T., Deformation of AR glass roving embedded in the warp knitted structure, *The Journal of the Textile Institute*, 102(4), 308–314, April 2011.

Gao, S. L., Mäder, E., and Plonka, R., Coatings for glass fibres in a cementitious matrix, *Acta Materialia*, 52, 4745–4755, 2004.

Garmendia, L., San-Jos, J. T., Garcia, D., and Larrinaga, P., Rehabilitation of masonry arches with compatible advanced composite material, *Construction and Building Materials*, 25, 4374–4385, 2011.

Glowania, M. and Gries, T., Coating of high-performance fibers and textiles for textile reinforced concrete, in *Eighth fib PhD Symposium in Kgs*, Lyngby, Denmark, June 20–23, 2010.

Glowania, M. H., Linke, M., and Gries, T., Coating of AR-glass fibers with poly-urethane for textile-reinforced concrete, in *Ninth International Symposium on High Performance Concrete: Design, Verification and Utilization*, Energy Events Centre, Rotorua, New Zealand, August 2011.

Gries, T. and Roye, A., Three dimensional structures for thin walled concrete elements, in M. Curbach (ed.), *Textile Reinforced Structures: Proceedings of the Second Colloquium on Textile Reinforced Structures (CTRS2)*, Dresden, Germany, Sonderforschungsbereich 528, Technische, 2003, pp. 513–524.

Haim, E. and Peled, A., Impact behavior of textile and hybrid cement-based composites, *ACI Materials Journal*, 103(8), 235–243, 2011.

Han, F., Chen, H., Jiang, K., Zhang, W., Lv, T., and Yang, Y., Influences of geometric patterns of 3D spacer fabric on tensile behavior of concrete canvas, *Construction and Building Materials*, 65, 620–629, 2014.

Hartig, J., Jesse, F., Schicktanz, K., and Häußler-Combe, U., Influence of experimental setups on the apparent uniaxial tensile load-bearing capacity of Textile Reinforced Concrete specimens, *Materials and Structures*, 45, 433–446, 2012.

Hausding, J., Engler, T., Franzke, G., Köckritz, U., and Offermann, P., Concrete reinforced with stitch bonded multi-plies: A review, in C. Aldea (ed.), *Thin Fiber and Textile Reinforced Cementitious Systems*, ACI Symposium Publication 244, Farmington Hills, MI, 2007, pp. 1–16.

Hausding, J., Lorenz, E., Ortlepp, R., Lundahl, A., and Cherif, C., Application of stitch-bonded multi-plies made by using the extended warp knitting process: Reinforcements with symmetrical layer arrangement for concrete, *The Journal of the Textile Institute*, 102(8), 726–738, August 2011.

Hegger, J., Bruckermann, O., and Voss, S., *ACI Convention on AR-Glass and Carbon Fibres in Textile Reinforced Concrete—Simulation and Design*, SP-244-4, Farmington Hills, MI, 2007, pp. 57–75.

Hegger, J. and Voss, S., Textile reinforced concrete under biaxial loading, in *Sixth RILEM Symposium on Fibre Reinforced Concrete (FRC), BEFIB*, Varenna, Italy, September 2004, pp. 1463–1472, 20–22.

Hegger, J. and Voss, S., Investigations on the bearing behaviour and application potential of textile reinforced concrete, *Engineering Structures*, 30, 2050–2056, 2008.

Hegger, J., Will, N., Bruckermann, O., and Voss, S., Load–bearing behaviour and simulation of textile reinforced concrete, *Materials and Structures*, 39, 765–776, 2006.

Hinzen, M. and Brameshuber, W., Hybrid short fibres in fine grained concrete, in J. Hegger, W. Brameshuber, N. Will (eds.), *Textile Reinforced Concrete: Proceedings of the First International RILEM Symposium*, Aachen, Germany, RILEM, Paris, France, September 6–7, 2006, pp. 23–32.

Hinzen, M. and Brameshuber, W., Influence of short fibres on strength, ductility and crack development of textile reinforced concrete, in H. W. Reinhardt and A. E. Naaman (eds.), *High Performance Fibre Reinforced Cement Composites (HPFRCC5): Proceedings of the Fifth International RILEM Workshop*, Mainz, Germany, 2007, pp. 105–112.

Hinzen, M. and Brameshuber, W., Improvement of serviceability and strength of textile-reinforced concrete elements with short fiber mixes, in C. M. Aldea (ed.), *ACI Convention on Design and Applications of Textile-Reinforced Concrete*, SP-251-1, Farmington Hills, MI, 2008, pp. 7–18.

Hinzen, M. and Brameshuber, W., Improvement of serviceability and strength of textile reinforced concrete by using short fibres, in *Fourth Colloquium on Textile Reinforced Structures (CTRS4)*, Dresden, Germany, 2009, pp. 261–272.

Hinzen, M. and Brameshuber, W., Influence of matrix composition and short fibres on the workability of fine grained fibre concrete, in *International RILEM, Conference on Material Science (MATSCI)*, Aachen, Germany, ICTRC, Vol. 1, 2010, pp. 131–140.

Hull, D. and Clyne, T. W., *An Introduction to Composite Materials*, 2nd edn., Cambridge Solid State Science Series, Cambridge University Press, U.K., 1996.

Kakemi, M., Hannant, D. J., and Mulheron, M., Filament fracture within glass fibre strands in hybrid fibre cement composites, *Journal of Materials Science*, 33, 5375–5382, 1998.

Kanda, T. and Li, V. C., Interface property and apparent strength of high-strength hydrophilic fibre in cement matrix, *Journal of Materials in Civil Engineering*, 10(1), 5–13, 1998.

Ko, F. K., Weaving its role with advanced composite materials, *Automotive Engineering*, 5, 6–66, May 1987.

Kobayashi, K. and Cho, R., Flexural characteristics of steel fibre and polyethylene fiber hybrid: Reinforced concrete, *Composites*, 13, 164–168, 1982.

Koeckritz, U., Cherif, C., Weiland, S., and Curbach, M., In-situ polymer coating of open grid warp knitted fabrics for textile reinforced concrete application, *Journal of Industrial Textiles*, 40(2), 157–169, October 2010.

Krüger, M., Prestressed textile reinforced concrete, PhD thesis, Institute of Construction Materials, University of Stuttgart, Stuttgart, Germany, 2004.

Krüger, M., Reinhardt, H. W., and Yong, X., Sulphoaluminate cement matrices used for textile and glass fibre reinforced concrete elements, in A. E. Naaman and H. W. Reinhardt (eds.), *International Workshop High Performance Fibre Reinforced Cement Composites, (HPFRCC4)*, Ann Arbor, MI, RILEM, 2003, pp. 349–360.

Larrinaga, P., Chastre, C., Biscaia, H. C., and San-Jose, J. T., Experimental and numerical modeling of basalt textile reinforced mortar behavior under uniaxial tensile stress, *Materials and Design*, 55, 66–74, 2014.

Larrinaga, P., Chastre, C., San-José, J. T., and Garmendia, L., Non-linear analytical model of composites based on basalt textile reinforced mortar under uniaxial tension, *Composites Part B: Engineering*, 55, 518–527, 2013.

Lawler, J. S., Wilhelm, T., Zampini, D., and Shah, S. P., Fracture processes of hybrid fiber-reinforced mortar, *Materials and Structures*, 36(257), 197–208, 2003.

Leung, C. K. and Li, V. C., Effect of fibre inclination on crack bridging stress in brittle matrix composites, *Journal of the Mechanics and Physics of Solids*, 40, 1333–1362, 1992.

Leung, C. K. and Ybanez, N., Pullout of inclined flexible fiber in cementitious composite, *Journal of Engineering Mechanics*, 123, 239–246, 1997.

Li, Q. and Xu, S., Experimental research on mechanical performance of hybrid fibre reinforced cementitious composites with polyvinyl alcohol short fiber and carbon textile, *Journal of Composite Materials*, 45(1), 5–28, January 2011.

Mecit, D. and Roye, A., Investigation of a testing method for compression behavior of spacer fabrics designed for concrete applications, *Textile Research Journal*, 79(10), 867–875, 2009.

Meyer, C. and Vilkner, G., Glass concrete thin sheets prestressed with aramid fiber mesh, in *Proceedings: High Performance Fibre Reinforced Cement Composites (HPFRCC4)*, Ann Arbor, MI, RILEM, 2003, pp. 325–336.

Mobasher, B., Dey, V., Peled, A., and Cohen, Z., Correlation of constitutive response of hybrid textile reinforced concrete from tensile and flexural tests, *Cement and Concrete Composites*, 53, 148–161, 2014.

Mobasher, B., Jain, N., Aldea, C. M., and Soranakom, C., Mechanical properties of alkali resistant glass fabric composites for retrofitting unreinforced masonry walls, *American Concrete Institute*, SP-244-8, 124–140, 2007.

Mobasher, B. and Li, C. Y., Mechanical properties of hybrid cement based composites, *ACI Materials Journal*, 93(3), 284–293, 1996.

Mobasher, B., Pahilajani, J., and Peled, A., Analytical simulation of tensile response of fabric reinforced cement based composites, *Cement and Concrete Composites*, 28, 77–89, 2006a.

Mobasher, B., Peled, A., and Pahilajani, J., Pultrusion of fabric reinforced high flyash blended cement composites, in *Sixth RILEM Symposium on Fibre-Reinforced Concretes (FRC), BEFIB*, Varenna, Italy, September 2004, pp. 1473–1482.

Mobasher, B., Peled, A., and Pahilajani, J., Distributed cracking and stiffness degradation in fabric-cement composites, *Materials and Structures*, 39(3), 317–331, 2006b.

Mu, B. and Meyer, C., Flexural behavior of fibre mesh-reinforced concrete with glass aggregate, *ACI Materials Journal*, 99, 425–434, September–October 2002.

Mu, B., Meyer, C., and Shimanovich, S., Improving the interface bond between fiber mesh and cementitious matrix, *Cement and Concrete Research*, 32, 783–787, 2002.

Naaman, A. E., Evolution in ferrocement and thin reinforced cementitious composites, *Arabian Journal for Science and Engineering*, 37(2), 421–441, 2012.

OPERHA, Open and fully compatible next generation of strengthening system for the rehabilitation of Mediterranean building heritage, Contract no: 517765 (INCO), 6th FP, 2006–2008.

Orlowsky, J., Raupach, M., Cuypers, H., and Wastiels, J., Durability modelling of glass fibre reinforcement in cementitious environment, *Material and Structures*, 38, 155–162, 2005.

Pekmezci, B. Y., Kayaoglu, B. K., Pourdeyhimi, B., and Karadeniz, A. C., Utility of polyvinyl alcohol fibre-based needle punched nonwoven fabric as potential reinforcement in cementitious composites, *Journal of Composite Materials*, 48(25), 3129–3140, 2014.

Peled, A., Reinforcement with textile fabrics of cement materials, DSc thesis, Technion, Israel Institute of Technology, Haifa, Israel, 1995.

Peled, A., Pre-tensioning of fabrics in cement-based composites, *Cement and Concrete Research*, 37(5), 805–813, May 2007.

Peled, A. and Bentur, A., Reinforcement of cementitious matrices by warp knitted fabrics, *Materials and Structures*, 31, 543–550, October 1998.

Peled, A. and Bentur, A., Geometrical characteristics and efficiency of textile fabrics for reinforcing composites, *Cement and Concrete Research*, 30, 781–790, 2000.

Peled, A. and Bentur, A., Fabric structure and its reinforcing efficiency in textile reinforced cement composites, *Composites: Part A*, 34, 107–118, 2003.

Peled, A., Bentur, A., and Yankelevsky, D., Woven fabric reinforcement of cement matrix, *Advanced Cement Based Materials Journal*, 1, 216–223, 1994.

Peled, A., Bentur, A., and Yankelevsky, D., Effect of fabric geometry on bonding, *Advanced Cement Based Materials*, 7, 20–27, 1998a.

Peled, A., Bentur, A., and Yankelevsky, D., The nature of bonding between monofilament polyethylene yarns and cement matrices, *Cement and Concrete Composites*, 20, 319–327, 1998b.

Peled, A., Bentur, A., and Yankelevsky, D., Flexural performance of cementitious composites reinforced with woven fabrics, *Journal of Materials in Civil Engineering*, 11(4), 325–330, 1999.

Peled, A., Cohen, Z., Janetzko, S., and Gries, T., Hybrid fabrics as cement matrix reinforcement, in M. Curbach and R. Ortlepp (eds.), *Sixth Colloquium on Textile Reinforced Structures (CTRS6)*, Berlin, Germany, September 2011a, pp. 1–14.

Peled, A., Cohen, Z., Pasder, Y., Roye, A., and Gries, T., Influences of textile characteristics on the tensile properties of warp knitted cement based composites, *Cement and Concrete Composites*, 30(3), 174–183, 2008a.

Peled, A., Cyr, M., and Shah, S. P., Hybrid fibers in high performances extruded cement composites, in M. D. I. Prisco, R. Felicetti, and G. A. Plizzari (eds.), *Proceedings RILEM Symposium, BEFIB*, Varenna, Italy, PRO 39, 2004, pp. 139–148.

Peled, A. and Mobasher, B., Pultruded fabric-cement composites, *ACI Materials Journal*, 102(1), 15–23, January–February 2005.

Peled, A. and Mobasher, B., Properties of fabric-cement composites made by pultrusion, *Materials and Structures*, 39, 787–797, 2006.

Peled, A. and Mobasher, B., Tensile behavior of fabric cement-based composites: Pultruded and cast, *Journal of Materials in Civil Engineering*, 19(4), 340–348, 2007.

Peled, A., Mobasher, B., and Cohen, Z., Mechanical properties of hybrid fabrics in pultruded cement composites, *Cement and Concrete Composites*, 31(9), 647–657, 2009.

Peled, A., Sueki, S., and Mobasher, B., Bonding in fabric-cement systems: Effects of fabrication methods, *Cement and Concrete Research Journal*, 36(9), 1661–1671, 2006.

Peled, A., Zaguri, E., and Marom, G., Bonding characteristics of multifilament polymer yarns and cement matrices, *Composites: Part A*, 39(6), 930–939, 2008b.

Peled, A., Zhu, D., and Mobasher, B., Impact behavior of 3D fabric reinforced cementitious composites, in H. W. Reinhardt and G. Parra-Montesinos (eds.), *Workshop on High Performance Fibre-Reinforced Cement Composites (RILEM) HPFRCC-6*, Ann Arbor, MI, June 19–22, 2011b, pp. 543–551.

Perez-Pena, M. and Mobasher, B., Mechanical properties of fibre reinforced lightweight concrete composites, *Cement and Concrete Research*, 24(6), 1121–1132, 1994.

Rambo, D. A. S., Silva, F. A., Toledo Filho, R. D., Fonseca, O., and Gomes, M., Effect of elevated temperatures on the mechanical behavior of basalt textile reinforced refractory concrete, *Materials and Design*, 65, 24–33, 2015.

Reinhardt, H. W. and Kruger, M., Prestressed concrete plates with high strength fabrics, in M. Di Prisco, R. Felicetti, and G. A. Plizzari (eds.), *Proceedings of the RILEM Symposium on Fibre-Reinforced Concrete, BEFIB*, Varenna, Italy, RILEM PRO 39, 2004, pp. 187–196.

Reinhardt, H. W., Kruger, M., and Grosse, U., Concrete prestressed with textile fabric, *Journal of Advanced Concrete Technology*, 1(3), 231–239, 2003.

Roye, A., Hochleistungsdoppelraschelprozess für Textilbetonanwendungen, Doctoral dissertation, RWTH Aachen University, Aachen, Germany, 2007.

Roye, A. and Gries, T., Design by application: Customized warp knitted 3-D textiles for concrete applications, *Kettenwirk-Praxis 39*, 4, 20–21, 2004.

Roye, A. and Gries, T., Technical spacer fabrics: Textiles with advanced possibilities, in *13th International Techtextil-Symposium "Focusing on Innovation"*, Frankfurt am Main, Germany, June 6–9, 2005.

Roye, A. and Gries, T., 3-D textiles for advanced cement based matrix reinforcement, *Journal of Industrial Textiles*, 37(2), 163–173, 2007a.

Roye, A. and Gries, T., Three-dimensional and online-shaped textile production with double needle bar raschel machines and weft insertion for concrete applications, *American Concrete Institute*, SP-244-5, 77, 2007b.

Roye, A., Gries, T., Engler, T., Franzke, G., and Cherif, C., Possibilities of textile manufacturing for load, *Adapted Concrete Reinforcements*, SP-250-2, 23–34, 2008.

Roye, A., Gries, T., and Peled, A., Spacer fabric for thin walled concrete elements, in M. Di Prisco, R. Felicetti, and G. A. Plizzari (eds.), *Fibre Reinforced Concrete: BEFIB*, RILEM PRO 39, Varenna, Italy, 2004, pp. 1505–1514.

Scheffler, C., Gao, S. L., Plonka, R., Mader, E., Hempel, S., Butler, M., and Mechtcherine, V., Interphase modification of alkali-resistant glass fibres and carbon fibres for textile reinforced concrete II: Water adsorption and composite interphases, *Composites Science and Technology*, 69, 905–912, 2009.

Schleser, M., Einsatz polymerimprägnierter, alkaliresistenter glastextilien zur bewehrung zementgebundener matrices ("Use of polymer-impregnated, alkali-resistant glass fabrics for reinforcement of cement-based"), Doctoral dissertation, Institut für Schweißtechnikund Fügetechnik der RWTH Aachen University, Aachen, Germany, 2008.

Scholzen, A., Chudoba, R., and Hegger, J., Thin-walled shell structure made of textile reinforced concrete Part I: Structural design and realization, *Structural Concrete*, 16(1), 106–114, 2015.

Schorn, H., Hempel, R., and Butler, M., Mechanismen des Verbundes von textilbewehrtem Beton ("Mechanism of TRC composite"), Tagungsband 15. Internationalen Baustofftagung ibausil, Bauhaus-Universität Weimar, Weimar, Germany, September 2003, pp. 24–27.

Shibata, N. A., Nishimum, A., and Norita, T., Graphite fiber's fabric design and composite properties, *SAMPE Quarterly*, 7(4), 25–33, July 1976.

Silva, F. A., Melo Filho, J. A., Toledo Filho, R. D., and Fairbairn, E. M. R., Mechanical behavior and durability of compression moulded sisal fibre cement mortar laminates (SFCML), in *First International RILEM Conference on Textile Reinforced Concrete (ICTRC)*, Aachen, Germany, 2006, pp. 171–180.

Silva, F. A., Mobasher, B., and Toledo Filho, R. D., Cracking mechanisms in durable sisal fibre reinforced cement composites, *Cement and Concrete Composites*, 31, 721–730, 2009.

Silva, F. A., Toledo Filho, R. D., Melo Filho, J. A., and Fairbairn, E. M. R., Physical and mechanical properties of durable sisal fibre cement composites, *Construction and Building Materials*, 24, 777–785, 2010.

Sim, J., Park, C., and Moon, D. Y., Characteristics of basalt fibre as a strengthening material for concrete structures, *Composites: Part B*, 36, 504–512, 2005.

Stang, H., Significance of shrinkage-induced clamping pressure in fibre–matrix bonding in cementitious composite materials, *Advanced Cement Based Materials*, 4, 106–115, 1996.

Toledo Filho, R. D., Silva, F. A., Fairbairn, E. M. R., and Melo Filho, J. A., Durability of compression molded sisal fibre reinforced mortar laminates, *Construction and Building Materials*, 23, 2409–2420, 2009.

Tsesarsky, M., Katz, A., Peled, A., and Sadot, O., Textile reinforced concrete (TRC) shells for strengthening and retrofitting of concrete elements—Influence of admixtures, *Materials and Structures*, 48, 471–484, 2015.

Tsesarsky, M., Peled, A., Katz, A., and Antebi, I., Strengthening concrete elements by confinement within textile reinforced concrete (TRC) shells: Static and impact properties, *Construction and Building Materials*, 44, 514–523, 2013.

Tysmans, T., Adriaenssens, S., Cuypers, H., and Wastiels, J., Structural analysis of small span textile reinforced concrete shells with double curvature, *Composites Science and Technology*, 69, 1790–1796, 2009.

Vilkner, G., Glass concrete thin sheets reinforced with prestressed aramid fabrics, Dissertation, Columbia University, New York, 2003.

Wastiels, J., Sandwich panels in construction with HPFRCC-faces: New possibilities and adequate modeling, in H. W. Reinhardt and A. E. Naaman (eds.), *High Performance Fibre Reinforced Cement Composites*, RILEM Publications, Cachan, France, 1999, pp. 143–151.

Weichold, O., Preparation and properties of hybrid cement-in-polymer coatings used for the improvement of fiber-matrix adhesion in textile reinforced concrete, *Journal of Applied Polymer Science*, 116, 3303–3309, 2010.

Xu, G., Magnani, S., and Hannant, D. J., Tensile behavior of fibre–cement hybrid composites containing polyvinyl alcohol fibre yarns, *ACI Materials Journal*, 95(6), 667–674, 1998.

Xu, S., Krueger, M., Reinhardt, H. W., and Ozbolt, J., Bond characteristics of carbon, alkali resistant glass and aramid textiles in mortar, *Journal of Materials in Civil Engineering*, 16(4), 356–364, 2004.

Zamir, M., Dvorkin, D., and Peled, A., Fabric cement-based composites with nanoparticles filler, interfacial characteristics, in E. Schlangen, M. G. Sierra Beltram, M. Lukovic, and G. Ye (eds.), *Third International RILEM Conference on Strain Hardening Cementitious Composites (SHCC3)*, Dordrecht, the Netherlands, November 3–5, 2014, pp. 171–178.

Zhu, D., Mobasher, B., and Peled, A., Experimental study of dynamic behavior of cement-based composites, *Journal of Sustainable Cement-Based Materials*, 2(1), 1–12, 2013.

Zhu, D., Peled, A., and Mobasher, B., Dynamic tensile testing of fabric–cement composites, *Construction and Building Materials*, 25, 385–395, 2011.

Zhu, W. and Bartos, P., Assessment of interfacial microstructure and bond properties in aged GRC using novel microindentation method, *Cement and Concrete Research*, 27(11), 1701–1711, 1997.

Chapter 6

Mechanics of TRC composite

6.1 Introduction

Textile-reinforced concrete (TRC) materials exhibit strain-hardening cement composite (SHCC) characteristics and are therefore well suited for applications that may involve large energy absorption for thin sections, high strain capacity, fatigue, impact resistance, or for structures in seismic regions where high ductility is desired or reduction of conventional reinforcement is needed. In addition, strain-hardening materials are attractive for use in industrial structures, highways, bridges, earthquake, hurricane, and high-wind loading conditions. The design and implementation of these systems requires applications that go beyond the elastic response and into the strain-hardening range. This range is attributed to multiple cracking under tensile stresses and the postcrack response that exceeds the first-crack stress over a large strain range.

Classes of strain-hardening fiber-reinforced concrete (FRC) materials such as TRC, as discussed by Naaman and Reinhardt (2006), are unlike conventional FRC where fracture localization occurs immediately after the first crack is formed. In the strain-hardening composites, distribution of cracking throughout the specimen is facilitated by the fiber bridging mechanism. Since a substantial amount of energy is required to extend existing cracks, secondary parallel cracks form. Single crack localization is therefore shifted to multiple distributed cracking mechanisms, leading to macroscopic pseudo-strain-hardening behaviors such as those shown in Figure 6.1. When used as a continuous reinforcement in cement matrices, the enhanced mechanical bond strength presented by the textiles results in a composite with strain-hardening and multiple-cracking behavior. The simulation of reduced stiffness may be based on either empirical approaches obtained from experimental data (Mobasher et al., 2006a) or analytical models that simulate the debonding of textiles (Sueki et al., 2007).

The Aveston, Cooper, and Kelly model (ACK model as discussed in Chapter 4) addresses the increased strain capacity of the matrix and the multiple-cracking mechanism in the presence of fibers in unidirectional

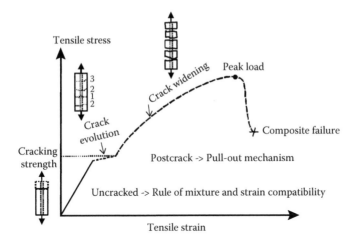

Figure 6.1 Tensile response of a strain-hardening fiber cement-based composite. (From Soranakom, C., Multi scale modeling of fibre and fabric reinforced cement based composites, PhD dissertation, Arizona State University, Tempe, AZ, 2008.)

composites (Aveston et al., 1971). This method, however, does not address the tension-stiffening effect, which is defined as the ability of the uncracked segments in between the two parallel cracks to carry tensile force. The parameter ε_{mu} may theoretically be obtained using the ACK approach or other methods (Mobasher and Li, 1996) that can predict the strain capacity of the matrix phase in the presence of fibers as shown in Equation 6.1:

$$\varepsilon_{mu} = \left[\frac{12\tau\gamma_m E_f V_f^2}{E_c E_m^2 r V_m} \right]^{\frac{1}{3}} \tag{6.1}$$

where
   τ is the shear strength of the matrix
   γ_m is the fracture toughness of the matrix
   E_c is the composite modulus
   r is the fiber radius
   V_m is the volume fraction of the matrix

This equation indicates that the strength of the matrix phase increases in the presence of fibers. It further shows that the significant energy absorption due to the frictional bond between the fibers and matrix also contributes to additional crack formation and parallel microcracking. The ACK model has been verified for cement-based materials by results that show the increasing strength of the matrix in the presence of fibers; however, the assumptions of perfect bond parameters have been modified to take into account the effect of debonding and reduction in stiffness due to sequential cracking in cement-based materials (Mobasher and Shah, 1990).

6.2 Experimental observations of mechanical response

6.2.1 Nonlinear stress–strain response

The uniaxial tension test captures the various modes of failure that take place in a TRC specimen. By using this method, one can obtain a tensile stress–strain response as well as gain insight into nonlinear modes of behavior such as distributed cracking, fiber debonding, and pullout mechanisms. Tension testing is normally conducted using a grip system to allow for the deformation of the specimen. Both dog-bone and straight prismatic specimens have been used. Figure 6.2a shows the schematic drawing of

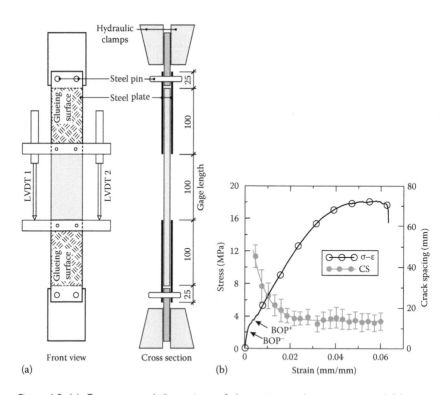

Figure 6.2 (a) Geometry and dimensions of the static tensile test setup and (b) stress and crack spacing as a function of applied strain in a tensile specimen. (From Mobasher, B. et al., *J. Cement Concr. Compos.*, 28(1), 77, 2006a; Mobasher, B. et al., *Mater. Struct.*, 39, 317, 2006b.)

the specimen and fixture assembly for a uniaxial tension specimen that utilizes prismatic specimens with pins added in for transferring the load (Mechtcherine et al., 2011). Note that the elongation needs to be measured using displacement-measuring linear variable differential transformers (LVDTs) so that the inclusion of spurious deformations in the strain values is minimized. Comprehensive experimental programs have been conducted in studying the effects of varied fabric types, matrix formulations, and processing parameters on the mechanical response of cement-based composites (Peled and Mobasher, 2005).

Figure 6.2b represents a typical tensile stress–strain response of an AR glass fabric reinforced cement composite. The tensile response shows a linear behavior up to about 3 MPa; beyond this level, the stress–strain response becomes nonlinear, while a major change in the stiffness of the sample occurs at the bend-over point (BOP), or around 3–4 MPa. This is characterized by a knee in the stress–strain curve. The specimen continues to carry load at a significantly lower stiffness up to an ultimate strain level of 6%. In the region between the BOP and the sample's ultimate strength, crack activity is visible in the formation of a distributed crack and later on, the widening of the crack. The experimental data consist of a range of parameters in terms of the initial stiffness, BOP stress (with initial and final levels indicated by – and + signs), BOP strain, post-BOP stiffness, ultimate strength capacity, toughness, and pullout stiffness (Mobasher et al., 2006b).

6.2.2 Effect of specimen thickness and fabric orientation

This section presents the effect of sample processing parameters on the TRC composites. A cast process was used to study the effect of fabric content and ply orientation. In the first set of experiments, two layers of AR glass fabric were used and the thickness was adjusted as a variable. The fabric was freely laid while casting the specimens. The effect of specimen thickness on the tensile stress–strain response using the same number of fabric layers is shown in Figure 6.3a (Singla, 2004). In this case, reducing the thickness of the composite increases the effective fiber fraction and allows the fabric to distribute the cracks throughout, thus improving the ductility of the system. Also, a specimen with a thinner cross section has a higher volume fraction for the same number of fabric layers, which results in the higher strength of the specimen due to multiple-cracking response. (The thinner sample is 6.7 MPa, whereas the strength of thicker sample is 2.9 MPa.) Reducing the thickness by 33.33% can increase tensile strength by 56.7%.

Changing the orientation of the fabric affects the tensile response as shown in Figure 6.3b. Specimens made with fabric at 0°, 6°, and 12° offset with respect to the machine direction show a marked decrease in the strength and stiffness values. The initial stiffness of 6.3 GPa measured for 0° orientation drops to 4.0 and 2.7 GPa for 6° and 12°, respectively.

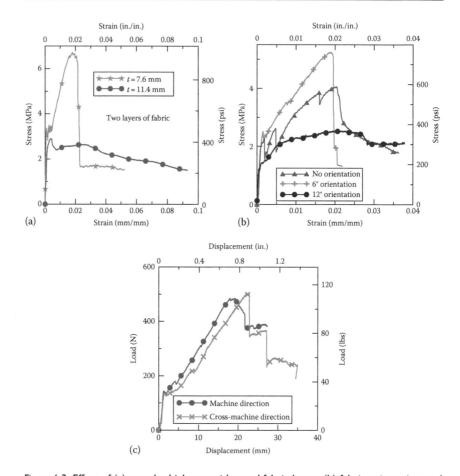

Figure 6.3 Effect of (a) sample thickness with equal fabric layers, (b) fabric orientation, and (c) machine and cross-machine direction on the tensile stress–strain response. (From Singla, N., Experimental and theoretical study of fabric cement composites for retrofitting masonry structures, MS thesis, Arizona State University, Tempe, AZ, 2004.)

The tensile strength increased marginally for 6° from 4.56 to 4.9 MPa but decreased to 2.7 MPa for the 12° oriented samples. The machine and cross-machine directions are compared in Figure 6.3c. Due to the nature of bonding of the fabric at machine and cross-machine direction (MD and XM, respectively) junctions, the MD yarns are placed over and under the XMD yarns that are inserted in an aligned fashion. This results in better anchorage of the MD yarns as compared to the XMD yarns. The strength of the XM yarns are, however, higher due to the reduction in variation of the load along the yarn length.

6.2.3 Distributed cracking and spacing evolution

An automated procedure based on image analysis was developed by Mobasher et al. to quantitatively measure the distribution of crack spacing (Peled et al., 2004). This approach results in a statistically viable sampled set of data at each strain level. Figure 6.4a and b show the stress–strain and crack spacing evolution in the glass and PE samples. The strength and stiffness retention of both samples are different due to the postcrack stiffness of AR glass (ARG) samples being as much as an order of magnitude higher than the PE composites. The strain capacities of the two samples are comparable, but the crack densities in the PE composites are nearly twice that of ARG samples. Figure 6.4c compares the crack spacing distributions of the ARG and the PE samples. The strain associated with the completion of cracking phase is also referred to as the point of saturation of crack density and occurs over a large range of strains for both samples. These distributions differ in their cumulative distribution functions as shown in Figure 6.4c.

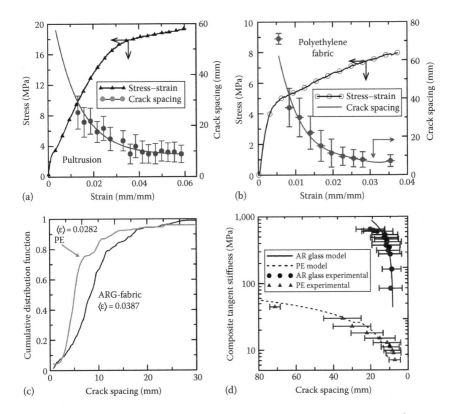

Figure 6.4 (a–d) Comparison of two fabric systems in stiffness degradation as a function of crack spacing. (From Peled, A. and Mobasher, B., *ASCE J. Mater. Civil Eng.*, 19(4), 340, 2007; Mobasher, B. et al., *J. Mater. Struct.*, 39, 317, 2006b.)

In fact, the ARG-TRC at 50% higher strain has a less dense crack spacing than the PE composites, which indicates that the delamination between the glass and the matrix phase is higher than in the PE composites.

Using an exponential decay representation, the crack spacing parameter can be incorporated through utilizing a damage evolution law. The model parameters are the experimental results of using average crack spacing as a function of applied strain. The form of the decay function representing the crack spacing versus strain is given by

$$S(\varepsilon_i) = S_1 + S_0 e^{-\alpha(\varepsilon_i - \varepsilon_{mu})} \quad \varepsilon_i > \varepsilon_{mu} \tag{6.2}$$

where
  S is the average crack spacing
  S_1 is a parameter describing saturation crack spacing
  S_0 and α are the decay rate parameters
  ε_i is variable strain
  ε_{mu} is strain at the BOP(+) level

The function representing experimentally obtained crack spacing is also plotted in Figure 6.4c and d along with the crack spacing. This function indicates an inverse relationship between applied strain and crack spacing until a constant level referred to as the crack saturation density, "CSD" is reached. The CSD in the ARG-TRC samples is indicated by the flattening of the crack spacing curve at about 10 and 8 mm in PE-TRC samples. Beyond this point, a reduction in crack spacing is not observed, as further increase in the strain causes fabric debonding, pullout, and widening of existing cracks. Typical values of S_1, S_0, ε_{mu}, and α for different matrix and fabric combinations are given by Mobasher et al. (2006a). Parameters for representing the decay in stiffness, which has a similar form as Equation 6.2, are also provided. The numbers represent an average value for the representation of crack spacing functions and relate properties governing the cracking, debonding, and reduction of stiffness in the composite.

The crack spacing density can be correlated with the change in the stiffness of the sample. Figure 6.4d represents the relationship between the crack spacing and the stiffness degradation of AR glass and PE fabric composites. As the imposed strain is increased, both the crack spacing and the composite stiffness decrease. The rate of degradation is a function of bond characteristics, delamination, cracking strength of the matrix, and stiffness of the fabric. The ultimate level of composite response can be determined from responses shown in Figure 6.4d.

The first crack strength defined as BOP is a function of the matrix, interface, number of lamina per unit thickness, as well modification to the matrix phase such as addition to short reinforcement. Figure 6.5 is a representative stress–strain curve obtained from the quasistatic tensile test

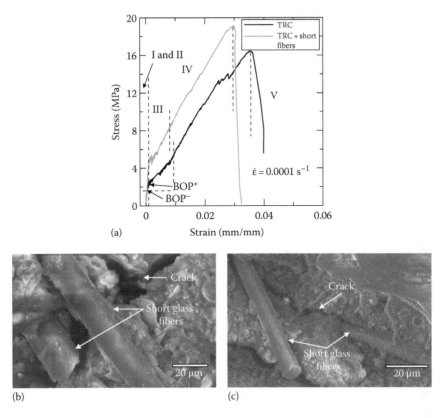

Figure 6.5 Quasistatic tensile behavior of TRC and TRC reinforced with short glass fibers: (a) stress versus strain behavior, (b) and (c) ESEM micrographs showing short glass fibers bridging microcracks. (From Silva, F.A. et al., *Mater. Sci. Eng. A*, 528(3), 1727, 2011a.)

(strain rate of 0.0001 s⁻¹) of TRC with a plain matrix as compared with a TRC with the matrix reinforced with short glass fibers. From a macroscopic perspective, the BOP of each of the curves corresponds to the formation of the first matrix crack crossing the entire cross section. Five distinct zones are identified using Roman numerals, with two zones prior to and three zones after the BOP. Zone I corresponds to the elastic-linear range in which both matrix and fiber behave linearly. Due to the low volume fraction of the fibers, the stiffness of the composite is influenced predominantly by Young's modulus of the matrix. The linear zone is terminated by the initial crack formation in the matrix phase (reported as BOP⁻ from experiments), and as shown in Figure 6.2, the BOP⁻ is determined at the point where its linearity is lost. The strain range within Zone II is associated with the formation of matrix microcracks; however, in this case, no single crack traverses the entire width. The term defined as BOP$^+$ corresponds to the stress level at which the first matrix crack completely propagates across the width of specimen (as shown in Figure 6.4). The addition of short glass fibers increased the σ_{BOP}^- from the average value of 2.72 MPa (0.53 MPa) to 3.63 MPa (0.67 MPa) and σ_{BOP}^+ from 3.55 MPa (0.86 MPa) to 4.71 MPa (0.62 MPa) (the values in the parentheses give the standard deviation). This behavior is attributed to the capacity of short glass fibers to bridge microcracks, as can be observed in Figure 6.5b and c.

The post-BOP stage is characterized by the formation of distributed cracking (Zone III). After the initiation of cracks in the matrix, the load-carrying capacity of TRC does not reduce to zero since the cracks are bridged by fabrics. Immediately after the initiation of the first matrix crack, other matrix cracks also propagate throughout the specimen at approximately regular intervals. In this phase, as the applied strain increases, more cracks form. Zone IV comes after the completion of the cracking phase. This zone is dominated by progressive damage and is characterized by a crack-widening phase ultimately leading to the failure of the specimen.

An increase in average tensile strength from 16 to 17 MPa was noticed when adding 0.5% of short fibers to TRC. The addition of short fibers decreased the energy absorption capacity (work-to-fracture) from 18 to 16 J when tested at a rate of 0.0001 s^{-1}.

6.3 Modeling of tension response using experimental crack spacing results

In Figures 6.1 and 6.2, the schematic representation of the damage evolution as a function of applied strain are compared. In these cases, the post-BOP stage is characterized by the formation of distributed cracking. However, the stiffness of the fabric cement system is sufficiently high to keep the newly formed cracks from widening and thus promoting additional cracking. This stiffness additionally reduces the rate of crack spacing. The gradual reduction of matrix stress levels in the vicinity of the cracked matrix is referred to as the softening zone. In this zone, the matrix cracks widen, and while there may be no localization in the strain-softening zone, the response is based on contributions from a softening matrix and the fabric pullout force (Sueki et al., 2007). The final range of response is dominated by progressive damage and characterized by crack widening, which ultimately leads to failure by fabric pullout, failure, or delamination. Automated procedures to measure spacing between the cracks formed during testing have been discussed in detail in several papers by Peled, Mobasher, and coworkers (Mobasher et al., 2006b; Peled et al., 2007; Sueki et al., 2007). Tensile tests can further be used to address the effects of (1) fabric type and geometry, (2) matrix formulations, and (3) processing parameters.

6.3.1 Model representation

A schematic representation of various stages of cracking in Figure 6.6a shows the stress strain model. Figure 6.6b shows the schematic interaction of the various damage mechanisms in terms of distributed damage and crack spacing evolution as a function of strain. The standard model proposed for the interaction between stress–strain responses of the matrix, fabric, and composite is shown in Figure 6.6. Using a composite laminate analysis, the model for tension can be developed by considering four distinct stages defined in the context of the extent of damage as a function of applied strain. The four distinct zones are identified using roman numerals with two zones prior to the BOP and two zones after the BOP range. These distinct zones of response are used to break down the tension-stiffening model into the initial stages elastic response and microcrack development in the matrix. These two regions identified as I and II address the formation of first major crack. In these zones, the stiffness offered by the reinforcement phase that bridges the microcracks acts to hinder their coalescence during the transition from the elastic range to cracking. The higher the fiber stiffness, the

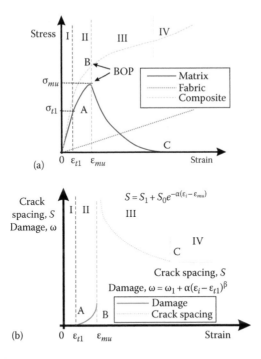

Figure 6.6 Schematic representation of various stages of cracking in strain-hardening cement composite systems (a) stress strain model and (b) the damage and crack spacing evolution as a function of strain. (From Mobasher, B. et al., *J. Cement Concr. Compos.*, 28(1), 77, 2006a; Mobasher, B. et al., *Mater. Struct.*, 39, 317, 2006b.)

more delayed the formation of the first main crack in the matrix, and hence an increase in the BOP level has been addressed by many researchers. This leads to a gradual transition of the bond-slip mechanism, which then can be simplified to a linear interfacial delamination model.

Zone I is modeled as the elastic-linear range where both the matrix and the fabric behave linearly. Best addressed using the composite laminate theory with an iso-strain formulation, this zone relates the properties of fabric and matrix to the composite response. Zone I is terminated by initial crack formation in the matrix phase at a point labeled "A" and is designated as σ_{t1} at the strain level ε_{t1} (Agarwal and Broutman, 1990) (reported as of σ_{BOP}^- from experiments). After the initiation of cracks in the matrix, its load-carrying capacity does not vanish as the newly formed microcracks are immediately bridged by the longitudinal yarns. Zone II occurs between the two stress levels of σ_{BOP}^- and σ_{BOP}^+. During this stage, as matrix cracks begin to form and propagate across the width of the specimen, load is transferred to the fibers. However, no single crack has traversed the entire length of the specimen. This range (Zone II) is defined by the formation of the first complete crack across the cross section between points "A" and "B." The isolated crack system is modeled as a dilute concentration of cracks in a medium, and the stiffness degrades up to the BOP+ level according to the single scalar damage parameter "ω." The evolution of the damage parameter proposed by Karihaloo (1995) is expressed as a power law:

$$\omega(\varepsilon_i) = \omega_1 + \alpha(\varepsilon_i - \varepsilon_{t1})^\beta \quad \varepsilon_{t1} < \varepsilon_i < \varepsilon_{mu} \tag{6.3}$$

where
  ε_i is the strain at which the damage parameter is computed
  ε_{t1} is the strain at first crack
  ω_1, α, and β are constant terms in the power law model

The values of these constants (taken from Mobasher et al., 2006a) are $\alpha = 1.0$, $\beta = 0.3$, and $\omega_1 = \varepsilon_{t1}H$, where H is the gage length of the specimen used. Parameter σ_{t1} is the tensile strength of the matrix in the absence of fibers, and $\varepsilon_{t1} = \sigma_{t1}/E_{m0}$ is defined as the strain at failure under uniaxial tension for the matrix in an unreinforced condition. A damage model by Horii et al. (1989) and Nemat-Nasser and Hori (1993) has been used to simulate the decrease in the stiffness of the cracked matrix as the strain increases and cracking saturates, as the stiffness decrease is directly related to these variables. The stiffness parameter $E_m(\omega)$, as a function of damage and initial matrix elastic modulus E_{m0} in Zone II, is defined by the authors as

$$E_m(\omega) = \frac{E_{m0}}{1 + \dfrac{16}{3}\omega(1 - \upsilon_m^2)} \tag{6.4}$$

where v_m is the Poisson ratio of the matrix. This value of matrix stiffness is used in the rule of mixtures to obtain the longitudinal stiffness of the lamina $E_1(\omega)$. A modified rule of mixture is used in modeling the stiffness computation of a lamina according to Equation 6.5:

$$E_1(\omega) = E_f V_f + E_m(\omega)(1 - V_f) \tag{6.5}$$

where
E_f is the stiffness of the fiber
V_f is the volume fraction of the fiber

The stress is computed using an incremental approach by adding the products of strain increments by the effective stiffness at that level. The stress in the matrix phase beyond the elastic range is calculated incrementally as

$$\sigma_1^i(\omega) = \sigma^{i-1} + \Delta\sigma_t^i = \sigma_{t1} + \sum_{n=1}^{i} E_m(\omega)(\varepsilon_n - \varepsilon_{n-1}) < \sigma_{mu} \quad \varepsilon_{t1} < \varepsilon_i < \varepsilon_{mu} \tag{6.6}$$

The maximum stress in the matrix phase is achieved at a strain level of ε_{mu} described in the next section using Equation 6.6 and is referred to as σ_{mu}.

The model for an estimation of tension response is shown in Figure 6.7a and b. The effect of the fiber volume fraction is considered in Figure 6.7a. As the fiber content increases, the overall stiffness increases and the composite carries an additional load since the fiber loading is in excess of the critical volume fraction. The gradual decay in the stiffness of the matrix is shown in the change in the stiffness of the composite.

Figure 6.7b simulates the sequential failure of the layers in the composite consisting of three identical layers subjected to tension. Three different cases are shown representing the stiffness reduction due to failure of the 1/3, 2/3, and finally 3/3 layers of the matrix within the lamina. The response of the composite changes drastically as the overall composite stiffness reduces. These results are compared with the experimental tension data gained from earlier work by Mobasher and Pivacek (1998) and show the effective stiffness of the unidirectional lamina systems.

6.3.2 Stresses and deformations in the distributed cracking zone

The post-BOP stage is characterized by the formation of distributed cracking in Zone III and is represented by parameters S_0, S_1, α, and ε_{mu} (Equation 6.2). Since the stiffness of the fabric cement system is sufficiently high, it may keep the newly formed cracks from widening and thus promote additional cracking instead. This stiffness directly affects the rate of reduction of the

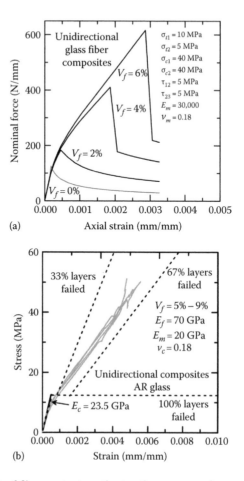

Figure 6.7 (a) Effect of fiber content on the tensile response of a composite and (b) the effect of ply failure on the overall response of the brittle matrix-brittle fiber systems. (From Mobasher, B., *Mechanics of Fibre and Textile Reinforced Cement Composites*, CRC Press, September 2011, p. 480.)

crack spacing, or α parameter. The damage level corresponding to the stress at the BOP $(\sigma_{BOP}{}^+)$ is also the ultimate strength of the matrix in the presence of fibers σ_{mu} (as shown by the ACK model and represented in Equation 6.1) when Zone II terminates. Similar to the incremental approach in Zone II, the degraded stiffness at each strain value from ε_{t1} up to the BOP strain level (ε_{mu}) is computed and used to calculate the stress.

The experimental values of $\varepsilon_{BOP(+)}$ are in the range of 400–1000 microstrains and as much as 300% higher than $\varepsilon_{BOP(-)}$. These values correspond to the ACK Model parameter for ε_{mu} and correlate with microcrack growth over a relatively large strain range. Depending on the fabric cement system used, the utilization of parameter γ_m/r in the range 0.5–5.0 N mm/mm yields an accurate correlation between experimental and theoretical simulations of the stress and strain at BOP level.

Zone III is dominated by the formation of parallel microcracking, which occurs beyond the BOP level. The gradual reduction of matrix stress levels in the vicinity of the cracked matrix is referred to as the softening phenomenon; however, as long as there is a bond between the fabric and the matrix, a portion of the load is carried by the matrix. This fractional load carried by the matrix, when averaged over the entire length of the specimen, can be viewed in the context of the strain-softening component. The strain softening is therefore a structural response and different than the conventional definition of softening that is observed as the widening of a single crack in an unreinforced specimen. In this zone, the cracks in the matrix widen, and while there may be no localization in the strain-softening zone, the overall composite response is modeled empirically by contributions from a softening matrix and the fabric pullout force. The stress–strain response of the matrix in the postpeak region is assumed to be an exponentially decaying function of the maximum stress and its asymptotic approach to zero:

$$\sigma_1^i\left(\varepsilon_i\right) = \sigma_{mu}e^{-q\left(\varepsilon_i - \varepsilon_{mu}\right)} \quad \varepsilon_i > \varepsilon_{mu} \tag{6.7}$$

In the previous equation, q represents the exponent coefficient affecting the rate of decay in stress from the peak composite stress, which represents the rate of decay. The typical value of q between 0.1 and 5.0 gives a reasonable rate of stress decay in the postpeak region of the matrix phase. The definition of strain in this region is gage length dependent, so one has to use the mean strain over the length of several cracks in the matrix in order to capture an averaged response. As the specimen undergoes strain softening, an exponential decaying stiffness similar to Equation 6.7 is used, utilizing the stiffness at peak. The modulus $E_m(\varepsilon_i)$, computed for each strain level ε_i, can be taken proportional to the reduction of the stress from the peak value.

Zone IV is dominated by progressive damage and is characterized by crack widening, which ultimately leads to failure by fabric pullout. This zone is asymptotically terminated at the saturation crack spacing and is represented by parameter S_1. The behavior of both the matrix and the fabric in addition to the interaction between the two is studied in each of these four ranges, and the formulations are compiled together to present a comprehensive material simulation model (Mobasher et al., 2006a).

6.3.3 Comparison with experimental results

Analytical model parameters can be calibrated with the experimental results of stress, crack spacing, and stiffness decay versus strain response of several composite systems, as demonstrated for three fabric types: AR glass, PE, and PP; two matrix formulations, with and without addition of fly ash; and two processing parameters, with low and high pressures applied on top of the composite at its fresh stage (Mobasher et al., 2006a).

The experimental stress–strain response with the model predictions is shown in Figure 6.8a for low-modulus PP fabric cement composites.

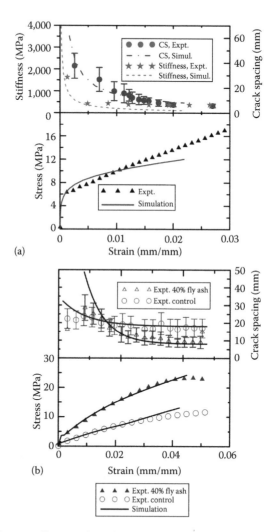

Figure 6.8 (a) Stress, stiffness, and crack spacing plotted versus strain for a sample with PP fabric compared with the theoretical model response and (b) stress and crack spacing plots versus strain for samples with glass fabric (with and without fly ash) with the theoretical model. (From Mobasher, B. et al., *J. Cement Concr. Compos.*, 28(1), 77, 2006a; Mobasher, B. et al., *Mater. Struct.*, 39, 317, 2006b.)

The three responses of stress–strain, stiffness decay, and the crack spacing simulations correlate well with the experimental measurements. While the saturation crack spacing of these specimens is much smaller than glass fabric specimens, the postcracking stiffness values are low due to the low stiffness of the polymeric fibers. The general agreement of both the crack spacing and stress versus strain responses clearly shows the applicability of the proposed methodology to a range of composites. In this simulation, a nonlinear approach for the fabric pullout model was essential in developing a gradual change in the stiffness (Mobasher et al., 2006a).

The influence of matrix rheology and composition on the packing and bond enhancements was studied by comparing the control mixture with another mixture with 40% by volume of Portland cement replaced with Class F fly ash. Both systems use AR glass fabrics. Results of this simulation are shown in Figure 6.8b. In order to simulate these values, a bond strength of 2.62 MPa was used for the matrix with 40% fly ash as compared to a bond strength of 1.76 MPa for the control sample. These bond values resulted in different fabric pullout slip responses, directly affecting the tensile response. The fly ash use modifies the overall response of the system and the crack spacing parameters, resulting in much higher postcrack stiffness as compared to the control sample. The improved bond strength and anchorage of the fabric is a dominant parameter that resists crack opening and debonding to take place. The crack saturation density in the control specimens is 20 mm as compared to 12 mm for fly ash–modified mixtures.

6.3.4 Distributed cracking and tension stiffening

A finite difference tension-stiffening model by Soranakom (2008) and Soranakom and Mobasher (2009, 2010) simulates the crack spacing and stress–strain response of SHCC materials under static and dynamic loads. Based on a finite difference formulation, the model takes into account the nonlinear bond-slip model and separates the various fiber, matrix, interface, and junction bond regions into their own constitutive properties. The cracking criterion and constitutive material equations are illustrated in Figure 6.9 representing three distinct failure mechanisms of matrix strength cracking criterion, interface bond-slip characteristics, and tensile stress–strain of the continuous fibers. In the case of textiles, an additional parameter representing the stiffness and strength of the transverse yarn junction forces is also included.

A cracked specimen under uniform tensile stress is idealized as a series of 1D segments that consist of fiber, matrix, and interface elements (for more details, see Chapter 4). The individual pullout segments continue to transfer load at cracked locations. The load is carried by the longitudinal fibers at a cracked plane solely but is transferred back to the matrix by means of

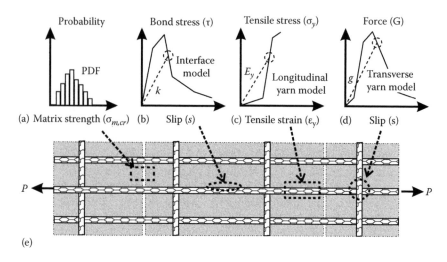

Figure 6.9 Schematic drawing of four distinct mechanisms in TRC: (a) matrix strength cracking criterion, (b) interface bond-slip response, (c) tensile stress-strain of fiber, (d) force-slip response of the transverse yarn junction bond, and (e) composite diagram with potential matrix, bond, fiber and transverse junction failure locations. (From Soranakom, C. and Mobasher, B., *Theor. Model. Mater. Struct.*, 43(2010), 1217, 2010.)

interface elements. As the load on the composite increases such that the cracking stress of the matrix is reached, additional crack planes form sequentially. An iterative solution algorithm based on nonlinear analysis enforces that a load–deformation response follows the material constitutive laws. The slip distributions are solved to obtain the corresponding stress and strain responses.

By treating the matrix as a brittle material with no strain softening, only the cracking strength, $\sigma_{m,cr}$, as shown in Figure 6.9a is used. The bond between fiber and matrix is described by a generalized free-form bond stress $\tau = \tau(s)$ and expressed as a function of slip (s) (Figure 6.9b). Several linear segments define the pre- and postpeak behaviors of the bond characteristics. At each load step, a secant modulus k enforces the local bond stress and slip at each node in the finite difference model to follow the prescribed bond-slip relation. The third aspect (Figure 6.9c) addresses fiber properties.

The equilibrium equations are derived from free body diagrams of the nodes and expressed as a coefficient, and the unknown variable slip (s) is defined as the relative difference between the elongation of the continuous fibers and matrix. A finite length between two consecutive nodes i and $i + 1$ along the longitudinal x-axis is used. The embedded length L is discretized into "n" nodes with equal spacing of "h" as shown in Figure 6.10a through c.

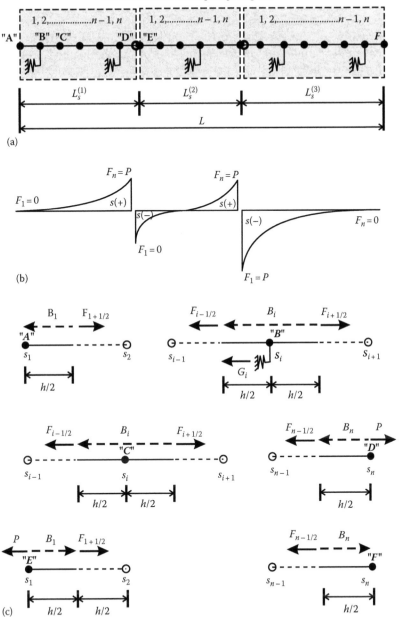

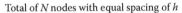

Figure 6.10 Finite difference model: (a) discretized pullout model, (b) convention for slip and boundary conditions for force in fiber, and (c) free body diagram of five representative nodes labeled as "A"–"E." (From Soranakom, C. and Mobasher, B., *Theor. Model. Mater. Struct.*, 43(2010), 1217, 2010.)

The bond stress is assumed constant over the small spacing h for each node within each linear domain. At the left end, force in the fiber is imposed to be zero, simulating a stress-free condition, implying that the fiber strain or derivative of slip vanishes. At the right end, the nodal slip is prescribed incrementally, simulating displacement control. As the loading progresses, the part of the fiber that slips out of the matrix has no frictional bond resistance; thus, fiber elongation is the only term in that section. The extruding part can be easily implemented by checking the amount of slip versus the embedded length of each node. If the slip is greater than the embedded length, zero bond stress is applied to that node.

Both deterministic and stochastic crack patterns can be used. Either uniform matrix strengths with predetermined sequential equidistant cracking locations can be specified or random matrix strengths at nodes along the length may be generated and used as a local cracking criterion. Fiber efficiency factor is introduced to take into account the imperfect nature of the bond and inability of the fiber stiffness to be utilized within the test. This efficiency factor can be directly related to the measurements of inner and outer sleeve bond characteristics (Brameshuber, 2006).

The pullout model similar to Figure 6.11a can be applied to experimental data addressing single-fiber or textile systems (Sueki et al., 2007). Figure 6.11b represents both the fiber and interface model plotted on the same scale and used in the simulation. As long as the applied load is less than the first-cracking limit, tensile response is calculated by the rule of mixtures and strain compatibility. Once the first strength criterion is met, the section is divided into two parts and each segment is modeled as a pullout problem and solved independently. As the load increases, additional cracks form at new locations in between existing cracks, or a crack and specimen boundaries. The cracked specimen, represented by a number of independently solved pullout segments that are combined to represent the entire specimen. The analysis terminates as the stress in the yarn reaches its ultimate tensile strength (UTS) or a solution is not found due to slip instability (very large slip values).

It is observed from Figure 6.12a and b that by decreasing the efficiency of the interfacial bond, expressed as η, the total strain of the composite increases up to a point that the fiber starts to fracture. Fiber failure occurs when η equals 0.3. An efficiency factor of 0.6 corresponds to an approximate ultimate strain of 1.5% (close to experimental results). The crack spacing continues to extend to smaller values due to significant debonding and slip as the efficiency factor η decreases. The influence of the matrix first-crack strength on the tensile and crack spacing is shown in Figure 6.12b. Both the ultimate strain and crack spacing increase as the matrix first-crack strength is increased, ranging from 3.5 to 6.5 MPa. There is no effect on the UTS.

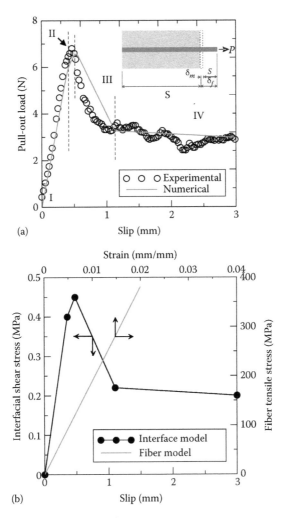

Figure 6.11 Fiber pullout test results of a sample tested at 3 days of curing L = 20 mm: (a) comparison of the experimental and numerical result from the finite difference model and (b) interface constitutive relation used in the finite difference simulation. (From Silva, F.A. et al., *Cement Concr. Compos.*, 33(8), 814, 2011b.)

Figure 6.13 compares the experimental with the predicted tensile response. The formation of crack spacing behavior is also shown.

Using an efficiency factor of $\eta = 0.6$, a matrix first-crack strength of 7 MPa, and the interfacial bond model presented in Figure 6.11a and b, the predicted stress–strain response shows a good fit with experimental values

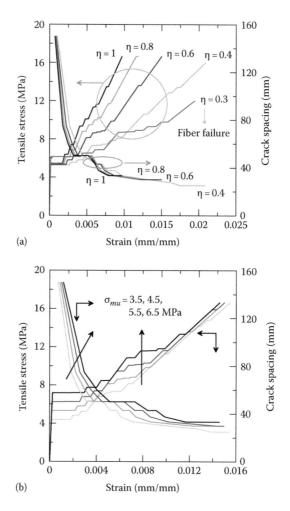

Figure 6.12 Simulation of tensile response and crack spacing: (a) effect of efficiency factor of fiber modulus and strength and (b) effect of matrix first-crack strength. (From Silva, F.A. et al., *Cement Concr. Compos.*, 33(8), 814, 2011b.)

and validates the model with the lower and upper bound experimental curves up to a strain of 1.0%. Beyond that point, the model follows the upper bound experimental curve and overestimates the UTS. The predicted crack spacing response correlates well with the experimental results as shown in Figure 6.13 and accurately predicts the saturated crack spacing up to a strain value of 0.005%. A crack saturation of approximately 30 mm is obtained, which compares with 23 mm for the experiments.

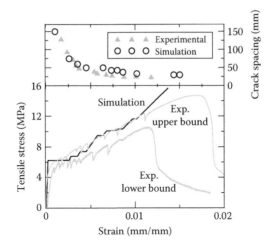

Figure 6.13 Comparison of experimental and numerical results of the composite tensile response and crack spacing showing the lower and upper bound experimental data. (From Silva, F.A. et al., *Cement Concr. Compos.*, 33(8), 814, 2011b.)

References

Agarwal, B. D. and Broutman, L. J., *Analysis and Performance of Fibre Composites*, 2nd edn., Wiley, New York, 1990.

Aveston, J., Cooper, G., and Kelly, A., The properties of fibre composites, in *Conference Proceedings of the National Physical Laboratory*, IPC Science and Technology Press Ltd., Surrey, England, 1971, pp. 15–26.

Brameshuber, W., *Textile Reinforced Concrete: State of the Art Report of RILEM Technical Committee, TC 201-TRC*, RILEM, Bagneux, France, 2006, p. 292.

Horii, H., Hasegawa, A., and Nishino, F., Fracture process and bridging zone model and influencing factors in fracture of concrete, in Shah, S.P. and Swartz S.E. (eds.), *Fracture of Concrete and Rock*, Springer, New York, NY, 1989.

Karihaloo, B. L., *Fracture Mechanics and Structural Concrete*, Longman Scientific & Technical, Harlow, England, 1995.

Mechtcherine, V., Silva, F. A., Butler, M., Zhu, D., Mobasher, B., Gao, S.-L., and Mäder, E., Behaviour of strain-hardening cement-based composites under high strain rates, *Journal of Advanced Concrete Technology*, 9(1), 51–61, 2011.

Mobasher, B., *Mechanics of Fibre and Textile Reinforced Cement Composites*, CRC Press, September 2011, p. 480.

Mobasher, B. and Li, C. Y., Effect of interfacial properties on the crack propagation in cementitious composites, *Journal of Advanced Cement Based Materials*, 4(3), November–December 93–106, 1996.

Mobasher, B. and Pivacek, A., A filament winding technique for manufacturing cement based cross-ply laminates, *Cement and Concrete Composites*, 20(5), 405–415, 1998.

Mobasher, B. and Shah, S. P., Interaction between fibres and the cement matrix in glass fibre reinforced concrete, *American Concrete Institute, ACI*, SP-124(1990), 137–156, 1990.

Mobasher, B., Pahilajani, J., and Peled, A., Analytical simulation of tensile response of fabric reinforced cement based composites, *Journal of Cement and Concrete Composites*, 28(1), 77–89, 2006a.

Mobasher, B., Peled, A., and Pahalijani, J., Distributed cracking and stiffness degradation in fabric-cement composites, *Materials and Structures*, 39, 317–331, 2006b.

Naaman, A. E. and Reinhardt, H. W., Proposed classification of HPFRC composites based on their tensile response, *Materials and Structures*, 39(289), 547–555, 2006.

Nemat-Nasser, S. and Hori, M., *Micromechanics: Overall Properties of Heterogeneous Materials*, 2nd edn., Elsevier, 1999.

Peled, A. and Mobasher, B., Pultruded fabric-cement composites, *ACI Materials Journal*, 102(1), 15–23, 2005.

Peled, A., Mobasher, B., and Sueki, S., Technology methods in textile cement-based composites, in K. Kovler, J. Marchand, S. Mindess, and J. Weiss (eds.), *Concrete Science and Engineering: A Tribute to Arnon Bentur*, RILEM Proceedings PRO 36, March 2004, pp. 187–202.

Peled, A. and Mobasher, B., Tensile behavior of fabric cement-based composites: Pultruded and cast, *ASCE Journal of Materials in Civil Engineering*, 19(4), 340–348, 2007.

Silva, F. A., Zhu, D., Mobasher, B., Soranakom, C., and Toledo Filho, R. D., High speed tensile behavior of sisal fibre cement composites, *Materials Science and Engineering: A*, 527, 544–552, 2010.

Silva, F. A., Butler, M., Mechtcherine, V., Zhu, D., and Mobasher, B., Strain rate effect on the tensile behaviour of textile-reinforced concrete under static and dynamic loading, *Materials Science and Engineering: A*, 528(3), 1727–1734, 2011a.

Silva, F. A., Mobasher, B., Soranakom, C., and Toledo Filho, R. D., Effect of fibre shape and morphology on interfacial bond and cracking behaviors of sisal fibre cement based composites, *Cement and Concrete Composites*, 33(8), 814–823, 2011b.

Singla, N., Experimental and theoretical study of fabric cement composites for retrofitting masonry structures, MS thesis, Arizona State University, Tempe, AZ, 2004.

Soranakom, C., Multi scale modeling of fibre and fabric reinforced cement based composites, PhD dissertation, Arizona State University, Tempe, AZ, 2008.

Soranakom, C. and Mobasher, B., Geometrical and mechanical aspects of fabric bonding and pullout in cement composites, *Materials and Structures*, 42, 765–777, 2009.

Soranakom, C. and Mobasher, B., Modeling of tension stiffening in reinforced cement composites: Part I. Theoretical Modeling, *Materials and Structures*, 43(2010), 1217–1230, 2010.

Sueki, S., Soranakom, C., Peled, A., and Mobasher, B., Pullout-slip response of fabrics embedded in a cement paste matrix, *ASCE Journal of Civil Engineering Materials*, 19, 9, 2007.

Chapter 7

Flexural modeling and design

7.1 Introduction

The high level of strength, ductility, and versatility attained in the general field of strain-hardening cement composites (SHCC) has been observed for a broad range of high-performance composites such as UHPFRC (ultra-high-performance cement-based composites) (Russell and Graybeal, 2013), textile-reinforced concrete (TRC), and engineered cementitious composites (ECC) (Naaman and Reinhardt, 2003). While these composites are manufactured differently, their mechanical response can be under the general category of strain hardening and/or deflection hardening. For example, with as much as one order of magnitude higher strength and two orders of magnitude higher in ductility than fiber-reinforced concrete (FRC), TRC's development has utilized innovative fabrics, matrices, and manufacturing processes.

A variety of fiber and fabric systems such as alkali-resistant (AR) glass fibers, polypropylene (PP), polyethylene (PE), and polyvinyl alcohol (PVA) have been utilized (Mobasher, 2011; Peled and Mobasher, 2005). These composites undergo a large strain capacity and exhibit a well-formed distributed crack system. Mechanical properties under uniaxial tensile, flexural, and shear tests indicate superior performance such as tensile strength as high as 25 MPa and strain capacity of 1%–8%. In order to fully utilize these materials, design procedures are needed to determine the dimensions and expected load-carrying capacity of structural systems. In many applications, the loading is of the flexural type; therefore, there is a need for developing design procedures for this case that are different than those used for tensile behavior covered in Chapter 6. This chapter presents an approach for analysis, simulation, back-calculation, and design of SHCC systems and is applicable to all classes of TRC, ECC, UHPFRC, and SHCC composites.

The general strain-hardening behavior of cement composites is discussed in Chapter 6, and typical tensile behavior is presented in Figure 7.1 for

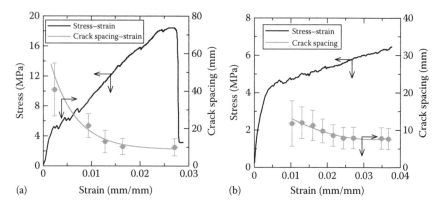

Figure 7.1 Tensile stress–strain response of (a) AR glass and (b) PE fabric composites. (From Peled, A. and Mobasher, B., *ASCE J. Mater. Civil Eng.*, 19(4), 340, 2007.)

AR glass and PE TRC, showing the stress–strain curves and crack spacing behavior (Mobasher et al., 2006). Note that the slope in the postcrack phase is significantly lower than the AR glass composites; however, the strain capacity is higher and the crack spacing is much smaller than the AR glass composites.

Tensile modeling was discussed in Chapter 6, and the essentials of this modeling are presented in Figure 7.2 in terms of a trilinear model representing tensile strain-hardening response and an elastic perfectly plastic compression model (Mobasher, 2011; Soranakom and Mobasher, 2008). Normalizing all parameters with respect to the minimum number of variables, tensile response is defined by stiffness E, first-crack tensile strain ε_{cr}, cracking tensile strength $\sigma_{cr} = E\varepsilon_{cr}$, ultimate tensile capacity ε_{peak}

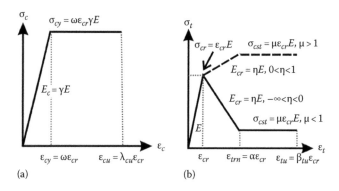

Figure 7.2 Material model for SHCC and SSCC FRC: (a) compression and (b) tension model. (From Soranakom, C. and Mobasher, B., *Cem. Concr. Compos.*, 30, 465, 2008.)

(same as ε_{trn}), and postcrack modulus E_{cr}. The softening range is shown as a constant stress level $\mu E \varepsilon_{cr}$. The compression response is defined by the compressive strength σ_{cy} defined as $\omega \gamma E \varepsilon_{cr}$. Material parameters required for the strain softening and hardening (SSCC and SHCC) are summarized as follows. The normalized parameters α, μ, η, and ω are defined to represent tensile strain at maximum stress, post-peak residual strength, post-crack modulus, and compressive yield strain, respectively. Material parameters required for the simplified models are summarized as follows:

$$\text{Cracking tensile strain,} \quad \varepsilon_{cr} = \frac{\sigma_{cr}}{E} \tag{7.1}$$

$$\text{Normalized tensile strain at peak strength,} \quad \alpha = \frac{\varepsilon_{peak}}{\varepsilon_{cr}} \tag{7.2}$$

$$\text{Normalized post-crack modulus,} \quad \eta = \frac{E_{cr}}{E} \tag{7.3}$$

$$\text{Normalized yield compressive strain,} \quad \omega = \frac{\sigma_{cy}}{E \varepsilon_{cr}} = \frac{\sigma_{cy}}{\sigma_{cr}} \tag{7.4}$$

The only variables defined in terms of the applied tensile strain at the extreme fiber, β, which can be correlated to the extreme compressive strains at extreme fiber, λ, are defined as

$$\text{Normalized tensile strain at the bottom fiber,} \quad \beta = \frac{\varepsilon_t}{\varepsilon_{cr}} \tag{7.5}$$

$$\text{Normalized compressive strain at the top fiber,} \quad \lambda = \frac{\varepsilon_c}{\varepsilon_{cr}} \tag{7.6}$$

The ratio of compressive and tensile modulus, parameter γ, has negligible effect on the ultimate moment capacity. In typical SHCC, the compressive strength is several times higher than the tensile strength; hence, the flexural capacity is controlled by the tensile component.

7.2 Quantification of flexural behavior

7.2.1 Derivation of moment–curvature relationship

Moment capacity of a beam section according to the imposed tensile strain at the bottom fiber ($\varepsilon_t = \beta \varepsilon_{cr}$) can be derived based on the assumed linear strain distribution as shown in Figure 7.3a. By using parameterized material models described in Equations 7.1 through 7.6, Figure 7.2a and b,

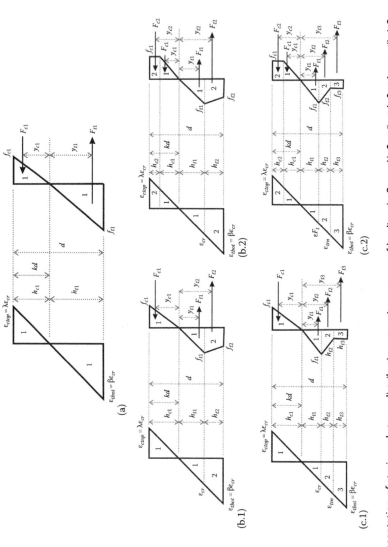

Figure 7.3 Representation of strain and stress distribution at various stages of loading. (a. Stage 1): $0 < \beta \leq 1, 0 < \lambda \leq \omega$, (b.1 Stage 2.1): $1 < \beta \leq \alpha$, $0 < \lambda \leq \omega$, (b.2 Stage 2.2): $1 < \beta \leq \alpha$, $\omega < \lambda \leq \lambda_{cu}$, (c.1 Stage 3.1): $1 < \beta \leq \alpha$, $\omega < \lambda \leq \lambda_{cu}$, (c.2 Stage 3.2): $\alpha < \beta \leq \beta_{tu}$, $0 < \lambda \leq \omega$, (c.2 Stage 3.2): $\alpha < \beta \leq \beta_{tu}$, $\omega < \lambda \leq \lambda_{cu}$. (From Soranakom, C. and Mobasher, B., Cem. Concr. Compos., 30, 465, 2008.)

the corresponding stress diagram for a linear distribution of strain across the cross section for a case of flexural loading is obtained as shown in Figure 7.3b, in which the stress distribution is subdivided into a compression zone 1, tension zone 1 and 2 force components, their centroidal distance to the neutral axis, the moment, and curvature distributions are obtained. Force components and their centroidal distance to the neutral axis in each zone can be expressed as:

$$\frac{F_{c1}}{bh\sigma_{cr}} = \frac{\beta\gamma k^2}{2(1-k)}; \quad \frac{y_{c1}}{h} = \frac{2}{3}k \tag{7.7}$$

$$\frac{F_{t1}}{bh\sigma_{cr}} = \frac{(1-k)}{2\beta}; \quad \frac{y_{t1}}{h} = \frac{2}{3}\frac{(1-k)}{\beta} \tag{7.8}$$

$$\frac{F_{t2}}{bh\sigma_{cr}} = \frac{(1-k)(\beta-1)(\eta\beta-\eta+2)}{2\beta}; \quad \frac{y_{t2}}{h} = \frac{2\eta\beta^2 - \eta\beta - \eta + 3\beta + 3}{3\beta(\eta\beta-\eta+2)}(1-k) \tag{7.9}$$

where
    F and y are the force and its centroid, respectively
    subscripts $c1$, $t1$, and $t2$ designate compression zone 1, tension zone 1, and tension zone 2, respectively
    b and h are the width and the height of the beam, respectively

The neutral axis parameter k is found by solving the equilibrium of net internal forces equal to zero, $F_{c1} + F_{t1} + F_{t2} = 0$.

The depth of the neutral axis and the nominal moment capacity M_n are also obtained and expressed as a product of the normalized nominal moment m_n and the cracking moment M_{cr}. The neutral axis parameter k is found by solving the equilibrium of net internal forces, and the nominal moment capacity M_n is obtained by taking the first moment of force about the neutral axis as shown in Equations 7.7 through 7.9. This procedure can also be done for every combination of tension and compression (three tensile modes and two compression modes) and is shown in Table 7.1. The location of neutral axis and moment capacity are presented in Table 7.1 with all potential combinations for the interaction of tensile and compressive failure responses. In case 2.1 in Table 7.1, which represents cracking tension and elastic compression response, the parameters for neutral axis and bending moment are expressed as:

$$k = \frac{C_1 - \sqrt{\beta^2 C_1}}{C_1 - \beta^2}, \quad \text{where } C_1 = \eta(\beta^2 - 2\beta + 1) + 2\beta - 1 \tag{7.10}$$

Table 7.1 Neutral axis k, normalized moment m, and curvature φ' for stages of normalized tensile strain at the bottom fiber (β), $\phi = \Phi/\Phi_{cr}$ where $\phi'_n = \beta/\left(2\left(1-k_n\right)\right)$

Stage, n		K	$m = M/M_{cr}$
1	$0 < \beta \leq 1$	$k_1 = \dfrac{1}{1+\sqrt{\gamma}}$	$m_1 = \dfrac{2\beta\left[(\gamma-1)k_1^3 + 3k_1^2 - 3k_1 + 1\right]}{1-k_1}$
2.1	$1 < \beta \leq \alpha$	$k_{21} = \dfrac{D_{21} - \sqrt{D_{21}\gamma\beta^2}}{D_{21} - \gamma\beta^2}$	$M'_{21} = \dfrac{1}{1-k_{21}}\left[\left(2\gamma\beta^3 - C_{21}\right)k_{21}^3 + 3C_{21}k_{21}^2 - 3C_{21}k_{21} + C_{21}\right]$
	$0 < \lambda \leq \omega$	$D_{21} = \eta(\beta_2 - 2\beta + 1) + 2\beta - 1$	$C_{21} = \dfrac{1}{\beta^2}\left[\left(2\beta^3 - 3\beta^2 + 1\right)\eta + 3\beta^2 - 1\right]$
2.2	$1 < \beta \leq \alpha$	$k_{22} = \dfrac{D_{22}}{D_{22} + 2\omega\gamma\beta}$	$M'_{22} = \left(3\gamma\omega\beta^2 + C_{22}\right)k_{22}^2 - 2C_{22}k_{22} + C_{22}$
	$\omega < \lambda \leq \lambda_{cu}$	$D_{22} = D_{21} + \gamma\omega_2$	$C_{22} = C_{21} - \dfrac{\gamma\omega^3}{\beta^2}$
3.1	$\alpha < \beta \leq \beta_{tu}$	$k_{31} = \dfrac{D_{31} - \sqrt{D_{31}\gamma\beta^2}}{D_{31} - \gamma\beta^2}$	$M'_{31} = \dfrac{1}{1-k_{31}}\left[\left(2\gamma\beta^3 - C_{31}\right)k_{31}^3 + 3C_{31}k_{31}^2 - 3C_{31}k_{31} + C_{31}\right]$
	$0 < \lambda \leq \omega$	$D_{31} = \eta(\alpha_2 - 2\alpha + 1) + 2\mu(\beta - \alpha) + 2\alpha - 1$	$C_{31} = \dfrac{1}{\beta^2}\left[\left(2\alpha^3 - 3\alpha^2 + 1\right)\eta - 3\mu\left(\alpha^2 - \beta^2\right) + 3\alpha^2 - 1\right]$
3.2	$\alpha < \beta \leq \beta_{tu}$	$k_{32} = \dfrac{D_{32}}{D_{32} + 2\omega\gamma\beta}$	$M'_{32} = \left(3\gamma\omega\beta^2 + C_{32}\right)k_{32}^2 - 2C_{32}k_{32} + C_{32}$
	$\omega < \lambda \leq \lambda_{cu}$	$D_{32} = D_{31} + \gamma\omega_2$	$C_{32} = C_{31} - \dfrac{\gamma\omega^3}{\beta^2}$

The nominal moment capacity M_n is obtained by taking the first moment of force about the neutral axis, $M_n = F_{c1}y_{c1} + F_{t1}y_{t1} + F_{t2}y_{t2}$, and it is expressed as a product of the normalized nominal moment m_n and the cracking moment M_{cr} as follows:

$$m_n = C_2 \frac{k^2 - 2k + 1}{\beta^2} + \frac{2\beta k^3}{1-k}, \quad \text{where } C_2 = C_1 + 2C_1\beta - \beta^2 \tag{7.11}$$

$$M_n = m_n M_{cr} \quad M_{cr} = \frac{\sigma_{cr}bh^2}{6} \tag{7.12}$$

The maximum moment capacity is obtained when the normalized tensile strain at the bottom fiber ($\beta = \varepsilon_t/\varepsilon_{cr}$) reaches the tensile strain at peak strength ($\alpha = \varepsilon_{peak}/\varepsilon_{cr}$) (Soranakom and Mobasher, 2009). If the full moment-curvature response is desired, then the location of the neutral axis and moment capacity are obtained under the definitions provided in Table 7.1 where the derivations of all potential combinations of interaction of tensile and compressive response are presented. Note that depending on the relationship among material parameters, any of the responses present in zones 2.a and 2.b or 3.a and 3.b are potentially possible. The general moment–curvature profile as shown in the parametric studies and the analysis of these equations indicates that the contribution of fibers is mostly apparent in the post-cracking tensile region, in which the response continues to increase after cracking, as shown in Figure 7.1a. The post-crack modulus E_{cr} is relatively flat, maintaining values of $\eta = 0.00$–0.40 for the majority of cement composites. The tensile strain at peak strength ε_{peak} is relatively large compared to the cracking tensile strain ε_{cr} and may be as high as $\alpha = 100$ for polymeric-based fiber systems. These unique characteristics cause the flexural strength to continue to increase after cracking.

Since typical strain-hardening FRC do not have significant postpeak tensile strength, the flexural strength drops after passing the tensile strain at peak strength. In the most basic sense, one needs to determine two parameters in terms of postcrack stiffness η and postcrack ultimate strain capacity α to estimate the maximum moment capacity for the design purposes (Mobasher, 2011). Furthermore, the effect of postcrack tensile response parameter μ can be ignored for a simplified analysis.

The steps in the calculation of load–deflection response from the moment–curvature have been discussed in detail in several publications dealing with SHCC and softening composites (Soranakom and Mobasher, 2007a,b), so we will not dwell on them here. The load–deflection response of a beam can be obtained by using the moment–curvature, crack localization rules, and moment–area method.

7.2.2 Simplified procedure for generation of moment–curvature response

According to the trilinear tension and elastic compression models shown in Figure 7.2a and b, the maximum moment capacity is obtained when the normalized tensile strain at the bottom fiber ($\beta = \varepsilon_t/\varepsilon_{cr}$) reaches the tensile strain at peak strength ($\alpha = \varepsilon_{peak}/\varepsilon_{cr}$). However, the simplified equations (7.10 through 7.12) for moment capacity are applicable for the compressive stress in the elastic region only. The elastic condition must be checked by computing the normalized compressive strain developed at the top fiber λ and comparing it to the normalized yield compressive strain ω. The general solutions for all the cases are presented in Table 7.1. Using the strain diagram in Figure 7.3a, the relationship between the top compressive strain and the bottom tensile strain is:

$$\frac{\varepsilon_c}{kh} = \frac{\varepsilon_t}{(1-k)h} \tag{7.13}$$

By substituting $\varepsilon_c = \lambda\varepsilon_{cr}$ and $\varepsilon_t = \beta\varepsilon_{cr}$ in Equation 7.13 and limiting the maximum compressive strain to its yield value $\varepsilon_{cy} = \omega\varepsilon_{cr}$, we can express the compression failure in a normalized form as:

$$\lambda = \frac{k}{1-k}\beta \leq \omega \tag{7.14}$$

In this section, the case represented by 2.1 in Table 7.1, where the tensile behavior is elastic–plastic while the compressive behavior is still elastic, is studied. Equations for other cases can also be developed through this example. The general solution presented in Table 7.1 can be simplified by representing the location of neutral axis as a function of applied tensile strain β as:

$$k = \frac{\sqrt{A}}{\sqrt{A} + \beta\sqrt{\gamma}} \quad A = \eta\left(\beta^2 + 1 - 2\beta\right) + 2\beta - 1 \tag{7.15}$$

This equation can be easily simplified by assuming equal tension and compression stiffness ($\gamma = 1$). For an elastic, perfectly plastic tension material ($\eta = 0$), the equation reduces to

$$k = \frac{\sqrt{2\beta - 1}}{\sqrt{2\beta - 1} + \beta} \tag{7.16}$$

Table 7.2 presents the case of ($\gamma = 1$) for different values of postcrack stiffness $\eta = 0.5, 0.2, 0.1, 0.05, 0.01$, and 0.001. One can conveniently

Table 7.2 Location of neutral axis, moment, and moment–curvature response of a strain-hardening composite material with $\gamma = 1 \ \eta = 0.0001{-}0.5$

η	$A, \left(k = \dfrac{\sqrt{A}}{\sqrt{A}+\beta}\right)$	$M'(k)$	$M'(\varphi)$
0.5	$0.5(\beta^2 + 1 - 2\beta) + 2\beta - 1$	$-0.773 + 0.108 \times 10^{-1}k^{-6}$	$0.507 + 0.686\varphi$
0.2	$0.2(\beta^2 + 1 - 2\beta) + 2\beta - 1$	$0.654 + 0.516 \times 10^{-2}k^{-6}$	$1.105 + 0.383\varphi$
0.1	$0.1(\beta^2 + 1 - 2\beta) + 2\beta - 1$	$1.276 + 0.289 \times 10^{-2}k^{-6}$	$1.461 + .234\varphi$
0.05	$0.05(\beta^2 + 1 - 2\beta) + 2\beta - 1$	$1.645 + .1632 \times 10^{-2}k^{-6}$	$1.720 + .1401\varphi$
0.01	$0.01(\beta^2 + 1 - 2\beta) + 2\beta - 1$	$0.852 + 0.456k^{-1}$	$1.342 + 0.371\sqrt{\varphi}$
0.0001	$0.0001(\beta^2 + 1 - 2\beta) + 2\beta - 1$	$3.177 - 3.068k$	$3.021 - 2.047/\sqrt{\varphi}$

interpolate between any of these two parameters. The neutral axis is a function β and can be used in the calculation of the moment, or the moment–curvature relationship. These general responses are shown in Figure 7.4a and b and validate the premise that as the applied tensile strain increases, the neutral axis moves toward the compression zone. Having said that, this change is a function of the postcrack tensile stiffness factor. The moment–curvature relationship in this range is ascending, but its rate is a function of the postcrack tensile stiffness. The parameter-based

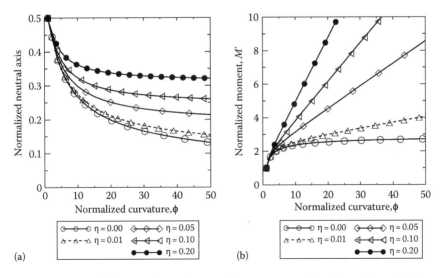

Figure 7.4 Effect of (a) depth of neutral axis on the moment capacity of a section and (b) the moment–curvature response in the range 2.1. (From Mobasher, B., *Mechanics of Fibre and Textile Reinforced Cement Composites*, CRC Press, September 2011, 480pp.)

fit equations in the third and fourth columns are obtained by curve fitting the simulated response from the closed-form derivations and are applicable within 1% accuracy to the closed-form results. Using these equations, one can generate the moment capacity and moment–curvature response for any cross section using basic tensile material parameters in the 2.1 range, as defined in Table 7.1.

7.2.3 Algorithm to predict load–deflection response

The steps in the calculation of load–deflection response from the moment–curvature were developed in detail in recent publications dealing with strain-hardening- and strain-softening-type composites (Soranakom and Mobasher, 2007b). These calculations tell us that the load–deflection response of a beam can be obtained by using the moment–curvature (Figure 7.4b), crack localization rules, and moment–area method as follows:

1. For a given cross section and material properties, the normalized tensile strain at the bottom fiber β is incrementally imposed to generate the moment–curvature response using the expressions given in Table 7.1. For each value of β in stages 2 and 3, the condition for compressive stress $\lambda < \omega$ or $\lambda > \omega$ is verified in advance of moment–curvature calculation, as shown in Figures 7.3 and 7.5.
2. The beam is segmented into finite sections. For a given load step, static equilibrium is used to calculate moment distribution along the beam and the moment–curvature relationship along with crack localization rules in order to identify the curvature, as shown in Figure 7.5.

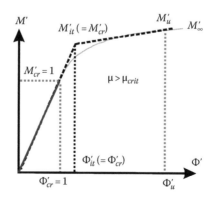

Figure 7.5 Moment–curvature relationship of strain-hardening FRC and its bilinear idealization. (From Soranakom, C. and Mobasher, B., *J. Eng. Mech.*, 133(8), 933, 2007a.)

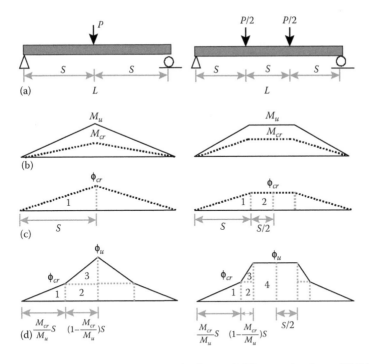

Figure 7.6 Moment–curvature distribution for 3PB and 4PB strain-hardening FRC beams. (From Soranakom, C. and Mobasher, B., *J. Eng. Mech.*, 133(8), 933, 2007a.)

3. Since a moment–curvature diagram determines the maximum load allowed on a beam section, the discrete moments along the diagram are used to calculate the applied load vector $P = 2M/S$, where S is the spacing between the support and loading point; hence $S = L/2$ for three-point bending and $S = L/3$ for four-point bending, as shown in Figure 7.6.
4. The deflection at midspan is calculated by using the numerical moment–area method of discrete curvature between the support and midspan. This procedure is applied at each load step until a complete load–deflection response is obtained.

7.2.4 Deflection computation using a bilinear moment–curvature assumption

With the moment–area method, the curvature diagram of a half-model (due to symmetry) is divided into several areas as shown in Figure 7.6. By assuming a bilinear moment–curvature relationship and with the application of moment–area method, the midspan deflection is directly obtained by using the double integration of curvature distribution resulting in closed-form analytical solutions. Testing of flexural samples may indicate formation of diagonal tension cracks due to shear failure, but provisions for shear cracking are not accounted for in the present approach. Many of the calculations generated from these methods are, therefore, limited to the response up to the maximum load. Having said that, a set of equations for calculating the midspan deflection δ of the three-point bending at the first cracking (δ_{cr}) and at ultimate (δ_u) under the condition of $\mu > \mu_{crit}$ are presented in Equations 7.17 and 7.18:

$$\delta_{cr} = \frac{1}{12}L^2\phi_{cr} \qquad (7.17)$$

$$\delta_u = \frac{L^2}{24M_u^2}\left[\left(2M_u^2 - M_uM_{cr} - M_{cr}^2\right)\phi_u + \left(M_u^2 + M_uM_{cr}\right)\phi_{cr}\right] \qquad (7.18)$$

Similarly, for four-point bending, the deflection is obtained as in Equation 7.20:

$$\delta_{cr} = \frac{23}{216}L^2\phi_{cr} \qquad (7.19)$$

$$\delta_u = \frac{L^2}{216M_u^2}\left[\left(23M_u^2 - 4M_uM_{cr} - 4M_{cr}^2\right)\phi_u + \left(4M_u^2 + 4M_uM_{cr}\right)\phi_{cr}\right] \quad \mu > \mu_{crit} \qquad (7.20)$$

Alternatively, one can use the nonlinear description of the moment–curvature relationship as developed in Table 7.2 in order to compute the load–deflection response.

7.2.5 Parametric studies of load–deflection response

Parametric studies evaluate the effect of different parameters on the moment–curvature and load–deflection response. Due to the nature of modeling, a unique set of properties from the flexural tests cannot be obtained as long as there are a range of tensile properties that may result in similar flexural responses. It is, therefore, essential to measure and match both tension and flexural responses in the back-calculation processes. Figures 7.7 through 7.9 show the effect of model parameters μ, α, and η on the simulated response. The key parameters in these examples have been changed to best fit the experimental load–deflection and tensile stress–strain curves, although it should be noted that simulations that use direct tension data underestimate the equivalent flexural stress. This may be due to several factors including size effect, uniformity in tension loading versus the linear strain distribution in flexure, and variation in lamina orientation, which may lead to a wider variation in flexural samples. However, the underestimation of flexural capacity can be reduced by applying scaling parameters to the tensile capacity (Soranakom and Mobasher, 2007a,b) [9]. Therefore,

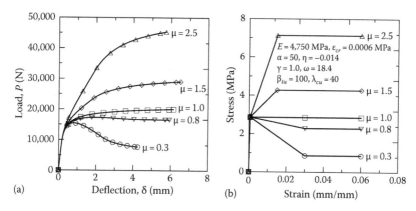

Figure 7.7 Parametric study of parameter μ as shown in (a) load–deflection and (b) stress–strain.

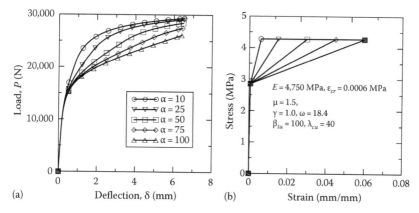

Figure 7.8 Parametric study of parameter α in (a) load–deflection and (b) stress–strain. (From Mobasher, B., *Mechanics of Fibre and Textile Reinforced Cement Composites*, CRC Press, September 2011, 480pp.)

this procedure still has potential for use of the flexural data to develop the moment–curvature response for the design of flexural load cases.

Figure 7.7a is a parametric study of parameter μ the residual tensile strength as shown in load–deflection. The stress–strain parameters for the response are shown in Figure 7.7b. Figure 7.7 shows that as the elastic–plastic yield limit is increased, the load–deflection capacity increases almost in proportion to the value of parameter μ, which represents the yield stress to the first-cracking stress magnitude. Figure 7.8a is a parametric study of parameter "alpha" the ultimate strain capacity as shown in load–deflection curve. The stress–strain parameters for the response are shown in Figure 7.8b. Figure 7.8 shows that as the ultimate tensile strain capacity increases for the same elastic–plastic yield strength, the rate of ascent of

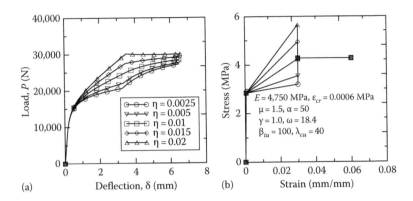

Figure 7.9 Parametric study of parameter η in (a) load–deflection and (b) stress–strain. (From Mobasher, B., *Mechanics of Fibre and Textile Reinforced Cement Composites*, CRC Press, September 2011, 480pp.)

the load–deflection response after the first cracking is affected. However, the maximum load is affected in a limited manner.

Figure 7.9 shows that the change in the postcracking stiffness by an order of magnitude affects the load–deflection response in the postcrack phase marginally and that the ultimate elastic–plastic load level remains static. This figure supports the hypothesis that while the bond strength affects the postcrack stiffness, it is the fiber frictional stress that is ultimately a more important parameter. As the cracks widen, the fiber pullout results in the maintaining of the stiffness and ductile deformation. The tensile ductility is, therefore, as important as the tensile strength of the material in studying the flexural response.

7.2.6 Inverse analysis of the load–deflection response of TRC composites

The tensile properties can be back-calculated from the flexural response by means of an inverse analysis and can be measured with experimentally obtained data. Thus, it is sufficient to describe the material behavior with four parameters: Young's elastic modulus E, first-cracking strain ε_{cr}, post-crack stiffness η, and strain at peak stress ε_{trn} (or ultimate strain parameter α). Since the compressive strength is several times higher than the first-crack tensile strength, constant parameters of $\gamma = 1.0$ and $\omega = 8$–12 are used. Parameter γ represents the ratio of tensile modulus to compressive modulus, and ω represents the ratio of compressive modulus to tensile first-crack strength.

The inverse analysis was performed by first adjusting Young's modulus until the initial slope of the predicted, and experimental flexural stress deflection responses were fitted. Next, the first-cracking strain was adjusted until the predicted postcrack response matched the proportional limit (LOP) of the experiments. Finally, the strain at peak stress was adjusted until the predicted and experimental peak stresses were coincident. A relatively low value of postcrack stiffness η, in the range of 0.01%–0.5%, can be used, indicating that the residual stiffness after cracking is in the range of 1%–5% of the elastic stiffness. Examples of analysis based on these concepts are outlined in the following for AR glass TRC and PE fiber ECC.

7.2.6.1 AR glass TRC

In order to address the variations in the experimental results due to operational, processing, random, and mixture parameters, a study was conducted to simulate a range of experimental data by means of selecting upper and lower bounds of input parameters. Three different TRC composites consisting of AR glass with alternate 100 lb or 200 lb (4.2–8.4 kPa) with uniform applied pressure during casting (denoted as GNS100 and GNS 200) as well as samples with 40% cement substitution by fly ash denoted as (GFA40) were used (Peled and Mobasher, 2005; Soranakom et al., 2006). Composites were manufactured using a cement paste with a w/c = 0.45 and eight layers of AR glass manufactured by Nippon Electric Glass Co. (Peled and Mobasher, 2005). Both experimental data from a set of specimens under uniaxial tension and three-point bending tests were utilized, but no attempt was made to obtain a best fit curve. The tensile TRC composites were $10 \times 25 \times 200$ mm. Flexural three-point bending samples were $10 \times 25 \times 200$ mm with a clear span of 152 mm. Material parameters were determined by fitting the hardening model to both tension and flexural tests. Results are depicted here with the simulated upper and lower bounds encompassing the TRCs in Figure 7.10. Figure 7.10a shows the predicted flexural load–deflection response, and Figure 7.10b shows the tensile stress–strain responses compared with experimentally obtained results.

Representative properties for the simulation of upper bound values obtained from the GNS200 samples were α = 50, μ = 3.9, η = 0.06, γ = 5.0, and ω = 10 with the constants being ε_{cr} = 0.0002 and E = 20,000 MPa, while the limits of the modeling were β_{tu} = 135 and λ_{cu} = 40. Representative material properties for the lower bound values from the GFA40 samples were α = 32, μ = 2.0, η = 0.032, γ = 5.0, and ω = 10. The constants were ε_{cr} = 180 μ_{str} and E = 20 GPa, while the limits of the modeling were β_{tu} = 150 and λ_{cu} = 40. These values apply to a typical set of data, and proper optimization with upper and lower bound values are required.

7.2.6.2 PE ECC

An ECC mix utilizing 2.0% by volume of PE fibers recommended by previous studies (Li, 1994; Maalej and Li, 1994) was also modeled, as shown in Figure 7.11. The flexural specimens for the four-point bending test were $76 \times 101 \times 355$ mm with a clear span of 305 mm (Maalej, 2004). Figure 7.11 shows the stress–strain and stress–deflection response.

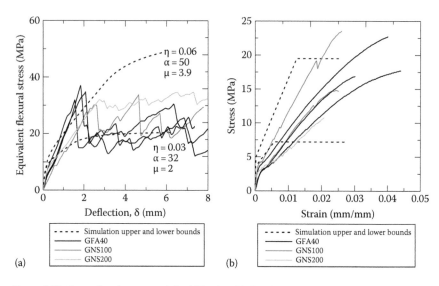

Figure 7.10 Strain-hardening model of AR glass TRC: (a) equivalent flexural stress deflection and (b) corresponding stress–strain response. (From Mobasher, B. and Barsby, C., Flexural design of strain hardening cement composites, *Proceedings of the Second International RILEM Conference on Strain Hardening Cementitious Composites (SHCC2-Rio)*, Rio de Janeiro, Brazil, 2012, pp. 53–60.)

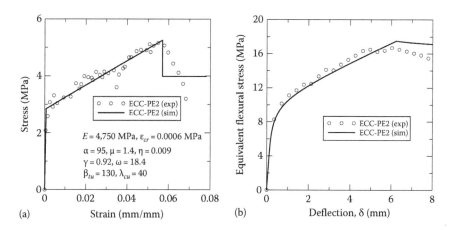

Figure 7.11 (a) Stress–strain response and (b) equivalent flexural stress deflection of ECC. (From Mobasher, B. and Barsby, C., Flexural design of strain hardening cement composites, *Proceedings of the Second International RILEM Conference on Strain Hardening Cementitious Composites (SHCC2-Rio)*, Rio de Janeiro, Brazil, 2012, pp. 53–60.)

7.3 Case studies

7.3.1 Prediction of load–deflection response

Two types of fabric systems are discussed in this section, including bonded AR glass and woven PE mesh. AR glass fabric-reinforced cement composites were manufactured for testing using a cement paste with a water-to-cement ratio of 0.45, while Saint-Gobain Technical Fabrics Inc. manufactured the AR glass fabrics were used. The grid size was 25.4 × 25.4 mm with two yarns in each of the longitudinal and transverse directions. Individual yarn fibers consisted of 1,579 filaments, with each being 19 μm in diameter. Two layers of fabric were placed at the top and bottom of the specimens to provide reinforcement in each direction: $V_L = V_T = 0.70\%$.

Figure 7.12a shows the tensile model and Figure 7.12b shows the predicted flexural load–deflection response of the cement composites. Note that in these systems, the high tensile stiffness and the strength of the composite lead to high values for the load, and distributed flexural cracking. An analysis of the samples indicates the formation of diagonal tension cracks in the samples due to the shear failure mechanism. No provisions for shear cracking were accounted for in the present approach, and no attempt was made to simulate the response beyond the first major flexural crack. Material properties for the simulation of PE fabric composites were defined in terms of α = 150, μ = 0.4, η = 0.008, γ = 1.0, and ω = 20.4. The constants were ε_{cr} = 0.0002 and E = 18,000 MPa, while the limits of the modeling were β_{tu} = 250 and λ_{cu} = 150. The results of the flexural tests are shown in Table 7.3, and the model parameters for the fit are shown in Table 7.4.

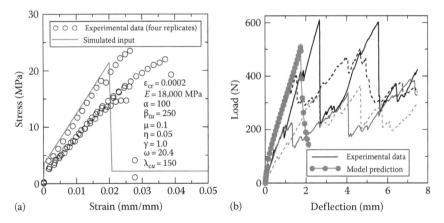

Figure 7.12 (a) Tensile stress–strain response input model and (b) predicted load–deflection response of AR glass fabric composites. (From Mobasher, B. and Barsby, C., Flexural design of strain hardening cement composites, *Proceedings of the Second International RILEM Conference on Strain Hardening Cementitious Composites (SHCC2-Rio)*, Rio de Janeiro, Brazil, 2012, pp. 53–60.)

Table 7.3 Data from experimental analysis of representative TRC and ECC samples

Sample ID	b (mm)	d (mm)	L (mm)	Flexural stiffness (N/mm)	Defl at max flex load (mm)	Max flex load (N)	Bending stress (MPa)	Flexural toughness (N mm/mm²)
GNS200	30	9	152	801	5.84	520	49	7.91
GNS100	34	7	152	565	6.25	138	20	2.88
GFA40	24	8	152	510	5.38	226	26	4.64
ECC-PE2	76	102	305	60,351	6.23	30,114	12	26.36

Source: Mobasher, B. and Barsby, C., Flexural design of strain hardening cement composites, *Proceedings of the Second International RILEM Conference on Strain Hardening Cementitious Composites (SHCC2-Rio)*, Rio de Janeiro, Brazil, 2012, pp. 53–60.

Table 7.4 Material model parameters from back-calculation of TRC and ECC samples

Sample ID	Young's modulus, E (GPa)	First-crack strength, σ_{cr} (MPa)	Post-crack tensile strength, μ	Transition tensile strain, α	Transitional tensile strain, ε_{trn} (%)	Ultimate strain, ε_{tu} (%)
GNS200	20	5	3.9	50	1.25	2.5
GNS100	20	3.6	2	32	0.576	2.7
GFA40	21	4.62	2.3	70	1.54	2.09
ECC-PE2	4.75	2.85	1.4	95	5.7	7.8

Source: Mobasher, B. and Barsby, C., Flexural design of strain hardening cement composites, *Proceedings of the Second International RILEM Conference on Strain Hardening Cementitious Composites (SHCC2-Rio)*, Rio de Janeiro, Brazil, 2012, pp. 53–60.

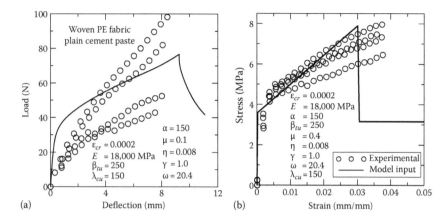

Figure 7.13 (a) Tension stress–strain response input model and (b) predicted load–deflection response of PE fabric composites. (From Mobasher, B. and Barsby, C., Flexural design of strain hardening cement composites, *Proceedings of the Second International RILEM Conference on Strain Hardening Cementitious Composites (SHCC2-Rio)*, Rio de Janeiro, Brazil, 2012, pp. 53–60.)

The overall tensile response for the PE composites is shown in Figure 7.13. The algorithm is able to predict the load–deflection response for strain-hardening material. The simulations developed by Mobasher and Barsby, 2012 are used to demonstrate the model with flexural specimens for the four-point bending test of 25 × 10 × 200 mm with a clear span of 152 mm. The results of the flexural fit are shown in Figure 7.13a. The material parameters for tension model were determined from the flexural test as shown by the solid line in Figure 7.13b and can also be compared by fitting the model to the uniaxial tension test results. The average material properties are characterized by compressive strength $f_c' = 73.4$ MPa. The initial tensile modulus $E = 18$ GPa and first cracking tensile strain $\varepsilon_{cr} = 200\mu str$ were used as they correlated with the uniaxial tensile test results. The ultimate compressive strain ε_{cu} was assumed to be $\alpha\varepsilon_{cr} = 150\,(0.0002) = 3\%$. All parameters used in the simulation are provided in the same figure.

The solid curves in Figure 7.13b show the predicted tension response obtained by the simulation process. Unfortunately, this prediction for the strain-hardening material during the precrack and postcrack stages overestimates the experimental results, partly because the tension data are from notched samples and the method of strain measurement is subjective (due to the reference of distributed cracking in flexure vs. a single dominant notch under tension). Note that the formation of the distributed crack system can be adequately simulated by the smeared pseudostrain model.

The predictions of the equivalent load–deflection response for AR glass and PE fabrics using the direct tension data underestimate the equivalent flexural stress. This may be due to several factors, including the effect of sample size and the uniformity in loading in tension versus the linear strain distribution in flexure. The underestimation of the flexural capacity of the beam can be addressed by increasing the apparent tensile strength as discussed in earlier publications. Alternatively, one can use the flexural response and develop a back-calculation procedure to calculate a direct fit to the experimental flexural data.

7.4 Flexural design

7.4.1 Design guidelines for 1D and 2D members

The present approach is applicable to the design and analysis of 1D beams and 2D members that include square, rectangular, or circular panels that may be simply supported, free, or fixed-edge plates. The development of yield line theory can be attributed to Johansen (1962) and Johansen and Pladeformler (1968), who were responsible in large part for covering a number of situations that are difficult to analyze without resorting to finite element methods. The only criticism of these derivations is that there are no specific guidelines in making accurate deflection calculations and that only the magnitude of the maximum load is measured. More recent publications addressing yield line analyses of slabs have been published by Kennedy and Goodchild (2003) and relate the section properties to the load-carrying capacity of the slabs. These formulations use Johansen's formulas for one-way, two-way, and flat slabs.

Since the postcrack flexural response of TRC is ductile, it can sustain large deflections after cracking. In many cases, the ultimate moment capacity can be used as a limit state design criteria for these materials. Therefore, the approaches developed for 2D slabs are very much applicable to the panels and to thin structural members. The application of yield line conditions results in obtaining an upper bound estimate to the ultimate load; however, this approach is sufficiently conservative so that the resulting failure modes are predictable and do not lead to catastrophic modes.

The methodology used in the design of concrete using a plastic analysis approach is adopted from the ACI 318 approach (2005). The nominal capacity of flexural member M_n must be decreased by the reduction factor ϕ_r to account for variability in materials, workmanship, and failure model assumed for strain-hardening composite as stipulated by ACI Sec. C.3.5 (ACI Committee 318, 2005). However, the reduced capacity must be greater than the ultimate moment M_u due to factored loading, as we can see in

$$\phi_r M_n \geq M_u \tag{7.21}$$

where ϕ_r is the reduction factor for strain-hardening cement composite and may be conservatively taken as 0.85, similar to the reduction factor for the compressive failure of plain concrete as stipulated by ACI Sec. C.3.5.

7.4.2 Capacity calculations based on section moment–curvature

The general strain-hardening tensile and elastic perfectly plastic compression model as derived by Soranakom and Mobasher (2008) and discussed in Chapter 6 and Section 7.1 (as shown in Figure 7.2) is used in these calculations. The tensile response is defined by tensile stiffness E, first-crack tensile strain ε_{cr}, ultimate tensile capacity ε_{peak}, and postcrack modulus E_{cr}. The softening range is shown as a constant stress level $\mu E \varepsilon_{cr}$. The compression response is defined by the compressive strength σ_{cy}, defined as $\omega \gamma E \varepsilon_{cr}$.

In order to set serviceability limits, ranges of allowable tensile and compressive strains are defined as limit states. Equations are then simplified to idealize bilinear tension and elastic compression models, as shown in Figure 7.14, by disregarding the postpeak ranges in tension. Furthermore, by disregarding the postpeak tensile responses and plasticity in the compression region, only one set of equations dominates the response. In addition, due to negligible differences in the compressive and tensile modulus E in compression and tension, they are assumed equal (Soranakom and Mobasher, 2008).

Tensile response is defined by tensile stiffness E, first-crack tensile strain ε_{cr}, ultimate tensile capacity ε_{peak}, and postcrack modulus E_{cr}, while the compression response is defined by the compressive strength $\sigma_{cy} = \omega E \varepsilon_{cr}$.

The contribution of reinforcement is mostly apparent in the postcracking tensile region, where the response continues to increase after cracking, as shown in Figure 7.14a. The postcrack modulus E_{cr} is relatively flat, and the tensile strain at peak strength ε_{peak} is large when compared to the cracking tensile strain ε_{cr}. These unique characteristics cause the flexural load carrying response to continue to increase after cracking. Since typical strain-hardening cement composite do not have significant postpeak tensile strength, the flexural strength decreases after passing the tensile strain at peak strength. It is therefore sufficient to use ε_{peak} to estimate the maximum

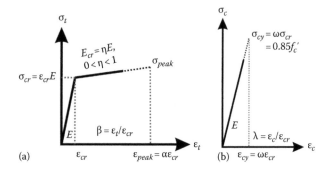

Figure 7.14 Simplified material models for strain-hardening cement composite: (a) tension and (b) compression. (From Mobasher, B., *Mechanics of Fibre and Textile Reinforced Cement Composites*, CRC Press, September 2011, 480pp.)

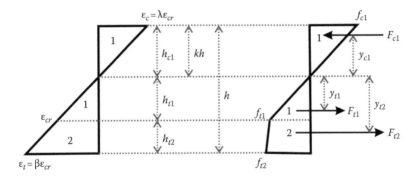

Figure 7.15 Strain and stress diagrams at the postcrack stage (Range 2.1 Table 7.1), (a) strain distribution and (b) stress distribution. (From Soranakom, C. and Mobasher, B., *Cem. Concr. Compos.*, 30, 465, 2008.)

moment capacity for design purposes. For typical strain-hardening FRC, the compressive strength is several times higher than the tensile strength. Thus, the flexural capacity is controlled by the weaker tension and the compressive stress is normally low in the elastic range. For this reason, the elastic compression model shown in Figure 7.14 is used and the compressive stress developed in a beam section is limited to the yield compressive stress $\sigma_{cy} = 0.85f'_c$ at compressive yield strain ε_{cy}, where f'_c is the uniaxial compressive strength. Range 2.1 as shown in Figure 7.15, and defined by parameters $(1 < \beta \le \alpha, 0 < \lambda \le \omega)$, is represented by the neutral axis location, moment, and curvature as shown in Figure 7.15 and defined as:

$$k_{21} = \frac{D_{21} - \sqrt{D_{21}\gamma\beta^2}}{D_{21} - \gamma\beta^2}, \quad D_{21} = \eta\left(\beta^2 - 2\beta + 1\right) + 2\beta - 1$$

$$M'_{21} = \frac{\left(2\gamma\beta^3 - C_{21}\right)k_{21}^3 + 3C_{21}k_{21}^2 - 3C_{21}k_{21} + C_{21}}{1 - k_{21}},$$

$$C_{21} = \frac{\left(2\beta^3 - 3\beta^2 + 1\right)\eta + 3\beta^2 - 1}{\beta^2} \tag{7.22}$$

7.4.3 Demand calculations using yield line analysis

The upper bound theory of limit analysis is utilized by applying the virtual work method. In this approach, a kinematically admissible displacement field that simulates the potential collapse mechanism of the structure is used. The internal and external work expressions are evaluated for each collapse mechanism and are set equal to one another in order to satisfy the conservation of energy criteria. In order to calculate the work done by the externally applied forces through a kinematically admissible displacement field, the expenditure of external loads and the dissipation of energy within the yield lines are carried out independently. The results are then made equal to each other and, from the resulting equation, the unknown, be it the ultimate moment m generated in the yield lines or the ultimate failure load q of the slab, is evaluated.

Simulation would require the slab to be divided into rigid regions that rotate about their respective axes of rotation along the support lines. If we give the point of maximum deflection a value of unity, then the vertical displacement of any point in the regions is thereby defined as a fractional value. The expenditure of external loads is evaluated by taking all external loads on each region, finding the equivalent resultant load and multiplying it by the distance it travels.

The dissipation of energy is quantified by projecting all the yield lines around a region onto, and at right angles to that region's axis of rotation. These projected lengths are multiplied by the moment acting on each length and by the angle of rotation of the region. At the small angles considered, the rotation is equated to the tangent of the angle produced by the deflection of the region. The sense of the rotations is immaterial. Two most common methods of approach, based on upper and lower bound theories of limit analysis, are shown in the following. In the simple case of the statically determinate beam, results are identical from both of them, since the solution is the exact solution to the problem.

7.4.3.1 Virtual work method (upper bound approach)

Using the principle of virtual work, it is stated that if the system is in equilibrium, and a kinematically admissible virtual displacement is imposed on the system, the work by the external forces on the virtual displacement is equal to the work done by the real internal stresses on the virtual strains. The virtual work expression can be written as a work balance equation using the virtual and real parameters. Indeed, one can use the principle of virtual work to equate the external and internal work measures. In the case of a simple beam subjected to force F, the internal plastic moment M developed can be used, as F and M represent the force and plastic bending moment, while $\delta\Delta$ and δn represent the virtual displacement and associated rotation as shown in Figure 7.23a:

$$W_{\text{int}} = W_{ext} \quad \sum F\delta\Delta = \int_0^L M_P \delta\phi dx \qquad (7.23a)$$

Flexural capacity of a simply supported beam subjected to concentrated load as shown in Figure 7.16 is developed first and then extended to a distributed load case. A plastic analysis methodology uses the principle of

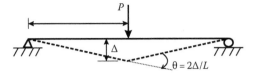

Figure 7.16 Collapse mechanism of a simply supported beam subjected to load at center point.

virtual work to equate the internal and external dissipated work to obtain the collapse load. The work equations are derived based on the concept presented in Figure 7.16:

$$\Sigma(F \times \Delta) = \Sigma(M \times L \times \theta) \qquad (7.23b)$$

$$P_u\Delta = M_P(2\theta) = M_P\left[2\left(\frac{\Delta}{0.5L}\right)\right] \quad P_u = \frac{4M_P}{L}$$

7.4.3.2 Equilibrium method (lower bound method)

This method is based on the concept presented in Figure 7.17:

$$\Sigma M = 0 \quad \frac{P_u}{2}\left(\frac{L}{2}\right) - M_P = 0$$

$$P_u = \frac{4M_P}{L} \qquad (7.24)$$

The flexural capacity of a simply supported beam subjected to distributed loads is developed next using the same approach, and the resultant four segments of N_R and rotation θ (from Figure 7.18, N_R acting at 1/3 of δ_{max}) and solving Equation 7.26 for the moments give us

$$N_R = \frac{qL}{2}, \quad \theta = \frac{2\delta_{max}}{L} \qquad (7.25)$$

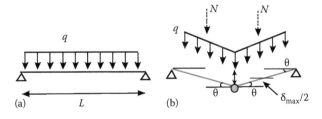

Figure 7.17 (a) Free body diagram and (b) the kinematic deformations.

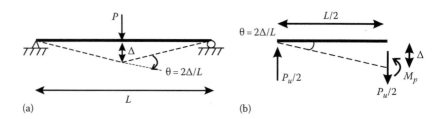

Figure 7.18 The collapse mechanism in a three-point bending beam with distributed load. (a) Geometry of the three point bending set up, and (b) the free body diagram and kinematics of rotation vs. deflection relationship.

$$W_{int} = W_{ext}, \quad 2M_p\theta = 2\left(\frac{qL}{2}\right)\frac{1}{2}\left(\frac{\theta L}{2}\right), \quad M_p = \frac{q_{ult}L^2}{8} \qquad (7.26)$$

Using the moment versus allowable load relationship, one can refer back to Equation 7.26, and depending on the applied load that determines the magnitude of q_{ult}, we can compute a required ultimate moment capacity. This then allows us to use this equation in conjunction with the work equation to compute the required section size to carry the given load.

7.4.4 Collapse mechanism in plastic analysis

Statically determinate beams fail upon the development of one plastic hinge. For indeterminate structures, as the load increases, the most highly stressed section yields locally and forms a plastic hinge. This plastic hinge will act as a real hinge insofar as increased loading is concerned, as seen in Figure 7.18. Additional loading causes additional plastic hinge formations. When the number and arrangement of actual and plastic hinges result in an unstable structure, then collapse occurs.

A case study of sequential failure pattern is demonstrated next. The loading history of the beam, from working load to collapse load, is traced in Figure 7.19. At working loads, before yielding begins anywhere, the distribution of bending moments will be as shown in Figure 7.19a, with maximum moment occurring at the fixed ends. As the load is gradually increased, yielding begins at the supports when the bending moment reaches the yielding moment (M_y). Further increase in the load will cause the simultaneous formation of plastic hinges at each end at a plastic moment (M_p). At this level of loading, the structure is still stable, the beam having been rendered statically determinate by the formation of two plastic hinges. Only when a third hinge forms will a failure mechanism be created. This happens when the maximum positive moment attains a value of M_p.

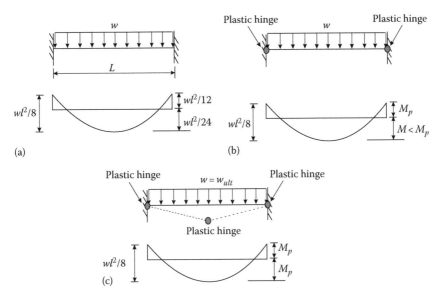

Figure 7.19 (a) Uniformly loaded beam and its bending moment diagram. (b) First plastic hinge formation at the supports. (c) Second plastic hinge formation at the center.

7.4.5 Analysis of 2D panels

7.4.5.1 Case study 1: Square panel with free edges

Plastic analysis theories use the principle of virtual work to measure the internal and external dissipated work to obtain the collapse load. For example, for a distributed load on a three-point bend flexural beam, as shown in Figure 7.20, the work equations are derived as:

$$W_{int} = W_{ext}, \quad 2M_P\theta = 2M_P\left(\frac{2\delta}{L}\right) = 2\left[q\left(\frac{L}{2}\right)\left(\frac{\delta}{2}\right)\right], \quad M_P = \frac{q_{ult}L^2}{8} \quad (7.27)$$

This yield pattern can be used to calculate the potential collapse mechanism of a plate supported along its two or four edges. If the panel has fixed edges, then the yielding along the edge also needs to be included in the calculations.

By requiring the equality of the energy expended by loads as the beam undergoes deflection, to the energy dissipated by rotations about the yield line, we get

$$\Sigma(N\delta) = \Sigma(ml\theta) \quad (7.28)$$

On the left-hand side, q is the uniformly distributed load and $L^2/4$ is the area of each wedge, so the equivalent point load is $(q \times L^2/4)$ and $\delta_{max}/3$ is

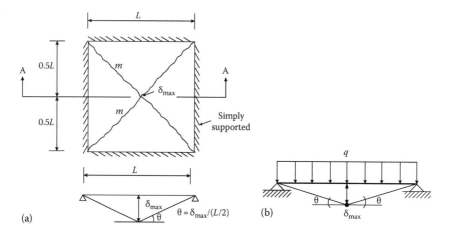

Figure 7.20 Simply supported square panel with (a) yield lines and (b) loading and rotation conditions through section A–A. (From Mobasher, B. and Barsby, C., Flexural design of strain hardening cement composites, *Proceedings of the Second International RILEM Conference on Strain Hardening Cementitious Composites (SHCC2-Rio),* Rio de Janeiro, Brazil, 2012, pp. 53–60.)

the deflection of the centroid. On the right-hand side, L is the length of the square as the rotations are projected onto the sides. The calculation of the rotation angle θ is shown in Figure 7.20, and upon simplifying the expression, we get

$$4\left(q\frac{L^2}{4}\frac{\delta_{max}}{3}\right) = 4\left(mL\frac{\delta_{max}}{0.5L}\right) \tag{7.29}$$

$$\frac{4L^2q}{12} = 8m, \quad m = \frac{qL^2}{24} \tag{7.30}$$

where
　　m is the moment along the yield lines
　　q is the uniformly distributed load
　　L is the length of the square side

7.4.5.2 Case study 2: Square panel with edges clamped

This case study is for a plate with clamped edges, as shown in Figure 7.21.
　　Energy expended is independent of the support conditions; therefore, it is the same as the expression obtained for the simply supported slab.

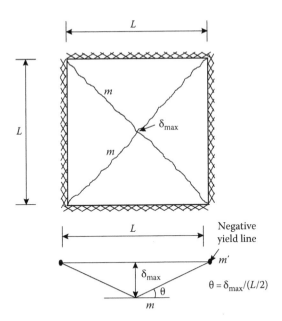

Figure 7.21 Continuous supported (clamped) square slab.

Energy dissipated in the case of clamped slabs will be higher due to rotation of the corner supports. All of the four wedges have a projection of their positive (sagging) yield line with a value m surrounding that region onto its axis of rotation with the length as L. The sides also have continuous supports (clamped) and negative (hogging) yield line causing a moment of value m' that forms along the support:

$$\Sigma(N\delta) = \Sigma(ml\theta)$$

$$4\left(q\frac{L^2}{4}\frac{\delta_{max}}{3}\right) = 4\left(mL\frac{\delta_{max}}{0.5L} + m'L\frac{\delta_{max}}{0.5L}\right) \quad (7.31)$$

If we assume $m = m'$ (i.e., that the sagging moment is equal to the hogging moment), we get

$$\frac{4L^2 q}{12} = 16m, \quad m = \frac{qL^2}{48} \quad (7.32)$$

7.4.5.3 Case study 3: Rectangular slab with clamped edges

The scheme of the rectangular slab with clamped edges is presented in Figure 7.22. The model assumes a two-way span slab that measures $a \times b$ and carries a load of q kN/m².

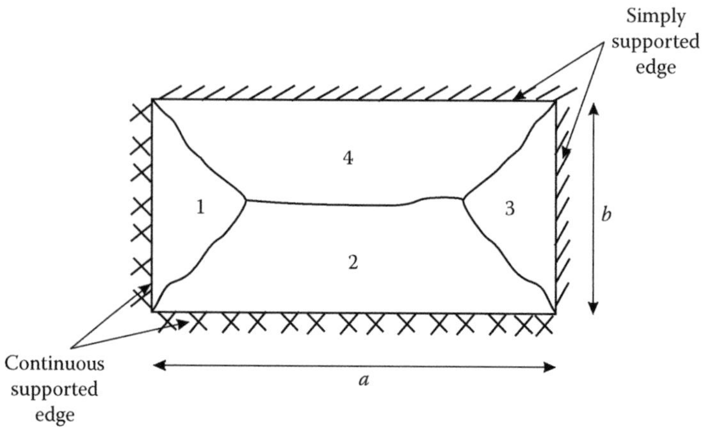

Figure 7.22 Continuous supported (clamped) and two-sided, simply supported rectangular slab.

The slab is divided into rigid regions that rotate about their respective axes of rotation along the support lines. If we give the point of maximum deflection a value of unity, then the vertical displacement of any point in the regions is thereby defined. The expenditure of external loads is evaluated by taking all external loads on each region, finding the center of gravity of each resultant load, and multiplying it by the distance it travels. Two groups are considered: (a) the triangles and (b) the trapezoidal sections representing the external work, W_e, as in:

$$W_e = \Sigma(N\delta) = \left(\frac{1}{3}qb^2\right) + \left(\frac{1}{2}q(a-b)b\right) = \frac{qb}{6}(3a-b) \qquad (7.33)$$

In this expression, the first half of the formulation consists of both of the triangles (regions 1 and 3 are completely triangular, with only parts of regions 2 and 4). Their area is b^2 and, therefore, the equivalent point load is expressed as qb^2. 1/3 is the deflection of the centroid when maximum deflection has been assumed as a unity. The second half of the expression is composed of the rectangle at the center, which consists of the remaining regions 2 and 4 (Figure 7.23).

The dissipation of energy is quantified by projecting all the yield lines around a region onto, and at right angles to, that region's axis of rotation. These projected lengths are multiplied by the moment acting on each length,

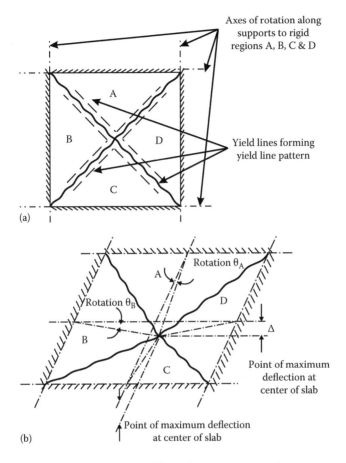

Figure 7.23 Schematic representation of the deformation pattern for a rectangular plate. (a) Plan view of the yield lines. (b) Rotation of the segments due to the rotation at the yield hinges.

and by the angle of rotation of the region. At the small angles considered, the angle of rotation is equated to the tangent of the angle produced by the deflection of the region. Assuming the direction of the moment with respect to the rotation of yield lines as m (positive/sagging) and moment caused due to the rotation about the clamped sides as m' (negative/hogging), the internal work will be:

$$W_i = \sum (ml\theta) = \left(mb\frac{2}{b} \right) + \left(m'b\frac{2}{b} \right) + \left(ma\frac{2}{b} \right) + \left(m'a\frac{2}{b} \right) + mb\frac{2}{b} + ma\frac{2}{b} \quad (7.34)$$

Terms 1 and 3 represent the triangular portion, terms 2 and 4 represent the negative moments in the triangular portions, and terms 5 and 6 represent the trapezoidal sections' contributions. Assuming that the moment acting about the yield line and the moment acting about the supports are equal to each other, that is, that the positive and negative moments are equal ($m = m'$), we get

$$W_i = 6m\left(1 + \frac{a}{b}\right) \tag{7.35}$$

Also, energy equilibrium between the internal and external work requires

$$m = \frac{qb^2(3a-b)}{36(a+b)} \tag{7.36}$$

As an example, the formula for one-way slabs supporting distributed loads as derived by Kennedy and Goodchild (2003) [23] is:

$$M_P = \frac{q_{ult}L^2}{2\left(\sqrt{1+i_1} + \sqrt{1+i_2}\right)^2} \tag{7.37}$$

In this case, M_P is the ultimate sagging moment along the yield line (kN m/m), M'_P is the ultimate support moment along the yield line (kNm/m), q is the ultimate load (kN/m²), and L is the span (m) with parameters i_1 and i_2 representing the ratio of support moments to yield midspan moments and should be chosen by the designer, $i_1 = m'_1/m$, $i_2 = m'_2/m$.

7.4.5.4 Case study 4: Circular slab with free edges

In Figure 7.24, we can see the flexural capacity of round slab simply supported subjected to a center-point loading. Note that depending on the number of yield lines, the internal energy dissipation changes. It is, however, shown that in the case of a simply supported round slab, the allowable applied load can be computed from the bending moment capacity determined through laboratory tests on flexural samples:

$$W_{int} = W_{ext}, \quad \theta = \frac{\delta}{R}, \quad dW_{int} = M_P R\theta d\alpha = M_P \delta d\alpha$$

$$W_{int} = W_{ext} = \int_0^{2\pi} M_P \delta d\alpha = 2\pi M_P \delta = P_{ult}\delta \tag{7.38}$$

$$P_{ult} = 2\pi M_P$$

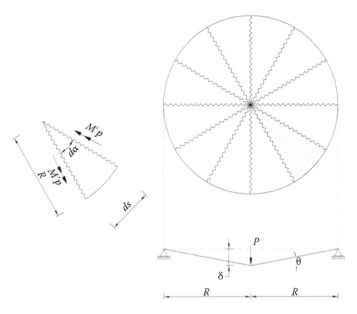

Figure 7.24 Principle of virtual work to determine the ultimate load-carrying capacity of a round panel test simply supported in its contour and subjected to center-point load.

If the support is fixed, the solution would yield

$$W_{int} = W_{ext} = \int_0^{2\pi} M_P \delta d\alpha + 2\pi R M_P \theta = 4\pi M_P \delta = P_{ult}\delta$$

$$(7.39)$$

$$P_{ult} = 4\pi M_P$$

7.4.6 Design of TRC members supported on a substrate

The design of a TRC member on a substrate can be drawn parallel to floor slabs, which are traditionally designed using either the Westergaard or Meyerhof method. While Westergaard uses elastic theory, the more up-to-date methods advocated by Meyerhof and his successors Losberg and Weisgerber depend on the yield line theory (Baumann and Weisgerber, 1983; Beckett, 2000; Meyerhof, 1962; Mobasher and Barsby, 2012; The Concrete Society, 1994).

Consider the following example, in which a concentrated load is applied to the top of a TRC member on a substrate. As the load increases, tensile stresses are induced and give rise to radial cracking emanating from the point of application of the load. These radial cracks increase in length until the bending stresses along a circumferential section become equal to the flexural strength, and a circumferential tension crack is formed on the top, at which point failure is assumed to have occurred. The Johansen formula (Johansen, 1962), which was later expanded by his successors, establishes the collapse load (P_u) in flexure as

$$P_u \left[1 - \sqrt[3]{\frac{\sigma_p}{p_c}} \right] = 2\pi(m + m') \tag{7.40}$$

In Equation 7.40, m and m' = the sagging and hogging flexural moments are the positive and negative flexural moments of resistance, respectively for a symmetric TRC composite, $m = m'$, and σ_p = the elastic or plastic modulus of the substrate reaction, and p_c is the stress on the slab under the contact area of the concentrated load. The term under the cube root in Equation 7.40 represents the substrate resistance. σ_p represents the resistance of the substrate, which if ignored simplifies the formula to:

$$P_u = 2\pi(m + m') \tag{7.41}$$

7.4.7 Design of simply supported TRC beam under distributed load

The objective of this section is to design a 300 mm span simply supported beam subjected to a uniformly distributed live load pressure of 550 kPa. The material being used here is a strain-hardening TRC with Young's modulus E of 15 GPa, a cracking tensile strength σ_{cr} of 5.85 MPa, and an ultimate compressive strength f_c' of 65 MPa. The ultimate tensile strength σ_{peak} of 9 MPa, tensile strain at peak strength ε_{peak} of 0.009, and density of 20.4 kN/m^3 are likewise used.

While the self-weight is negligible compared to the load applied, in this case, it is computed to illustrate the calculations of factored loads. By assuming a thickness between 100 and 150 mm, one can calculate the self-weight as

$$w_{sw} = 0.15 \times 20.4 = 3.06 \text{ kPa}$$

The ultimate factored load is calculated as

$$w_u = 1.2(DL) + 1.6(LL) = 1.2(3.06) + 1.6(550) = 884 \text{ kPa}$$

When we consider the beam over a strip 1 m in width, we can calculate the maximum moment at the midspan of the beam as the design load:

$$M_u = \frac{w_u L^2}{8} = \frac{884 \times 0.3^2}{8} = 9.94 \text{ kN m/m}$$

The next step is to calculate the normalized moment capacity used in this design, which requires the calculation of the material parameters according to Equations 7.1 through 7.4 as follows:

Cracking tensile strain, $\varepsilon_{cr} = \dfrac{\sigma_{cr}}{E} = \dfrac{5.85}{15,000} = 3.9 \times 10^{-4}$

Normalized tensile strain at peak strength, $\alpha = \dfrac{\varepsilon_{peak}}{\varepsilon_{cr}} = \dfrac{0.009}{3.9 \times 10^{-4}} = 23.1$

Normalized postcrack tensile modulus,
$$\eta = \frac{E_{cr}}{E} = \frac{9 - 5.85}{(0.009 - 0.00039) \times 15,000} = 0.0244$$

Normalized yield compressive strain, $\omega = \dfrac{\sigma_{cy}}{\sigma_{cr}} = \dfrac{0.85 \times f_c'}{\sigma_{cr}} = \dfrac{0.85 \times 65}{5.85} = 9.4$

The neutral axis and normalized moment capacity can be calculated by using the maximum value of $\beta = \alpha$ in Equations 7.8 and 7.10 as the ultimate tensile capacity:

$$C_1 = \eta(\beta^2 - 2\beta + 1) + 2\beta - 1$$

$$= 0.0244(23.1^2 - 2 \times 23.1 + 1) + 2 \times 23.1 - 1 = 57.12$$

Neutral axis depth parameter,

$$k = \frac{C_1 - \sqrt{\beta^2 C_1}}{C_1 - \beta^2} = \frac{57.12 - \sqrt{23.1^2 \times 57.12}}{57.12 - 23.1^2} = 0.246$$

$$C_2 = C_1 + 2C_1\beta - \beta^2 = 57.12 + 2 \times 57.12 \times 23.1 - 23.1^2 = 2,162$$

Therefore, the normalized moment capacity is obtained by

$$m_n = C_2 \frac{k^2 - 2k + 1}{\beta^2} + \frac{2\beta k^3}{1 - k}$$

$$= 2,162 \frac{0.246^2 - 2 \times 0.246 + 1}{23.1^2} + \frac{2 \times 23.1 \times 0.246^3}{1 - 0.246} = 3.21$$

At this stage in order to verify the validity of the equations used, one has to verify that the normalized compressive strain developed at the top fiber is still within the elastic range as indicated by Equation 7.14.

$$\lambda = \frac{k}{1-k}\beta = \frac{0.246}{1-0.246}23.1 = 7.54 <\omega = 9.5 \quad => OK$$

The compressive strain is in the elastic stage; therefore, the calculated normalized moment $m_n = 3.21$ is valid. By using the normalized moment capacity, one can determine the cracking moment required to carry the ultimate moment by Equations 7.9 and 7.12:

$$M_{cr} = \frac{M_n}{m_n} = \frac{M_u}{\phi_r m_n} = \frac{9.94}{0.65 \times 3.21} = 4.76 \text{ kNm/m}$$

Finally, the required thickness is determined by Equation 7.11:

$$h = \sqrt{\frac{6M_{cr}}{\sigma_{cr}b}} = \sqrt{\frac{6 \times 4,760}{5.85 \times 10^6 \times 1}} = 0.07 \text{ m}$$

At this stage, one can use a thickness of 75 mm, which is less than the assumed thickness of 150 mm in the conservative estimate of the self-weight. It may not be necessary to recalculate the new self-weight, since the weight is negligible compared to the live load.

Alternatively, one can use the design charts presented in Figure 7.25 to quickly estimate the normalized moment capacity. The use of the chart requires one to first draw a vertical line from $\beta = \alpha = 23.1$ to the curve at $\eta = 0.0244$ and obtain $m_n = 3.25$ from the m_n–β chart. This establishes the normalized moment capacity. In order to check for the range of applicability of the equation, one has to continue the vertical line and acquire $\lambda = 7.7$ from the λ–β chart. Since λ is lower than the normalized yield compressive strain $\omega = 9.4$, the assumption of failure in the elastic compression zone and the obtained value of $m_n = 3.25$ is valid. We can see that the values $\lambda = 7.7$ and $m_n = 3.25$ manually picked from the charts are very close to the exact values $\lambda = 7.54$ and $m_n = 3.21$, both of which are directly computed from the equations. Once m_n is identified, the required cracking moment M_{cr} and thickness h can be calculated using an elastically equivalent section, as shown before.

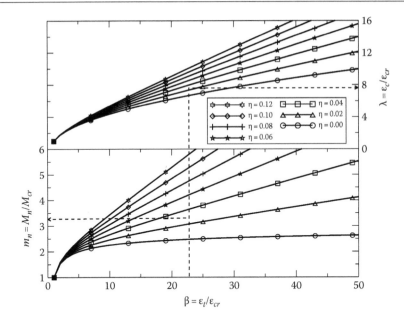

Figure 7.25 Design chart for the three-point beam design of strain hardening material using the closed-form equations. (From Soranakom and Mobasher, 2000.)

In the case of a serviceability-based design, the recommended approach is to choose a limiting or allowable maximum normalized tensile strain as input to get the normalized ultimate moment M'. This is done by using M'–β&η chart and drawing a vertical from $\beta = \alpha = 22.8$ to the curve at $\eta = 0.0232$. The moment capacity is obtained by drawing a horizontal line to the left, to $M' = 3.2$. This calculation requires a check for the normalized compressive stress to be less than the allowable value. In order to verify this check, using the λ_{dv}–β&η chart, one has to continue the vertical line to the curve at $\eta = 0.0243$ and then draw a horizontal line to the right to get $\lambda_{dv} = 7.8$. Since $\lambda_{dv} = 7.8 < \omega = 9.5$, the assumption regarding the elastic response in compression is validated. By doing that, we can see that the compressive strain is elastic and that the assumptions regarding the zone of failure are correct (tension cracking, compression elastic). Otherwise, equations for cracking tension–elastic plastic compression as defined in Table 7.1 should be used. The final stress strain response that was used for the design is plotted in Figure 7.26 based on the material properties assumed and used for design. This response should be evaluated against the potential candidate materials and used in the objective design of the composite in terms of reinforcement, strain capacity and tensile and compressive strength.

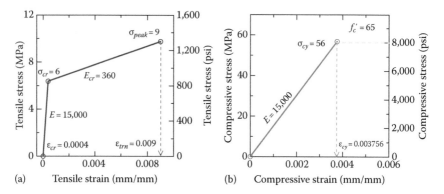

Figure 7.26 Stress–strain curves used for design based on assumed material properties. (a) The tensile stress–strain and (b) compressive stress–strain response. (From Mobasher, B. and Barsby, C., Flexural design of strain hardening cement composites, *Proceedings of the Second International RILEM Conference on Strain Hardening Cementitious Composites (SHCC2-Rio)*, Rio de Janeiro, Brazil, 2012, pp. 53–60.)

Based on the value retrieved from the table or calculated based on the equations provided, a value of $M' = 3.2$ for this material is used. The cracking moment and ultimate moment capacity are calculated as

$$M_{cr} = \frac{1}{6}\sigma_{cr}bh^2 = \frac{1}{6}6 \text{ MPa} \times 1 \times 0.075^2 \text{ m}^3 = 5.625 \text{ kN m/m}$$

$$\phi M_n = \phi M_{cr}M' = 0.85 \times 5.625 \times 3.2 = 15.23 \text{ kN m/m}$$

$$M_u = 9.9 \text{ kN m/m} < \phi M_n = 15.29 \text{ kN m/m}$$

Therefore, the proposed 75 mm thick section is sufficient assuming that the tensile and compressive stress strain response of Figure 7.26 could be achieved for the given composite material.

References

ACI Committee 318, *Building Code Requirements for Structural Concrete*, ACI Manual of Concrete Practice, American Institute, Detroit, MI, 2005.

Baumann, R. A. and Weisgerber, F. E., Yield-line analysis of slabs-on-grade, *ASCE Journal of Structural Engineering*, 109(7), 1553–1568, July 1983.

Beckett, D., Concrete industrial ground slabs, in J. A. Hemsley (ed.), *Design Application of Raft Foundations*, Thomas Telford, London, U.K., 2000, pp. 1–38.

Johansen, K. and Pladeformler, W., Polyteknisk Forlag, Copenhagen, Denmark, 1968, 240pp. (Yield line formulae for slabs, Translated by Cement and Concrete Association, London, U.K., 1972. Ref 12.044. p. 106).

Johansen, K. W., *Yield Line Theory*, Cement and Concrete Association, London, U.K., 1962.

Kennedy, G. and Goodchild, C., *Practical Yield Line Design*, Reinforced Concrete Council, British Cement Association, 2003.

Li, V. C., From micromechanics to structural engineering—The design of cementitious composites for civil engineering applications, *Structural Engineering/ Earthquake Engineering*, 10(2), 1–34, 1994.

Maalej, M., Fracture resistance of engineered fibre cementitious composites and implications to structural behavior, PhD thesis, University of Michigan, Ann Arbor, MI, 2004.

Maalej, M. and Li, V. C., Flexural/tensile-strength ratio in engineered cementitious composites, *Journal of Materials in Civil Engineering*, 6(4), 513–528, 1994.

Meyerhof, G. G., Load-carrying capacity of concrete pavements, *Journal of the Soil Mechanics and Foundations Division, Proceedings of the American Society of Civil Engineers*, 88(SM 3), 89–116, June 1962.

Mobasher, B., *Mechanics of Fibre and Textile Reinforced Cement Composites*, CRC Press, September 2011, 480pp.

Mobasher, B. and Barsby, C., Flexural design of strain hardening cement composites, in *Proceedings of the Second International RILEM Conference on Strain Hardening Cementitious Composites, (SHCC2-Rio)*, Rio de Janeiro, Brazil, 2012, pp. 53–60.

Mobasher, B., Peled, A., and Pahilajani, J., Distributed cracking and stiffness degradation in fabric-cement composites, *Materials and Structures*, 39(287), 317–331, 2006.

Naaman, A. E. and Reinhardt, H. W., Setting the stage: Toward performance based classification of FRC composites, in *Proceedings of Fourth International Workshop on High Performance Fibre Reinforced Cement Composites (HPFRCC-4)*, Ann Arbor, MI, June 15–18, 2003, pp. 1–4.

Peled, A. and Mobasher, B., Pultruded fabric-cement composites, *ACI Materials Journal*, 102(1), 15–23, 2005.

Peled, A. and Mobasher, B., Tensile behavior of fabric cement-based composites: Pultruded and cast, *ASCE Journal of Materials in Civil Engineering*, 19(4), 340–348, 2007.

Russell, H. G. and Graybeal, B. A., Ultra-high performance concrete: A state-of-the-art report, Bridge Community Publication no. FHWA-HRt-13-060, June 2013.

Soranakom, C. and Mobasher, B., Closed form solutions for flexural response of fibre reinforced concrete beams, *Journal of Engineering Mechanics*, 133(8), 933–941, 2007a.

Soranakom, C. and Mobasher, B., Closed-form moment-curvature expressions for homogenized fibre reinforced concrete, *ACI Material Journal*, 104(4), 351–359, 2007b.

Soranakom, C. and Mobasher, B., Correlation of tensile and flexural response of strain softening and strain hardening cement composites, *Cement & Concrete Composites, 2008*, 30, 465–477, 2008.

Soranakom, C. and Mobasher, B., Design flexural analysis and design of textile reinforced concrete textile reinforced structures, in *Proceedings of the Fourth Colloquium on Textile Reinforced Structures (CTRS4) und zur 1. Anwendertagung*, SFB 528, Technische Universität Dresden, Eigenverlag, Dresden, Germany, 2009, pp. 273–288.

Soranakom, C., Mobasher, B., and Bansal, S., Effect of material non-linearity on the flexural response of fibre reinforced concrete, in *Proceedings of the Eighth International Symposium on Brittle Matrix Composites BMC8*, Warsaw, Poland, 2006, pp. 85–98.

The Concrete Society, Concrete industrial ground floors: A guide to their design and construction, Concrete Society Technical Report 34, The Concrete Society, Slough, U.K., 1994, p. 148.

Chapter 8

High rate loading

8.1 Introduction

Textile-reinforced concrete (TRC) is a strain-hardening composite, and as such it is expected to have a unique and favorable response to dynamic loading, whether high strain rate or impact, which are dependent on the nature of the textile reinforcement and its geometrical characteristics in the composite.

TRC composites clearly demonstrate a significant improvement in the energy absorption capacity under static loading as compared to plain concrete materials and other fiber–cement composites (Häußler-Combe et al., 2004; Kruger et al., 2003; Mobasher et al., 2006; Peled and Bentur, 2003; Peled and Mobasher, 2007). These results show the potential of such components under high-speed loading (Butnariu et al., 2006; Peled, 2007).

The strain rate sensitivity for TRC is affected by the type and properties of textiles. Armenakas et al. (1970) studied the behavior of coated S-glass fibers at different rates of loading ranging from 0.00033 to 1.33 s^{-1}. While the strength of strong fibers (statistic average of 10 strongest fibers) was insensitive to strain rates, the weak fibers (statistic average of 10 weakest fibers) showed a considerable drop in the average strength from 1,379 MPa ($\dot{\varepsilon} = 0.00033$ s^{-1}) to 552 MPa ($\dot{\varepsilon} = 1.33$ s^{-1}). This suggests that fibers with less severe flaws behave differently to those with more severe flaws. Using a split Hopkinson bar (SHB), Wang and Xia (2000) found that E-glass fibers' strength and strain capacity in the range of 90–1,100 s^{-1} increased from 2,070 to 2,800 MPa and from 3.52% to 4.27%, respectively.

Differences in the behavior of fabrics under high-speed loading directly affect the behavior of composites made with the fabrics. Research on dynamic tensile strength under high strain rates of fibers and fabrics such as Kevlar® 49, aramid, Twaron, and Zylon has been reported by several authors and used as a basis for textiles applied in the construction industry (Amaniampong and Burgoyne, 1994; Cheng et al., 2005; Farsi et al., 2006; Xia and Wang, 1999; Wagner et al., 1990). These studies address rate dependence of Young's modulus, failure stress, and failure strain over a strain rate range of 10^{-4} to $1,350$ s^{-1}.

8.2 High-speed response and testing systems

Characterization of dynamic tensile properties of materials is challenging as the failure process is affected by the mode and manner of testing. Problems appear at high rate loading due to inertial effect, nonuniform loading, and improper measurement of mechanical properties. Lack of general agreement about the standards and methodology used to conduct dynamic tensile tests further complicates the merging of the databases (Xiao, 2008). Mechanical response of cement-based materials is well known to depend on the strain rate (Bharatkumar, Shah, 2004; Silva et al., 2010; Xu et al., 2004; Zhang et al., 2005). Most of the available literature on the dynamic tensile behavior of concrete are based on plain concrete in compression and only limited data support an increase in tensile strength due to strain rate effects (Candoni et al., 2001; Xiao et al., 2008). Dynamic tensile data on fiber- and fabric-reinforced concrete are even more limited.

A number of experimental techniques exist to investigate high–strain rate material properties: split Hopkinson pressure bar, falling weight devices, flywheel facilities, hydraulic machine, and so on (Kenneth, 1966; Meyers, 1994; Nicholas, 1981; Zabotkin et al., 2003). The use of servo-hydraulic machines in medium–strain rate tensile testing was reported for metals (Bastias et al., 1996; Bruce et al., 2004), plastics (Hill and Sjöblom, 1998; Xiao et al., 2008), composite materials (Fitoussi et al., 2005), and woven fabrics (Zhu et al., 2008a,b), but few applications exist in cement-based composites. Most of the experimental studies under high strain rates (above 1 s^{-1}) have been performed using SHB tests on cylindrical specimens. Instrumented SHB tests (Zhang et al., 2005) can be performed at strain rates ranging from 10 to 120 s^{-1}. Grote et al. (2001) reported that the compressive strength of mortar increases with increases in the strain rate. Experiments show that the behavior of mortar is significantly rate sensitive in the strain rate range of 10^{-3} to $1,700$ s^{-1} (Xu et al., 2004). The rate dependence is weaker for strain rates below 400 s^{-1} and at a strain rate of $1,500$ s^{-1}, the compressive strength is 160 MPa or approximately 3.5 times the quasistatic strength. SHB tensile tests on cylindrical specimens made of wet and dry concrete demonstrate a significant increase in tensile strength measured in the range of strain rates above 10 s^{-1}. This is attributed to microcracking inertia, microcrack shielding, and cleavage of aggregates.

The importance of specimen geometry and size in dynamic material testing has been recognized by the Society of Automotive Engineers, which coordinated the standardization of "High Strain Rate Tensile Test Techniques for Automotive Plastics" in order to develop guidelines for dynamic tensile testing at medium strain rates (Hill, 2004; SAE, 2006).

The International Iron and Steel Institute (IISI) also formed a consortium to develop a high–strain rate tensile test standard for sheet steel (Borsutzki et al., 2003), while European researchers have been working on an ISO standard (ISO Plastics, 2003). The SAE (2006) and IISI (Borsutzki et al., 2003) projects provide details of the relationship between specimen size and wave propagation, inertia effect, strain measurement technique, loading, and gripping devices.

Strain rate sensitivity has been studied by many investigators in relation to fiber–cement composites, and these are relevant to TRC. Kormeling and Reinhardt (1987) performed tensile tests on steel-fiber-reinforced concrete at low (static) and intermediate strain rates (1.25×10^{-6} s^{-1} and 2.5×10^{-3} s^{-1}) using a displacement-controlled servo-hydraulic testing rig. High–strain rate tests (from 1.5 to 20 s^{-1}) were performed with an SHB. Results indicated that high strain rates increase the tensile strength and fracture energy while Young's modulus and stress-free crack opening remain almost the same. Steel-fiber-reinforced concrete (for $V_f = 3\%$) showed an increase in tensile strength from 3.5 to 6.5 MPa when strain rate increased from 1.25×10^{-6} to 20 s^{-1} for specimens tested at 20°C.

Kim et al. (2009) investigated the strain rate effect on the tensile behavior of high-performance fiber-reinforced cement composites (HPFRCC) using hooked and twisted high-strength steel fibers at pseudostatic ($\dot{\varepsilon} = 1 \times 10^{-4}$ s^{-1}) and seismic ($\dot{\varepsilon} = 0.1$ s^{-1}) loading conditions. The tensile behavior of HPFRCC with twisted fibers was sensitive to the strain rate, while hooked fiber-reinforced specimens showed no rate sensitivity.

A setup of the dynamic tensile testing is presented in Figure 8.1. An environmental chamber, temperature controller, and liquid nitrogen can also be employed for testing under variable temperatures. Signals from the piezoelectric force-link transducer and actuator linear variable differential transformer (LVDT) as well as noncontact displacement measuring instrumentations can be recorded and test parameters including peak load, strain at peak, maximum strain, postcrack stiffness, and work-to-fracture reported. High-speed imaging can be accomplished with sampling rates of up to 100,000 fps, using a Phantom v7.3 high-speed digital camera in order to capture the cracking and failure behavior. Figure 8.1a shows the schematic diagram of the setup for high-speed tensile testing. The camera was placed in front of the specimen observing its full size in between the grips, as shown in Figure 8.1b (Yao, 2013). With such setup, dynamic tensile tests were conducted using an MTS high-speed servo-hydraulic testing machine with a load capacity of 25 kN operating under open loop at a maximum speed of 14 m/s. Development of this test equipment and discussions of its dynamic system were addressed in detail elsewhere (Zhu et al., 2011a). The speed of the actuator is controlled by controlling the servo-valve of hydraulic supply and the nominal strain rate is measured from the actuator speed and the gage length of the specimen.

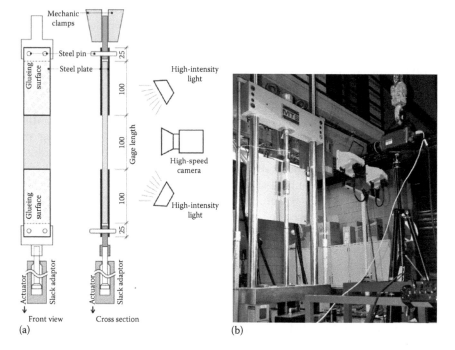

Figure 8.1 (a) Schematic diagram of the test setup for high-speed tensile testing. (b) Test setup using AR glass fabric–cement composite. (From Yao, Y., Application of 2-D digital image correlation (DIC) method to damage characterization of cementitious composites under dynamic tensile loads, MS thesis, Arizona State University, Tempe, AZ, 2013.)

8.2.1 Strain measurement techniques

Several parameters need to be addressed in the analysis of high-speed and impact test results. New challenges in the use and application of displacement measurement in the context of transducer-based (LVDT) measurement, strain-gage-based, and noncontact techniques such as laser extensometers and image analysis techniques are introduced in dynamic tensile testing.

Depending on the methodology used in the measurement of sample displacement, the values of strain rates may significantly depend on the test setup. Dynamic tensile tests have used the LVDT of the actuator in the servo-hydraulic high rate testing machine. The accuracy of the measurement is affected by the compliance and inertial effects of the testing fixtures and may differ from one testing machine to another. At high strain rates, Borsutzki et al. (2005) recommended strain measurement at high strain rates to be done by the relative displacement measured from points other than the gage section of the specimen, that is, the displacement between grips or the actuator displacement measured by LVDT. In order to ensure

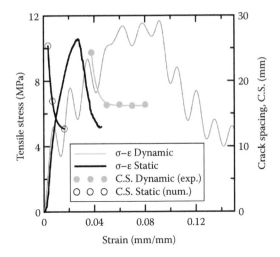

Figure 8.2 Dynamic versus static stress–strain and crack spacing behaviors of the sisal-fiber-reinforced cement composite. (From Silva, F.A. et al., *Mater. Sci. Eng. A*, 527(3), 544, 2010.)

the dynamic tensile testing accuracy, noncontacting devices and techniques should be used as an independent means for the verification of test results.

The strain-hardening composites exhibit a stiff response initially followed by distributed microcracking and formation of a single dominant crack that leads to final failure under both static and dynamic loadings. The progressive development of cracks during the loading process indicates sufficient stress transfer at the fabric–cement interface.

Silva et al. (2010) performed high-speed tensile tests on sisal-fiber-reinforced cement composites at $\dot{\varepsilon} = 24.6$ s^{-1}. Results are as shown in Figure 8.2. Pronounced strain rate dependence was noted for toughness and failure strain due to the pullout behavior. Mechanics of rate sensitivity under pullout were the dominant mechanisms affecting the response. High-speed tensile test results of Silva et al. (2010) on sisal-fiber-reinforced cement composites at $\dot{\varepsilon} = 24.6$ s^{-1} show a pronounced dependence on ultimate strain (average value of 10%) due to the pullout fracture of fibers.

Several studies have characterized TRC materials. Yao et al. (2015) studied four different types of specimens including plain mortar, mortar with addition of short glass fiber, glass TRC, and glass TRC with addition of short glass fibers. They were tested under high-speed tensile loads at nominal strain rates of 25, 50, and 100 s^{-1}. Results and parameters obtained from the experimental data include the stress–strain curves, tensile strength (peak stress), peak strain (strain at peak stress), maximum strain, and work-to-fracture. The work-to-fracture is evaluated using the total area under load versus displacement curve.

8.2.2 Strain measurement using digital image correlation (DIC) method

Noncontact instruments such as laser extensometer or DIC method, although labor intensive, present an accurate measure of displacement measurement for high strain rate testing. Image analysis techniques are also becoming commonplace as full-field approaches are widely used in displacement and strain field measurement. Digital image processing is a powerful tool to postprocess fringe images generated by the Moiré effect, photoelastic materials, or interferometric methods (Kobayashi, 1993; Silva et al., 2010). This method which was proposed by Peters and Ranson (1982) in the early 1980s, processes images of the test specimen before and after applying load and uses the observed shifts in the specimen surface speckle patterns to map the entire displacement field. Image correlation technique is a fundamental concept of computer vision recognition based on light-intensity pattern matching within a small area in the undeformed image and the same area in the deformed image (Poissant and Barthelat, 2010; Sutton et al., 1983). An approach based on DIC was applied to cement composites by Mobasher and Rajan (2003) and used to verify the accuracy of the displacement measurements. The light intensity patterns that are based on 256 gray levels are converted to continuous intensity patterns using a bilinear interpolation approach. Displacements that take place within in each subregion are obtained by solving an optimization algorithm to minimize the cross-correlation function between the twoimages obtained from the same subregion.

For dynamic tests, high-speed digital cameras such as the Phantom series are used to record sample deformation at a sampling rate of up to 100k fps for tests conducted at moderately high strain rates ($\dot{\varepsilon} = 25$ s^{-1}).

8.2.3 Noncontact laser-based strain extensometer

The noncontacting laser extensometer measures strain in dynamic tensile testing using the phase changes in a laser beam that is scattered from the surface of a specimen.

Figure 8.3a and b shows the optical layout of the laser extensometer. The apparent advantage of the laser extensometer over the stroke measurements is in eliminating the spurious deformations that occur at the grips; however, it falls short compared to the DIC method, which gives both full-field displacement and strain values and allows understanding the localized failure of cement-based composites.

The accuracy of laser extensometer is based on the measurement of a number of wavelengths of interferometry fringes that form on the surface of a specimen during deformation and depends on the quality of acquired signals that are filtered and fitted to construct the stress–strain curve. As a full-field measurement procedure, DIC-based image analysis is preferable

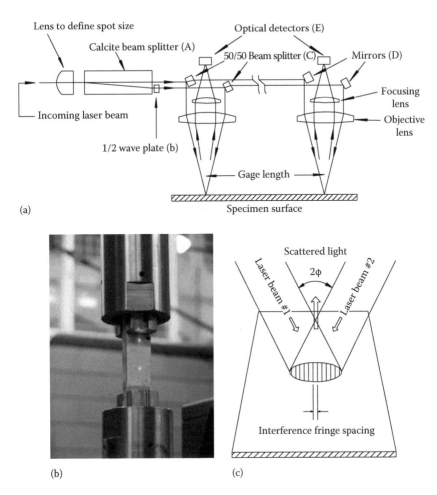

Figure 8.3 Optical layout of the laser extensometer and interference fringes measured from the surface refraction. (a) The set up of the specimen with (b) the laser markers used for displacement measurement, and (c) the development of interference fringes from two laser beams intersecting on the specimen surface. (From Zhu, D. et al., *Cement Concr. Compos.*, 34(2), 147, 2012.)

to stroke LVDT and laser measurements and provides information of local failure. The machine compliance does not affect the accuracy of the strain/displacement measurement for cement-based composites at the investigated strain rates as both the stroke LVDT and laser measurements provide consistent results. Zhu et al. developed a testing procedure for tensile tests with strain rates ranging from 0.0001 to 50 s^{-1}, thus covering both quasistatic and dynamic loading regimes. The speed of the actuator was controlled by the extent of the opening of the hydraulic supply servo-valve, which controls the hydraulic fluid flow and the actuator speed.

8.3 Failure mechanisms

Figure 8.4 presents the dynamic tensile stress versus time history of an AR glass TRC specimen. From a macroscopic perspective, the bend-over point (BOP) corresponds to the formation of matrix cracking. Five distinct stages are identified using Roman numerals with two stages prior to and three stages after the BOP. Stage I corresponds to the elastic–linear range as both matrix and textile behave linearly. Due to relatively low textile content, the stiffness of composite is dominated by matrix properties. The strain within Stage II is associated with the formation of matrix cracks, up until the BOP$^+$ point where a single crack traverses the entire width.

The macroscopic crack pattern is affected to a large extent by the geometry of reinforcing textile. Stage III which is the post-BOP stage, is characterized by the formation of distributed cracking since the load-carrying capacity does not vanish as the cracks are bridged by the textiles. With the initiation of the first matrix crack, other cracks also initiate at approximately regular intervals and begin to propagate across the width and homogenize the matrix. These stages are differentiated using the DIC approach. After completion of the cracking phase and the initiation of debonding in Stage IV, progressive damage leads to a crack-widening stage and ultimately leads to failure in Stage V.

Strain fields of a glass TRC specimen are shown in Figure 8.5a, where the longitudinal strain ε_{yy} based on the stroke displacement is computed with the DIC method. The time-lapse images illustrate the longitudinal strain distribution at varying stress levels. Per the legend, the various shades represent the range of low strain values to larger strain values indicating significant slip zones. At the beginning, a uniform distribution of strain throughout the specimen is in accordance with the elastic–linear stage (Stage I) until the tensile stress σ_t reached 3.7 MPa. The deformation was then localized in the regions of blue and green, identifying the strain concentration and crack initiation during Stage II. After the matrix cracks, the load-carrying capacity does not diminish in the uniform zones in purple where the tensile strains stay below 1,000 microstrains since the load is carried through intact interfaces. Increasing stresses result in additional transverse cracks and multiple localized zones with a rapid increase in tensile strain associated with Stage III. The consecutive crack formation is followed by crack widening, extensive debonding, and progressive damage, as shown in the final subimage of the specimen at failure with pronounced textile pullout. The far-field strains drop back to relatively lower levels due to the elastic recovery of textiles and crack closure.

The areas of the sections representing the opening width of the slip zones at different loading stages were measured and shown as percentage debonded

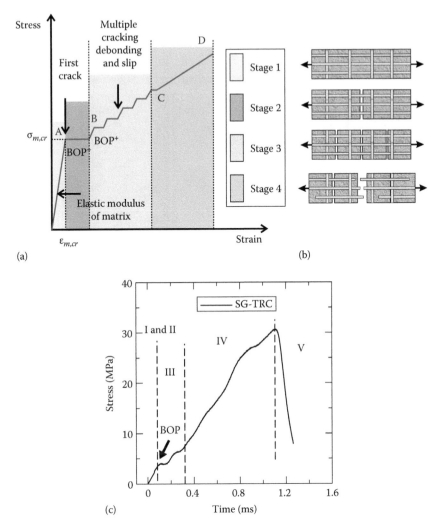

Figure 8.4 Schematic presentation of the tensile behavior of TRC including (a) tensile stress–strain evolution, (b) multiple-cracking mechanism, and (c) experimental stress versus time history of an SG-TRC specimen under dynamic tensile load. (From Yao, Y. et al., *Mater. Struct.*, 49, 2781, 2016; Yao, Y. et al., *Cement Concr. Compos.*, 64, 49, 2015.)

area or A_{slip}. The image analysis software (Ferreira and Rasband, 2012) was used in three steps: image region cropping, color threshold, and measurement of area and average length within region, as shown in Figure 8.5b. Comparative analysis shows that glass TRCs have the largest area and length of slip zones prior to failure compared to polypropylene (PP) TRCs.

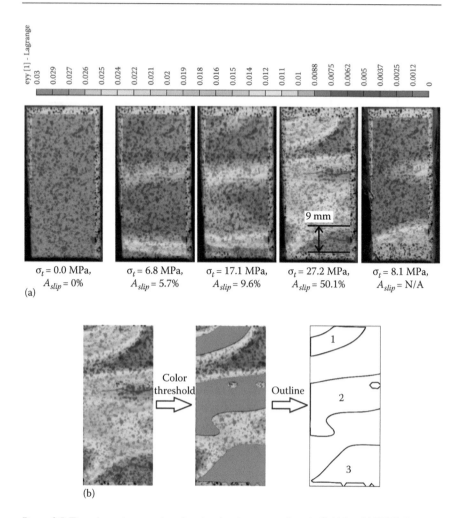

Figure 8.5 Time-lapse images showing the development of strain field for (a) PPTRC specimen tested at a strain rate of 100 s⁻¹, and (b) measurement of slip zone area.
(From Yao, Y. et al., *Mater. Struct.*, 49, 2781, 2016.)

The gradual increase of slip zones corresponds to pronounced textile
debonding and pullout, and result in energy dissipation. Once the shear
stress between fiber and matrix exceeds the interfacial bond strength, fiber
debonding starts and the slip zone is formed. Since the tensile stiffness and
strength of glass fibers are higher than those of PP, more load is transferred
from the matrix to the fiber during the cracking stages prior to complete
failure. However, due to the low tensile strength and stiffness of PP, failure
occurs at low stress level and bridging fibers elongate as another mechanism improving the energy absorption. The energy dissipated during the test

is related to multiple aspects including the length of frictional slip zones, which is about 9 mm. The width correlates with the distance between weft yarns of 10 mm, which indicates the contribution of weft yarns in resisting the slip in textiles.

8.4 Formation and characterization of distributed cracking

TRC composites are sensitive to strain rate as their Young's modulus, tensile strength, maximum strain, and toughness under dynamic loading are higher than those under quasistatic loading. Using the noncontact laser extensometer, the strain sensitivity has been measured for alkali-resistant (AR) glass fabric composite, and an average strain rate of 17 s^{-1} is used as an example to illustrate the deformation patterns, as shown in Figure 8.6, demonstrating the different stages of (a–c) multiple microcracking, (d) main crack widening and other microcracks closing, and (e) complete failure. These deformation patterns complicate the analysis of the data as each mode of failure is associated with change in local or overall stiffness of the sample under load. Therefore, parameters such as strain and stress need to be addressed in the context of the nature of damage at the time of measurement.

Zhu et al. (2011b) investigated three types of fabric-reinforced cement composites at strain rates ranging from 10–25 s^{-1} using a servo-hydraulic high-rate testing machine using carbon fabric, glass, and polyethylene (PE) fabric-reinforced composites. As shown in Figures 8.7a and b, for AR glass TRC, the tensile strength rose from 4.11 to 5.56 MPa when the strain rate increased from static (2.2 × 10^{-5} s^{-1}) to dynamic (18 s^{-1}) as shown in Figure 8.7a. No significant changes were noticed in the strain capacity response.

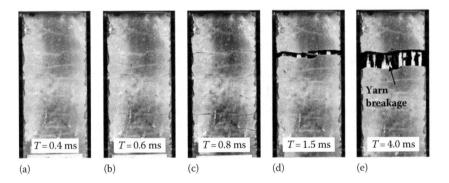

Figure 8.6 Images of AR glass composites under dynamic loading: (a–c) multiple microcracking, (d) main crack widening and other microcracks closing, and (e) complete failure; and under high-speed loading. (From Zhu, D. et al., *J. Mater. Civil Eng.*, 23(3), 230, 2011a; Zhu, D. et al., *Constr. Build. Mater.*, 25(1), 385, 2011b.)

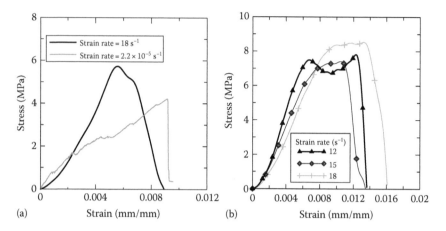

Figure 8.7 Response of AR glass TRC: (a) comparison of static and dynamic results and (b) stress–strain responses at varying strain rates. (From Zhu, D. et al., *J. Mater. Civil Eng.*, 23(3), 230, 2011a; Zhu, D. et al., *Constr. Build. Mater.*, 25(1), 385, 2011b.)

8.5 Hybrid systems: short fibers with TRC

The influence of matrix modifications by the addition of short fibers on the quasistatic tensile response of TRC was investigated by several researchers (Barhum and Mechtcherine, 2010; Hinzen and Brameshuber, 2009; Peled et al., 2009; Silva et al., 2011). Silva et al. (2011) subjected TRC with and without short glass fibers in the matrix to tensile loading ranging from 1 × 10^{-4} to 50 s^{-1}. An increase in tensile strength, strain capacity, and work-to-fracture was observed with increasing strain rates up to 0.1 s^{-1}.

Figure 8.8 shows the representative stress–strain curves for TRC specimens tested at different low strain rates, that is, 0.1 s^{-1} and below, with and without the addition of short fibers. TRC, with and without short fibers, displayed an increase in tensile strength, work-to-fracture, and strain capacity. The average tensile strength of TRC with plain matrix increased from 16.1 to 23.6 MPa, work-to-fracture rose from 18.5 to 33.5 J, and strain capacity increased from 3.08% to 3.92%. For TRC with the addition of short glass fibers, the tensile strength rose from 17.1 to 23.2 MPa, work-to-fracture from 16.1 to 31.7 J, and strain capacity from 2.65% to 3.58%. The σ_{BOP}^{+}, that is, the stress at which the first crack is formed, showed an increase with increasing strain rate for both materials under study, but with superior values for TRC with short fibers. Young's modulus showed a general tendency to increase; however, the scatter of the results was too high to draw firm conclusions.

Figure 8.9 shows typical stress–strain curves obtained at high strain rates (≥ 5 s^{-1}) on TRC specimens with and without short fibers. The shapes of the curves differ significantly from those obtained for lower strain rates; however, there was a similar multiple-cracking pattern in combination with

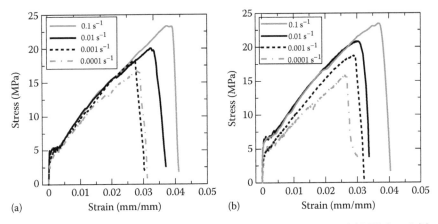

Figure 8.8 Effect of strain rate on the tensile stress–strain response of (a) TRC and (b) TRC with short glass fibers for rates up to 0.1 s⁻¹. (From Silva, F.A. et al., *Mater. Sci. Eng.A*, 528(3), 1727, 2011.)

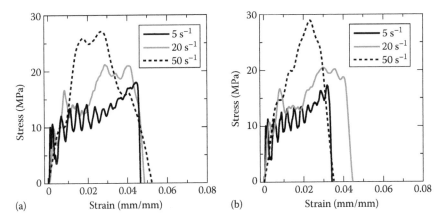

Figure 8.9 Effect of strain rate on tensile stress–strain response of (a) TRC and (b) TRC with short glass fibers for rates ranging from 5 to 50 s⁻¹. (From Silva, F.A. et al., *Mater. Sci. Eng.A*, 528(3), 1727, 2011.)

strain-hardening behavior. Oscillations of the system at its natural frequency are observed for specimens tested at 5 and 20 s⁻¹. These oscillations seemed to decrease as the strain rate increased. Composites tested at high strain rates showed an increase in tensile strength, work-to-fracture, and first-crack strength but a decrease in the strain capacity with increasing strain rate. The decrease in the strain capacity corresponds to the increase in the stiffness of the composite (measured as the slope of the stress vs. strain curve from the first-crack strength up to the ultimate strength) when increasing the strain rate after the first-crack formation.

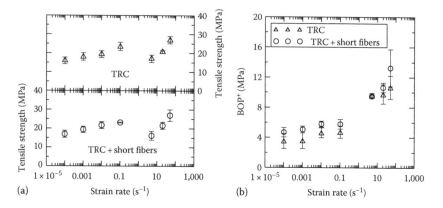

Figure 8.10 Effect of strain rate on (a) tensile strength and (b) on bend over point (BOP⁺) of TRC and TRC with short fibers. (From Silva, F.A. et al., *Mater. Sci. Eng. A*, 528(3), 1727, 2011.)

For strain rates above 5 s⁻¹, increase in the tensile strength and work-to-fracture was observed with a decrease in the strain capacity. High loading rate tests as shown in Figure 8.10 indicate a pronounced effect of the specimen length on the mechanical properties: with increasing gage length, the tensile strength and strain capacity decreased, while the work-to-fracture increased.

The overall effect of strain rate on the tensile strength and first-crack strength response of the samples with and without short fibers are shown in Figure 8.10a and b. Figure 8.10 also shows the increase in tensile strength as well as Young's modulus of the strain-hardening cement-based composites (SHCC).

The longitudinal strain (ε_{yy}) fields of a plain mortar and a TRC specimen (referred to as Mortar-ARG and TRC-ARG, respectively) reinforced with short glass fibers tested at 25 s⁻¹ are shown in Figure 8.11 using a color code, with purple representing the lowest strain values and red at 5.0% strain. Due to the inherent brittleness of matrix and the low fiber volume fraction (0.5%), only one macrocrack was observed for the Mortar-ARG sample (Figure 8.11a). Tensile strain concentrated in the vicinity of the crack while the far-field was uniformly deformed. Previous studies on the tensile behavior of SHCC containing 2% of short polyvinyl alcohol fiber, however, showed that in such a ductile material only very few cracks formed under high–strain rate loading (Mechtcherine et al., 2011a,b). Figure 8.11b illustrates the damage evolution of a TRC-ARG specimen such that at the beginning of the test (σ = 3.1 MPa), a relatively uniform strain distribution in accordance with linear–elastic stage (Stage 1) was obtained. As σ increased to 6.1 MPa, two bands in blue were formed, indicating matrix cracking and the onset of nonlinear behavior. Increasing tensile stress (σ = 16.7 MPa) resulted in additional transverse cracks into multiple fracture bands. Saturation of transverse cracks was coincident with maximum tensile stress (σ = 19.5 MPa) with three identified zones: (A) the localization zone

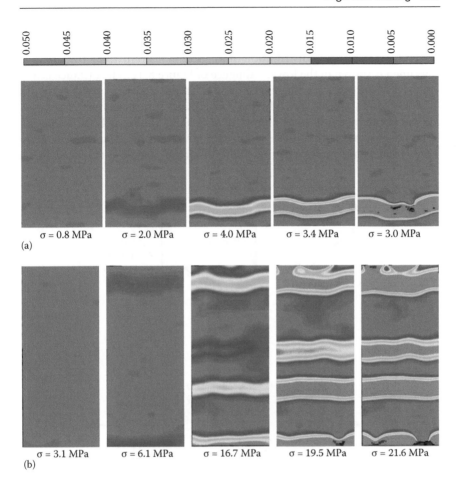

Figure 8.11 Strain fields (ε_{yy}) obtained by DIC for various specimens tested at a strain rate of 25 s^{-1}: (a) plain mortar with short fiber and (b) TRC with short fiber. (From Yao, Y. et al., *Cement Concr. Compos.*, 64, 49, 2015.)

initiating at a stress level of 2 MPa as shown in Figure 8.11a, which contains the transverse crack with the majority of the load carried by the textile phase; (B) the shear lag zone forming at a stress level of 4 MPa on both sides of the localized zone as shown in Figure 8.11a, where the slip between fiber and matrix cannot be ignored and the bond stress follows a shear lag pattern; and (C) the uniform zone where no crack is formed, composite is behaving linearly, and slip is negligible. This range is the constant stress range in between any two cracks by excluding the localized and shear lag regions. The fiber stress variation along the length reaches a maximum level in the bridge zone (A) and minimum value at the "perfectly bonded" zone (C). A similar pattern of strain map was also observed in the specimens tested at other strain rates.

The three distinct zones identified by DIC observations are shown in Figure 8.12. The strain map is selected from Figure 8.11b at σ = 19.5 MPa, and the corresponding distribution of longitudinal strain along the length of specimen is shown in the lower subfigure of Figure 8.12. The distance is normalized with respect to the length of AOI (in this case L_{AOI} = 44 mm), and different zones are separated by the dashed lines. The behaviors at zones A, B, and C can be modeled as an σ–ω relationship, nonlinear bond stress–slip relationship, and with a linear stress–strain relationship, respectively. These three models are integrated in a finite difference model introduced in the

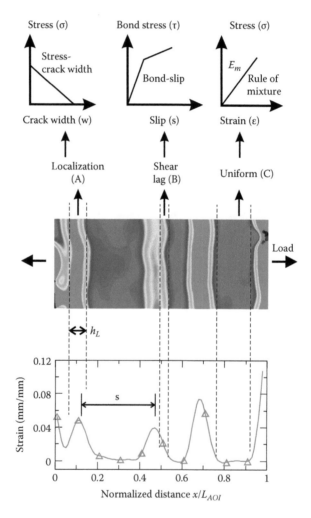

Figure 8.12 Identification of three zones: localization, shear lag, and uniform strain and corresponding modeling approaches. (From Yao, Y. et al., *Cement Concr. Compos.*, 64, 49, 2015.)

next section. Additionally, crack spacing (s) and the width of localization zone (h_L) were measured from the DIC data.

The width of the localization zone (h_L) is equivalent to a development length needed for the force transfer from fiber to matrix in order to reach the critical stress necessary for matrix cracking (see Figure 8.13a). The characteristic damage state (CDS) is a strain level where no more cracks in the matrix can develop due to the inability of the fibers in transferring sufficient load back into the matrix and correlates with the final crack spacing. The role of bond parameters in the formation of additional cracks and slip-related multiple cracks are expressed in terms of h_L and s representing the crack spacing. Parameters h_L and s for 15 individual TRC and TRC-ARG specimens were measured using DIC, and their probability distribution functions are expressed as a two-parameter Weibull distribution (Mobasher et al., 2014):

$$P(\sigma) = 1 - \exp\left[-\left(\frac{x}{\lambda}\right)^k\right] \qquad (8.1)$$

where
  x is the measured parameters (h_L or s)
  λ is the reference or scaling value related to the mean
  k is the Weibull modulus or shape parameter

The cumulative distribution function (CDF) of h_L and s as shown in Figure 8.13b and c indicates that the mean value of h_L decreased from 7.4 to 6.5 mm with the addition of short fibers (Figure 8.13b) as well as spacing s reduced from 10.3 to 8.4 mm (Figure 8.13c). This measurement confirms the role of short fibers in mitigating and bridging the microcracks in bond enhancement. At the microstructural level, short fibers improve the bond by means of active load transfer and cross-linking with hydration products, and thus a greater number of microcracks serve as nuclei for macrocrack formation (Butler et al., 2011). Addition of short fibers supports stress transfer across cracks as well as crack deflection mechanisms, both of which play a role in toughening. Therefore, stress relaxation of the matrix in the vicinity of cracks is less pronounced and a smaller development length is needed, hence cracks form more closely. As a result of narrower localization zones, finer crack pattern and smaller crack widths were obtained.

Hinzen and Brameshuber (2009) also investigated the effect of the addition of short glass, aramid, and carbon fibers on the tensile response of TRC. An increase of up to 40% was found in first-crack stress as well as a finer crack pattern, leading to an increase in ultimate strain. In studying the addition of glass and short carbon fibers, the first-crack stress was doubled due to the addition of 1.0% by volume of dispersed short glass fibers. The energy absorption capacity also increased but was not verified for high strain rates, since a significant increase in strain capacity was not gained.

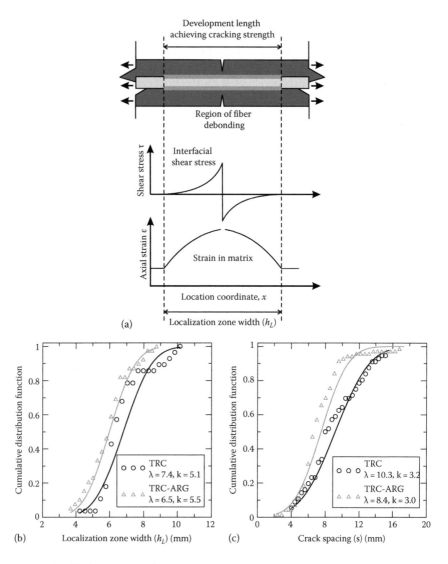

Figure 8.13 (a) Correlation of localization zone width (h_L) with development length to achieve cracking strength and curve fitting of Weibull CDF for (b) localization zone width (h_L) and (c) crack spacing (s). (From Yao, Y. et al., *Cement Concr. Compos.*, 64, 49, 2015.)

The decrease of failure strain probably can be traced back to a localization of crack widening to several scattered cracks.

Another possible explanation for some of the decrease in strain capacity due to the addition of short fibers can be the expected enhancement of the bond strength between textile and matrix. It is known that stronger bond leads to reduction of free fiber length in the vicinity of the cracked plane, causing a premature failure of multifilament yarns and consequently to TRC failure at lower strain values.

Similar to TRC, SHCC also shows a similar trend in properties affected by strain rate. Mechtcherine et al. (2011a,b) studied the dynamic behavior of SHCC reinforced with polyvinyl acetate (PVA) fibers under tensile load. In the case of tensile tests performed at strain rates up to 1×10^{-2} s^{-1}, as shown in Figure 8.14, SHCC exhibited an increase in tensile strength and a decrease in strain capacity with increasing strain rate. When loaded at high strain rates ranging from 10 to 50 s^{-1}, SHCC showed an increase in both tensile strength and strain capacity with increasing loading rate. It is observed that the tensile strength increases from 4.5 to 5.5 MPa while the elastic modulus rises from 19.7 to 23.9 GPa as the strain rate increases from 10^{-5} to 10^{-2} s^{-1}.

For strain rates higher than 10 s^{-1}, fiber pullout with an average free length of approximately 2.5 mm was the main failure mechanism, although a few ruptured fibers were observed as well. Figure 8.15 shows the capacity of PVA fiber to bridge cracks of approximately 200 µm of a specimen tested at a strain rate of 25 s^{-1}. Even though the SHCC does not show visible multiple cracking at strain rates of 10 s^{-1} and above, its increased pullout length and crack-bridging capacity provide high strain capacity and work-to-fracture

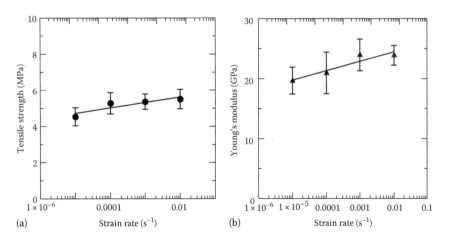

Figure 8.14 Effect of strain rate on (a) tensile strength and (b) elastic modulus of SHCC subjected to low strain rates (≤0.01 s^{-1}). (From Silva, F.A. et al., *Mater. Sci. Eng. A*, 528(3), 1727, 2011.)

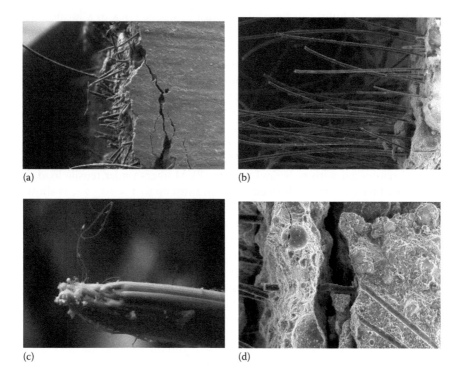

Figure 8.15 Failure mechanisms of SHCC at low and high strain rates: (a) fracture surface of a specimen tested at 10^{-3} s^{-1}, (b) fracture surface of a sample tested at 25 s^{-1}, (c) typical fiber fracture from (a), and (d) crack bridged by PVA fiber after being tested at 25 s^{-1}. (From Silva, F.A. et al., *Mater. Sci. Eng. A*, 528(3), 1727, 2011.)

in this range of strain rates. Besides the assumption that very fine, "invisible" cracks may contribute to the deformability of SHCC, plastic deformations of PVA fiber seem to play a much more important role in the case of high-speed loading. Fiber pullout from SHCC matrix at rates of 10^{-3} s^{-1} show a smoother surface in the fiber pulled out than at the higher strain rate of 25 s^{-1}. At strain rates higher than 10 s^{-1}, the fibers undergo during pullout a severe deformation process, which contributes considerably to the energy absorption capacity.

8.6 Modeling of behavior at high-speed using a tension-stiffening model

A finite difference method developed by Soranakom and Mobasher (2009, 2010a,b) was used to simulate the tension-hardening behavior in TRC and TRC-ARG specimens. The material model consists of three

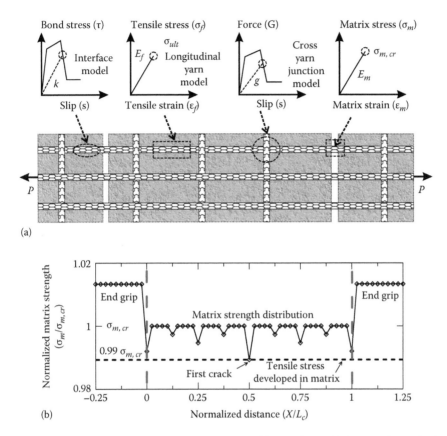

Figure 8.16 (a) Parameter identification in the model for mechanical behavior of a cracked composite specimen (matrix-cracking criterion, interface bond–slip model, longitudinal yarn tensile stress–strain relationship, mechanical anchorage provided by cross-yarn junctions as nonlinear spring model) and (b) prescription of a deterministic strength parameter distribution based on sequential cracking at midpoints of the segments.

homogenized phases: matrix, textile, and interface. Since a mean distribution of textile yarns in multiple layers is assumed, the locations of longitudinal yarns were not specified. A tension specimen is idealized as a series of 1D segments consisting of fiber, matrix, and interface elements with constitutive laws of each phase specified in Figure 8.16a, including matrix, longitudinal yarn stress–strain, and interface bond–slip models. The effect of transverse yarns through the mechanical anchorage is represented by a nonlinear spring model. The matrix stress–strain model is linear elastic and specified by its elastic modulus E_m and cracking strength $\sigma_{m,cr}$. Similarly, the tension model of textile is characterized by modulus E_f and ultimate tensile strength σ_{ult}. The bond–slip relationship is based on the fiber/textile pullout tests (Portala et al., 2014; Tuyan and Yazıcı, 2012) and associated follow-up models by Naaman et al. (1991) and Sueki et al. (2007). The dashed lines indicate the secant modulus k at the slip value s, which is used to compute the force applied at the node. A parameter representing the efficiency of the yarn stiffness ($\eta < 1$) is defined to represent the limitations in bonding that lead to telescopic or sleeve effect by Cohen and Peled (2010). This parameter has been quantified by experiments on sleeve filaments that are partially bonded to matrix and contribute to axial stiffness, while the core filaments provide marginal stiffness due to unbonded yarns (Banholzer et al., 2006). Using a uniform strength distribution along the length of the specimen and a deterministic sequential crack evolution, the first crack occurs at the center, then at the end grips, followed by 1/4, 1/8, and 1/16 points until crack saturation case is obtained (Figure 8.16b).

Figure 8.17a presents the discretized finite difference model of the cracked specimen with the total embedded length L discretized into N nodes of equal spacing, h. Transverse yarns are simulated by means of springs attached to the nodes at cross-yarn junction providing resistance to pullout force. Once cracking takes place, the specimen is divided into smaller segments $L_s^{(1)}, L_s^{(2)}, \dots L_s^{(q)}$ with each segment containing $n^{(q)}$ number of local nodes, where q is the segment index. An additional node is inserted at the crack location such that each cracked segment has its own end nodes and the problem can be solved independently. Free body diagrams of representative nodes are shown in Figure 8.17b, where s_i is the nodal slip, F_i is the nodal fiber force, B_i is the nodal bond force, and G_i is the nodal spring force. The equilibrium equations can be derived in terms of the primary unknown variable slip s, defined as the difference between the deformations of the longitudinal yarn with respect to the matrix:

$$s = \int_{x_i}^{x_{i+1}} \left(\varepsilon_y - \varepsilon_m \right) dx \tag{8.2}$$

where ε_y and ε_m are yarn and matrix strains distributed along the differential length dx. For typical low fiber volume fraction, the axial stiffness of the yarn $A_f E_f$ is considerably lower than the matrix term $A_m E_m$ and the contribution of matrix elongation to slip is ignored. Thus, the slip s and yarn strain ε_y are simplified to:

$$s = \int_{x_i}^{x_{i+1}} \varepsilon_y dx \quad \text{and} \quad \varepsilon_y = s' = \frac{ds}{dx} \tag{8.3}$$

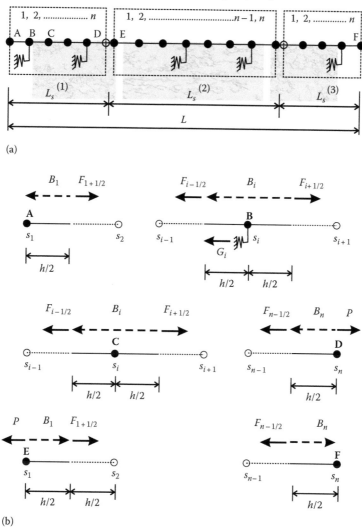

Total N number of nodes with equal spacing of h

(a)

(b)

Figure 8.17 Finite difference model: (a) discretized fabric pullout model and (b) free body diagram of six representative nodes labeled as "A"–"F." (Continued)

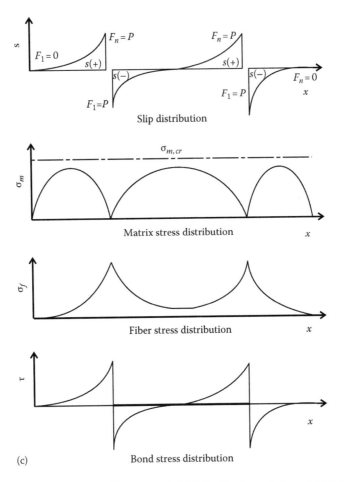

Figure 8.17 (Continued) Finite difference model: (c) distributions of slip, matrix stress, fiber stress, and bond stress.

Nodal equilibrium equations are constructed, and each nodal force is expressed as the product of slip by stiffness. A global system of equations including stiffness matrix $[C]$, nodal slip vector $\{S\}$, and force vector $\{T\}$ was subsequently obtained as follows:

$$[C]_{n,n}\{S\}_n = \{T\}_n \tag{8.4}$$

Once the solution of nodal slip values is obtained, the corresponding stress, strain, and crack spacing can be subsequently computed. The setup, assembly, and the solution algorithm of equilibrium equations based on several parametric studies were discussed in detail by Soranakom and Mobasher (2010).

Figure 8.17c schematically presents the distributions of slip, matrix stress (σ_m), fiber stress (σ_f), and bond stress (τ) in cracked segments. The tension force in both longitudinal yarns and matrix are positive values, while the distribution of the stress in matrix and fiber change in accordance with the placement of cracks. However, the load-carrying capacity of matrix in the uncracked segments does not diminish as a sign of tension-stiffening effect. The load carried by the fiber is transferred back to the matrix and σ_m is maximized at the center line of each cracked segment. As the load increases and σ_m reaches matrix-cracking strength $\sigma_{m,cr}$, new cracks form. Following a shear lag pattern, the bond stress varies from its maximum at the crack to a value of zero at the bonded region.

The crack saturation stage has not been discussed in significant detail in experiments and also in the simulation. The CDS is a strain level where no more cracks in the matrix can develop due to the inability of the fibers in transferring sufficient load back into it to result in cracking. The CDS level correlates with the final crack spacing. During the cracking phase, load transfer between the matrix and the bridging fibers indicates the continuity of total force across the crack; however, stress continuity can only be modeled by means of a shear lag approach. The bond parameters play an important role, leading to the formation of additional cracks and deformations associated with parallel cracking. The fiber stress variation along the length reaches a maximum level in the bridge zone and minimum value at the "perfectly bonded" zone. Three zones were defined: (A) the localization zone, where the transverse crack is located and the majority of the load is carried by the textile phase; (B) the shear lag zone, associated with an excessively higher strain in the fiber such that the fiber–matrix slip cannot be ignored; and (C) the uniform zone, where the slip is negligible and rule of mixtures is applicable.

8.7 Flexural impact loading

8.7.1 Introduction

Low-velocity impact behavior of cement-based materials has been the topic of several studies. Various impact tests include Charpy, Izod, drop-weight, and ballistic tests. These tests can be either instrumented or heuristically based and the resistance can be measured based on fracture energy, damage accumulation, and measurement of the number of drops to achieve a desired damage or stress level. Bindiganavile and Banthia (2001a,b) showed that flexural strength is higher under impact loading in comparison to quasistatic loading for fiber-reinforced concrete (FRC). Polymeric FRC showed an improvement in energy absorption under impact loading. Manolis et al. (1997) also showed that fibrillated PP fibers significantly improve the impact resistance of concrete slabs. Wang et al. (1996) identified two different damage mechanisms for FRC under drop-weight impact and showed that fiber fracture dominates the failure mechanism for fiber fractions lower than the critical fiber volume, whereas the fiber pullout mechanism was the dominant response under higher fiber fractions.

Zhu et al. (2009) studied the impact behavior of AR glass TRC. Maximum flexural stress and absorbed energy of beam specimen increased with the number of textile layers. Impact properties of PE TRC were investigated by Gencoglu and Mobasher (2007) and compared to AR glass textile. The PE textile composites showed a higher load-carrying capacity at large deflections and hence are more ductile than AR glass textile composites.

One of the main areas addressed in this section is the development of impact properties of sandwich composites (Bonakdar et al., 2013; Dey et al., 2015; Manalo et al., 2012; Memon et al., 2007; Tekalur, 2009; Uddin et al., 2006; Zhao et al., 2007). Low-velocity impact response of autoclaved aerated concrete (AAC) was investigated by Serrano-Perez et al. (2007), and the load-carrying capacity was stated to be limited due to its brittle nature. However, sandwich plates made with carbon fibers and AAC as core material are considerably more ductile when subjected to similar impact forces. Large-projectile-type impact tests (using wood lumber) and small-projectile, high-energy-type tests (using handguns) were conducted on fiber-reinforced cellular concrete (FRCC) by Zollo and Hays. Extent of penetration was evaluated, and the impact response of FRCC was documented to be influenced by its low-density void structure.

8.7.2 Impact test procedures

Impact test setup is frequently based on a free fall of an instrumented hammer that is dropped on a specimen placed in a three-point bending system. Such a system is shown in Figure 8.18. The drop heights can be adjusted within a range from 1 to 2,000 mm and released by means of a trigger switch and stopped by an electronic brake release mechanism. The impact force induced by the falling hammer is measured by a strain-gage-based load cell with a range of 90 kN mounted on the hammer behind the blunt-shaped impact head. Another load cell is mounted beneath the support plate that measures the force transmitted to the equipment base. An LVDT with a range of ± 10 mm was used to measure the sample deflection.

The raw data consisting of time, load, acceleration, and deflection are analyzed for each impact test. Figure 8.19 shows the analysis of a typical impact test of a representative specimen tested at a drop height of 25 mm. The time history for the impact load, deflection of midspan, and acceleration are shown. Flexural stress versus deflection of the same specimen is shown in Figure 8.19b. The latter can be categorized in five distinct zones. Zone 1 is the linear-elastic range that ends with the formation of the first crack. The stress corresponding to this point is defined as the limit of proportionality (LOP). This is followed by Zones 2 and 3, which are characterized by multiple

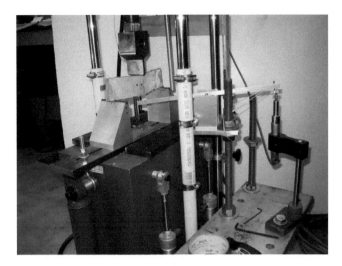

Figure 8.18 The impact test setup with a specimen after the test. The displacement measuring anchor is also shown. (From Zhu, D. et al., *Cement Concr. Compos.*, 31, 379, 2009.)

cracks associated with strain softening where fibers are bridging the crack. Zone 4 can be attributed to the fiber pullout and failure, followed by Zone 5, in which rebound takes place. Zones 4 and 5 can only be observed under special cases where the combination of specimen size and drop height allows for such behaviors. If there is sufficient ductility in the specimen to absorb the applied energy, some of the stored energy released causes a rebound that is characterized by a reduction in the deflection of the specimen as the load is decreased.

The response of TRC with AR glass fabric under flexural impact loading was investigated by Zhu et al. (2011b). It was reported that maximum stress increases with increasing the impact energy, sustaining high ranges of energy absorption capacity.

8.7.3 Impact response of TRC

The TRC specimens reinforced with six layers of fabrics (ARG6B) were tested at hammer drop heights of 50, 100, 200, and 250 mm, and the input energy varied from 7 to 35 J. Figure 8.20a shows the impact load–time curves of the ARG6B specimens tested for all the drop heights. At drop heights of 50, 100, and 200 mm, the rebound of specimens after impact is evident. In this case, it is shown that the ARG6B specimen responds to the impact loads as a stiff beam. The area under the impact load–deflection curve is the deformation energy that is initially progressively transferred from the hammer to the beam and then given back

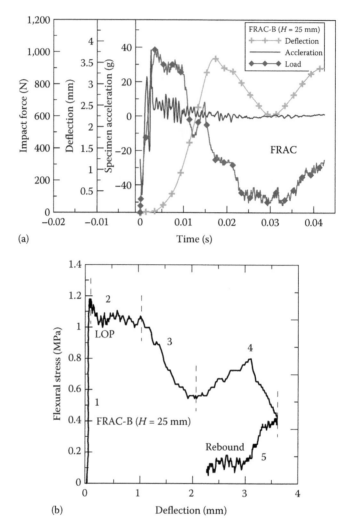

Figure 8.19 (a) Typical time history for impact event (post-analysis) and (b) typical flexural stress versus deflection for the same test. (From Dey, V. et al., *Cement Concr. Compos.*, 49, 100, 2014.)

from the beam to the rebounding hammer; the area included inside the curve refers to the energy absorbed by the test specimen during impact. At a drop height of 250 mm, the impact load–deflection curve after peak force was flat, indicating no rebound. The area under the load–deflection curve is the deformation energy that is progressively transferred from the hammer to the specimen and in this case of absence of rebound, is also the energy absorbed during the impact.

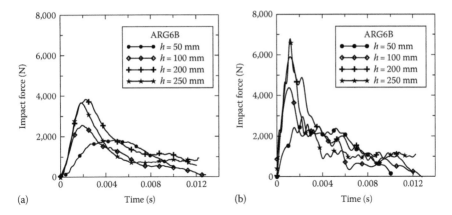

Figure 8.20 Impact force variation for ARG6B specimens at different drop heights: (a) measured by conventional load cell and (b) calculated from acceleration of hammer. (From Zhu, D. et al., *Cement Concr. Compos.*, 31, 379, 2009.)

8.7.4 *Effect of textile orientation*

Impact properties of 3D AR glass fabric cement-based composites compared to short AR glass fiber composites show that the 3D fabrics significantly improve the toughness and energy absorption to as much as 200-fold compared to short fiber composites. The setup of the 3D fabrics is shown in Figure 8.21.

In a series of instrumented drop-weight tests, the arrangement of the fabric within the composite was found to have a significant influence on the impact behavior. Improved toughness of about 50% was obtained when the fabric was placed vertically to the dropped hammer as compared to the horizontal arrangement. When energy absorption is considered, the thickness of the element and location of the 3D fabric faces relative to the impactor are important factors. In this work, the yarns in the z direction were spacer yarns with no real reinforcing abilities. For getting better performance, stronger yarns are suggested along the z direction. The performance is hence similar to two layers of textiles that are reinforcing the tension and compression zones of a beam.

The impact stress versus deflection curves of the horizontal and vertical arrangement specimens are presented in Figure 8.22a and b. A uniform impact behavior is observed; as for the horizontal arrangement this uniformity is even clearer. This indicates the uniformity of the composites and the ability of using this test to study the impact behavior of the TRC systems. It is also seen that the deflection of the TRC systems is very high, exhibiting values of up to about 16 mm. A rebound behavior of the specimens tested at vertical arrangement is clearly seen, that is, after reaching the maximum deflection, the specimen shifted back to some extent and no failure occurred.

(a) (b)

(c)

Figure 8.21 (a) 3D fabric structure and composite testing arrangements, (b) horizontal, and (c) vertical. (From Zhu, D. et al., *Construct. Build. Mater.*, 25(1), 385, 2011b.)

Comparison of the typical curves of each system is presented in Figure 8.22c and d, showing the impact stresses versus deflection and time. The rebound mechanism is obvious for the vertically tested system. On the other hand, no such rebound, that is, shifting back behavior, is seen for the composites tested at horizontal arrangement. Due to this rebound behavior, the entire duration of the test is greater, about twice as much for the composite tested vertically as compared to the composite tested horizontally. The toughness of the vertical system is also much greater than that of the horizontal system (Table 8.1), providing much better energy absorption of the vertically tested composite. However, the horizontal system exhibits greater maximum stress as compared to the vertical system. But the improved strength by the horizontal arrangement of about 15%, is much smaller than the improved toughness by the vertically system, of about 50%. So when energy absorption is considered, the thickness of the element and the location of the fabric faces relative to the impactor are important factors.

The crack pattern and development of these two systems are compared in Figure 8.23. When tested horizontally, at a duration of 0.75 ms, the developed crack is going through almost the entire thickness of the composite (Figure 8.23a). At later stages of the impact test, the crack is

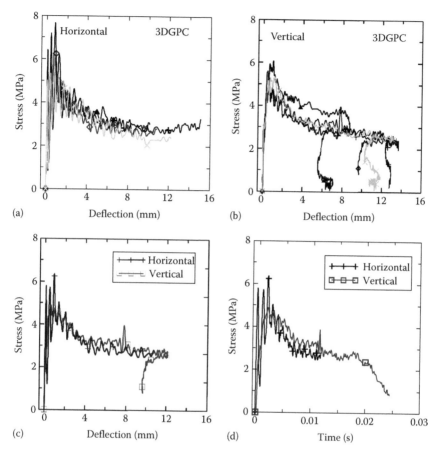

Figure 8.22 Impact behavior of 3D fabric composites tested at (a) horizontal, (b) vertical arrangements, (c) stress versus deflection, and (d) stress versus time. (From Zhu, D. et al., *Construct. Build. Mater.*, 25(1), 385, 2011b.)

widening, and the opening of the crack is relatively large at a duration of 13.25 ms (Figure 8.23b). The reinforcing yarns of the 3D fabric are pulling out and holding the specimen from a complete failure, as seen in Figure 8.24a, and major damage to the whole composite is obvious. Broken bundles and filaments at the tensile zone of the composite can be easily seen in Figure 8.24b. These last two images were taken after the impact test was ended. However, for the vertically tested system, the crack development is different; at an early test duration of 0.75 ms, a very fine crack is observed, which developed only up to the middle of the composite thickness (Figure 8.23c). For this composite, at later stages the crack developed through most of the composite thickness but the reinforcing yarns of the fabric bridged the crack through the specimen thickness and the composite

Table 8.1 Impact properties of the composites

Type	Sample direction	Rigidity (kN/mm)	Max. force (N)	Max. stress (MPa)	Max. defl. (mm)	Defl. at max. force (mm)	Toughness (kN * mm)
3DGFA	Horiz	27.2 ± 4.3	744 ± 73	5.2 ± 0.3	6.0 ± 0.53	0.140 ± 0.18	2.25 ± 0.38
3DGPC	Horiz	11.9 ± 1.1	876 ± 49	6.3 ± 0.9	12.0 ± 2.2	0.639 ± 0.34	5.46 ± 0.11
	Vert	11.0 ± 0.2	1,280 ± 1,24	5.4 ± 0.5	11.9 ± 2.2	0.824 ± 0.29	8.1 ± 0.96
GFI2	Horiz	29.7 ± 5	730 ± 67	5.7 ± 0.5	0.08 ± 0.0	0.037 ± 0.01	0.031 ± 0.00

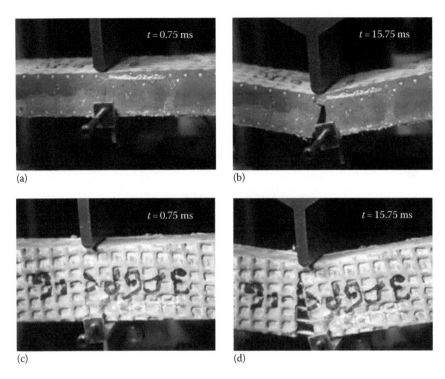

Figure 8.23 Comparison of 3D fabric composites tested at two arrangements at different durations. (a) Initiation of cracking at the moment of impact in a plate configuration. (b) Completion of crack propagation after impact in a plate specimen. (c) Initiation of impact at the moment of impact in abeam configuration. (d) Completion of crack propagation with fiber bridging mechanism.

did not fail (Figure 8.23d). When observing the vertically tested composite after the impact testing (Figure 8.24c), the crack is much finer, as part was closed due to the rebound mechanism, as discussed earlier. No significant damage to the fabric or yarn breakage was obtained with this composite at the end of testing, with no major failure of the composite. This composite can still hold loads and remain generally in its original shape after ending the test with the specific test parameters discussed earlier.

Comparison of the short fiber composites with the 3D fabric composites is shown in Figure 8.25. This figure compares typical curves of the 3D fabric composite with the two matrices: with fly ash (3DGFA) and without fly ash (3DGPC) and the short fibers (GF12), presenting stress versus deflection at log scales and stress versus time linear scale. The test arrangement in all is horizontal. The significant brittle behavior of short fiber composites is obvious as compared to the much more ductile behavior of the 3D fabric composites. Also, the duration of the test is much smaller for the

Figure 8.24 Comparison of 3D fabric composites at two test arrangements at the end of testing: (a) horizontal, (b) horizontal after impact test and (c) vertically after impact test. (From Zhu, D. et al., *Construct. Build. Mater.*, 25(1), 385, 2011b.)

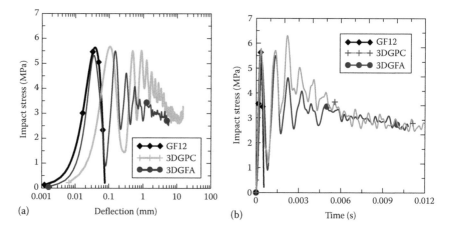

Figure 8.25 Comparison of composites' impact behavior with short glass fibers and 3D fabrics with plain cement and fly ash matrices: (a) stress versus deflection logarithm scale and (b) stress versus time. (From Zhu, D. et al., *Construct. Build. Mater.*, 25(1), 385, 2011b.)

fiber composites as compared with the 3D fabric composites. The short fiber composite duration lasts less than 0.1 ms with a maximum deflection of about 0.08 mm, as for the 3D fabric composites the test duration is about 0.01 ms with a deflection of about 10 mm.

References

Amaniampong, G. and Burgoyne, C. J., Statistical variability in the strength and failure strain of aramid and polyester yarns, *Journal of Material Science*, 29(19), 5141–5152, 1994.

Armenakas, A. E., Garg, S. K., and Sciammarella, C. A., Strength characteristics of glass fibers under dynamic loading, *Journal of Applied Physics*, 41(4), 1657–1664, 1970.

Banholzer, B., Brockmann, T., and Brameshuber, W., Material and bonding characteristics for dimensioning and modeling of textile reinforced concrete (TRC) elements, *Materials and Structures*, 39, 749–763, 2006.

Barhum, R. and Mechtcherine, V., Effect of short fibres on fracture behaviour of textile reinforced concrete, in B. H. Oh et al. (eds.), *Fracture Mechanics of Concrete and Concrete Structures*, Korea Concrete Institute, Seoul, Korea, 2010, pp. 1498–1503.

Bastias, P. C., Kulkarni, S. M., Kim, K. Y., and Gargas, J., Non-contacting strain measurements during tensile tests, *Experimental Mechanics*, 36(1), 78–83, 1996.

Bharatkumar, B. H. and Shah, S. P., Impact resistance of hybrid fiber reinforced mortar, in K. Kovler, J. Marchand, S. Mindess, and J. Weiss (eds.), *International RILEM Symposium on Concrete Science and Engineering: A Tribute to Arnon Bentur*, Evanston, IL, 2004.

Bindiganavile, V. and Banthia, N., Polymer and steel fiber-reinforced cementitious composites under impact loading. Part 1: Bond-slip response, *ACI Materials Journal*, 98(1), 10–16, 2001a.

Bindiganavile, V. and Banthia, N., Polymer and steel fiber-reinforced cementitious composites under impact loading. Part 2: Flexural toughness, *ACI Materials Journal*, 98(1), 17–24, 2001b.

Bonakdar, A., Mobasher, B., and Babbit, F., Physical and mechanical characterization of fiber-reinforced Aerated Concrete (FRAC), *Cement and Concrete Composites*, 38, 82–91, 2013.

Borsutzki, M., Cornette, D., Kuriyama, Y., Uenishi, A., Yan, B., and Opbroek, E., *Recommended Practice for Dynamic Tensile Testing for Sheet Steels*, International Iron and Steel Institute: High Strain Rate Experts Group, Brussels, Belgium, 2003.

Borsutzki, M., Cornette, D., Kuriyama, Y., Uenishi, A., Yan, B., and Opbroek, E., *Recommendations for Dynamic Tensile Testing of Sheet Steels*, International Iron and Steel Institute, Brussels, Belgium, 2005.

Bruce, D. M., Matlock, D. K., Speer, J. G., and De, A. K., Assessment of the strain-rate dependent tensile properties of automotive sheet steels, SAE technical paper 2004-01-0507, Troy, MI, 2004.

Butler, M., Hempel, S., and Mechtcherine, V., Modelling of ageing effects on crack-bridging behaviour of AR-glass multifilament yarns embedded in cement based matrix. *Cement and Concrete Research*, 41, 403–411, 2011.

Butnariu, E., Peled, A., and Mobasher, B., Impact behavior of fabric-cement based composites, in *Proceedings of the Eighth International Symposium on Brittle Matrix Composites (BMC8)*, Warsaw, Poland, 2006, pp. 293–302.

Candoni, E., Labibes, K., Albertini, C., Berra, M., and Giangrasso, M., Strain-rate effect on the tensile behaviour of concrete at different relative humidity levels, *Materials and Structures*, 34(1), 21–26, 2001.

Cheng, M., Chen, W., and Weerasooriya, T., Mechanical properties of Kevlar KM2 single fiber, *Journal of Engineering Materials and Technology*, 127(2), 197–204, 2005.

Cohen, Z. and Peled, A., Controlled telescopic reinforcement system of fabric–cement composites—Durability concerns, *Cement and Concrete Research*, 40, 1495–1506, 2010.

Dey, V., Bonakdar, A., and Mobasher, B., Low-velocity flexural impact response of fiber-reinforced aerated concrete, *Cement and Concrete Composites*, 49, 100–110, 2014.

Dey, V., Zani, G., Colombo, M., Prisco, D., and Mobasher, B., Flexural impact response of textile reinforced aerated concrete sandwich panels, *Journal of Materials of Materials and Design*, 86, 187–197, 2015.

Farsi, D. B., Nemes, J. A., and Bolduc, M., Study of parameters affecting the strength of yarns, *Journal of Physics*, 134, 1183–1188, 2006.

Ferreira, T. and Rasband, W., *ImageJ User Guide*, IJ 1.46r revised edn., National Institutes of Health, Bethesda, MD, 2012.

Fitoussi, J., Meraghni, F., Jendli, Z., Hug, G., and Baptiste, D., Experimental methodology for high strain rates tensile behavior analysis of polymer matrix composites, *Composite Science Technology*, 65(14), 2174–2188, 2005.

Gencoglu, M. and Mobasher, B., Static and impact behavior of fabric-reinforced cement composites in flexure, in *High-Performance Fiber-Reinforced Cement Composites (HPFRCC-5)*, Mainz, Germany, 2007, pp. 463-470.

Grote, D. L., Park, S. W., and Zhou, M., Dynamic behavior of concrete at high strain rates and pressures: I. Experimental characterization, *International Journal of Impact Engineering*, 25, 869–886, 2001.

Häußler-Combe, U., Jesse, F., and Curbach, M., Textile reinforced composites— Overview, experimental and theoretical investigations, in V. C. Li, C. K. Y. Leung, K. J. Willam, and S. L. Billington (eds.), *Fifth International Conference on Fracture Mechanics of Concrete and Concrete Structures, Ia-FraMCos 204*, Vail, CO, 2004, pp. 749–756.

Hill, S., Standardization of high strain rate test techniques for automotive plastics project, UDR-TR-2004-00016, UDRI: Structural Test Group, University of Dayton Research Institute, Dayton, OH, 2004.

Hill, S. and Sjöblom, P., Practical considerations in determining high strain rate material properties, SAE technical paper 1998-981136, Warrendale, PA, 1998.

Hinzen, M. and Brameshuber, W., Improvement of serviceability and strength of textile reinforced concrete by using short fibres, in M. Curbach and F. Jesse (eds.), *Textilbeton Theorie und Praxis, Fourth Colloquium on Textile Reinforced Structures (CSTR4)*, TU Dresden, Dresden, Germany, 2009, pp. 261–272.

ISO Plastics, Determination of tensile properties at high strain rates, A draft of ISO/CD 18872, Geneva, Switzerland, 2003. https://www.iso.org/.

Kenneth, G. H., Influence of strain rate on mechanical properties of 6061.T6 aluminum under uniaxial and biaxial states of stress, *Experimental Mechanics*, 6(4), 204–211, 1966.

Kim, D. J., El-Tawil, S., and Naaman, A. E., Rate-dependent tensile behavior of high performance fiber reinforced cementitious composites, *Materials and Structures*, 42(3), 399–414, 2009.

Kobayashi, A. S., *Handbook on Experimental Mechanics*, 2nd edn., Society for Experimental Mechanics, New York, 1993.

Kormeling, H. A. and Reinhardt, H. W., Strain rate effects on steel fiber concrete in uniaxial tension, *The International Journal of Cement Composites and Lightweight Concrete*, 9(4), 197–204, 1987.

Kruger, M., Ozbolt, J., and Reinhardt, H. W., A new 3D discrete bond model to study the influence of bond on structural performance of thin reinforced and prestressed concrete plates, in A. E. Naaman and H. W. Reinhardt (eds.), *Proceeding on High Performance Fiber Reinforced Cement Composites (HPFRCC4)*, Ann Arbor, MI, 2003, pp. 49–63.

Manalo, A. C., Aravinthan, T., and Karunasena, W., Mechanical properties characterization of the skin and core of a novel composite sandwich structure, *Journal of Composite Materials*, 47(14), 1785–1800, 2012.

Manolis, G. D. et al., Dynamic properties of polypropylene fiber-reinforced concrete slabs, *Cement and Concrete Composites*, 19, 341–349, 1997.

Mechtcherine, V., Millon, O., Butler, M., and Thoma, K., Mechanical behaviour of strain hardening cement-based composites under impact loading, *Cement and Concrete Composites*, 33, 1–11, 2011a.

Mechtcherine, V., Silva, F. A., Butler, M., Zhu, D., Mobasher, B., Gao, S.-L., and Mäder, E., Behaviour of strain-hardening cement-based composites under high strain rates, *Journal of Advanced Concrete Technology*, 9(1), 51–62, 2011b.

Memon, N. A., Sumadi, S. R., and Ramli, M., Ferrocement encased lightweight aerated concrete: A novel approach to produce sandwich composite, *Materials Letters*, 61, 4035–4038, 2007.

Meyers, M. A., *Dynamic Behavior of Materials*, John Wiley & Sons, New York, 1994.

Mobasher, B., Dey, V., Cohen, Z., and Peled, A., Correlation of constitutive response of hybrid textile reinforced concrete from tensile and flexural tests, *Cement and Concrete Composites*, 53, 148–161, 2014.

Mobasher, B., Peled, A., and Pahilajani, J., Distributed cracking and stiffness degradation in fabric-cement composites, *Materials and Structure (RILEM)*, 39(3), 317–331, 2006.

Mobasher, B. and Rajan, S. D., Image processing applications for the study of displacements and cracking in composite materials, in *16th ASCE Engineering Mechanics Conference (EM 2003)*, Seattle, WA, 2003.

Naaman, A., Namur, G., Alwan, J., and Najm, H., Fiber pullout and bond slip. I: Analytical study, *Journal of Structural Engineering*, 117(9), 2769–2790, 1991.

Nicholas, T., Tensile testing of material at high rates of strain, *Experimental Mechanics*, 21(5), 177–185, 1981.

Peled, A., Textiles as reinforcements for cement composites under impact loading, in H. W. Reinhardt and A. E. Naaman (eds.), *Workshop on High Performance Fiber Reinforced Cement Composites (HPFRCC-5)*, Mainz, Germany, 2007, pp. 455–462.

Peled, A. and Bentur, A., Fabric structure and its reinforcing efficiency in textile reinforced cement composites, *Composites Part A*, 34, 107–118, 2003.

Peled, A. and Mobasher, B., Tensile behavior of fabric cement-based composites: Pultruded and cast, *ASCE Journal of Materials in Civil Engineering*, 19(4), 340–348, 2007.

Peled, A., Mobasher, B., and Cohen, Z., Mechanical properties of hybrid fabrics in pultruded cement composites, *Journal of Cement and Concrete Composites*, 31, 647–657, 2009.

Peters, W. H. and Ranson, W. F., Digital imaging techniques on experimental stress analysis, *Optical Engineering*, 21(3), 427–431, 1982.

Poissant, J. and Barthelat, F., A novel subset splitting procedure for digital image correlation on discontinuous displacement fields, *Experimental Mechanics*, 50(3), 353–364, 2010.

Portala, N., Perezb, I., Thranec, L., and Lundgrena, K., Pull-out of textile reinforcement in concrete, *Construction and Building Materials*, 71, 63–71, 2014.

Serrano-Perez, J. C., Vaidya, U. K., and Uddin, N., Low velocity impact response of autoclaved aerated concrete/CFRP sandwich plates, *Composite Structures*, 80, 621–630, 2007.

Silva, F. A., Butler, M., Mechtcherine, V., Zhu, D., and Mobasher, B., Strain rate effect on the tensile behaviour of textile-reinforced concrete under static and dynamic loading, *Materials Science and Engineering: A*, 528(3), 1727–1734, 2011.

Silva, F. A., Zhu, D., Soranakom, C., Mobasher, B., and Toledo Filho, R. D., High speed tensile behavior of sisal fiber cement composites, *Materials Science and Engineering: A*, 527(3), 544–552, 2010.

Society of Automotive Engineers (SAE), High strain rate testing of polymers, J2749, Warrendale, PA, 2006. http://www.sae.org/.

Soranakom, C. and Mobasher, B., Geometrical and mechanical aspects of fabric bonding and pullout in cement composites, *Materials and Structures*, 42(6), 765–777, 2009.

Soranakom, C. and Mobasher, B., Modeling of tension stiffening in reinforced cement composites: Part II—Simulations vs experimental results, *Materials and Structures*, 43, 1231–1243, 2010a.

Soranakom, C. and Mobasher, B., Modeling of tension stiffening in reinforced cement composites: Part I—Theoretical modeling, *Materials and Structures*, 43, 1217–1230, 2010b.

Sueki, S., Soranakom, C., Mobasher, B., and Peled, A., Pullout-slip response of fabrics embedded in a cement paste matrix, *Journal of Materials in Civil Engineering*, 19(9), 718–727, 2007.

Sutton, M. A., Wolters, W. J., Peters, W. H., Ranson, W. F., and McNeill, S. R., Determination of displacement using an improved digital correlation method, *Image and Vision Computing*, 1(3), 133–139, 1983.

Tekalur, S. A., Shock loading response of sandwich panels with 3-D woven E-glass composite skins and stitched foam core, *Composites and Technology*, 69, 736–753, 2009.

Tuyan, M. and Yazıcı, H., Pull-out behavior of single steel fiber from SIFCON matrix, *Construction and Building Materials*, 35, 571–577, 2012.

Uddin, N., Fouad, F., Vaidya, U. K., Khotpal, A., and Perez, J. C. S., Structural characterization of hybrid fiber reinforced polymer (FRP)—Autoclaved aerated concrete (AAC) panels, *Journal of Reinforced Plastics and Composites*, 25, 981–999, 2006.

Wagner, H. D., Aronhime, J., and Marom, G., Dependence of tensile strength of pitch-based carbon and para-aramid fibers on the rate of strain, *Proceeding of the Royal Society of London, Mathematical and Physical Sciences*, A428, 493–510, 1990.

Wang, N. Z., Sidney, M., and Keith, K., Fiber reinforced concrete beams under impact loading, *Cement and Concrete Research*, 26(3), 363–376, 1996.

Wang, Y. and Xia, Y., Dynamic tensile properties of E-glass, Kevlar 49 and polyvinyl alcohol fiber bundles, *Journal of Materials Science Letters*, 19, 583–586, 2000.

Xia, Y. and Wang, Y., The effects of strain rate on the mechanical behavior of Kevlar fiber bundles: An experimental and theoretical study, *Composites Part A*, 29(11), 1411–1415, 1999.

Xiao, S., Li, H., and Lin, G., Dynamic behaviour and constitutive model of concrete at different strain rates, *Magazine of Concrete Research*, 60(4), 271–278, 2008.

Xiao, X. R., Dynamic tensile testing of plastic materials, *Polymer Testing*, 27(2), 164–178, 2008.

Xu, H., Mindess, S., and Duca, I. J., Performance of plain and fiber reinforced concrete panels subjected to low velocity impact loading, in M. di Frisco, R. Felicetti, and G. A. Plizzari (eds.), *Proceedings of RILEM Conference on Fiber reinforced Concrete, BEFIB*, Varenna, Italy, 2004, pp. 1257–1268.

Yao, Y., Application of 2-D digital image correlation (DIC) method to damage characterization of cementitious composites under dynamic tensile loads, MS thesis, Arizona State University, Tempe, AZ, 2013.

Yao, Y., Bonakdar, A., Faber, J., Gries, T., and Mobasher, B., Distributed cracking mechanisms in textile-reinforced concrete under high speed tensile tests, *Materials and Structures*, 49(7), 2781–2798, 2016.

Yao, Y., Silva, F. A., Butler, M., Mechtcherine, V., and Mobasher, B., Tension stiffening in textile-reinforced concrete under high speed tensile loads, *Cement and Concrete Composites*, 64, 49–61, 2015.

Zabotkin, K., O'Toole, B., and Trabia, M., Identification of the dynamic properties of materials under moderate strain rates, in *16th ASCE Engineering Mechanics Conference*, Seattle, WA, 2003.

Zhang, J., Maalej, M., Quek, S. T., and Teo, Y. Y., Drop weight impact on hybrid-fiber ECC blast / shelter panels, in N. Banthia, T. Uomoto, A. Bentur, and S. P. Shah (eds.), *Proceedings of Third International Conference on Construction Materials: Performance, Innovation and Structural Applications*, Vancouver, British Columbia, Canada, 2005.

Zhao, H., Elnasri, I., and Girard, Y., Perforation of aluminum foam core sandwich panels under impact loading—An experimental study, *International Journal of Impact Engineering*, 34, 1246–1257, 2007.

Zhu, D., Gencoglu, M., and Mobasher, B., Low velocity impact behavior of AR glass fabric reinforced cement composites in flexure, *Cement and Concrete Composites*, 31(6), 379–387, 2009.

Zhu, D., Mobasher, B., and Rajan, S. D., High strain rate testing of Kevlar 49 fabric, in *Society for Experimental Mechanics, 11th International Congress and Exhibition on Experimental and Applied Mechanics*, Vol. 1, 2008a, pp. 34–35.

Zhu, D., Mobasher, B., and Rajan, S. D., Image analysis of Kevlar 49 fabric at high strain rate, in *Society for Experimental Mechanics, 11th International Congress and Exhibition on Experimental and Applied Mechanics*, Vol. 2, 2008b, pp. 986–991.

Zhu, D., Mobasher, B., and Rajan, S. D., Dynamic tensile testing of Kevlar 49 fabrics, *Journal of Materials in Civil Engineering*, 23(3), 230–239, 2011a.

Zhu, D., Mobasher, B., and Rajan, S. D., Non-contacting strain measurement for cement-based composites in dynamic tensile testing, *Cement and Concrete Composites*, 34(2), 147–155, 2012.

Zhu, D., Peled, A., and Mobasher, B., Dynamic tensile testing of fabric-cement composites, *Construction and Building Materials*, 25(1), 385–395, 2011b.

Chapter 9

Durability of TRC

9.1 Introduction

The long term durability of textile-reinforced concrete (TRC) requires treatment and discussion at different levels, ranging from the durability of the composite itself in various exposure conditions and spanning its potential contribution to the long-term performance of structures in which TRC is being incorporated.

The durability of the composite itself requires addressing the chemical stability of the reinforcement in the alkaline cementitious matrix. In addition, it is essential to consider the implications of the processes that take place at the matrix–yarn interface, which are physical in nature, and may lead over time to changes in the nature of bonding. Such changes may result in variations in the overall mechanical response of the composite, which may be either beneficial or detrimental (Bentur and Mindess, 2007).

The long-term performance of TRC in structures is to a large extent dependent on the nature of cracking, either induced by environmental loads (drying shrinkage and temperature changes) or mechanical ones. The main characteristic to be considered is the special nature of TRC in which cracks are generated by multiple-cracking mechanism, leading to microcracks that are usually less than 100 μm in width. The ability to channel the mechanical behavior to achieve large deformation while keeping the cracks so small provides considerable serviceability, by reducing penetration of deleterious materials into the structure. This characteristic can potentially enhance long-term performance to an extent far greater than what might be achieved with conventional reinforced concrete. This is a common advantage of different types of strain-hardening cement composites and has been reviewed in recent years in several publications such as van Zijl et al. (2012) and Ahmed and Mihashi (2007).

9.2 Durability of the composite material

The most common components of TRC are portland cement matrix with various mineral additives such as fly ash, silica fume, and blast furnace slag and reinforcing fabrics made of glass, carbon, and various types of polymers. The yarn consists of multifilament bundles in the case of man-made high-modulus, high-strength reinforcements such as glass and carbon. The polymer yarns are in the form of either monofilaments or bundled filaments. The filament diameter can range from about 10 to 50 μm. The Portland-cement-based matrix is essentially durable in normal environments to which TRC is being exposed, and its long-term performance is an issue only in very extreme environmental conditions. Thus, the main aspects to be considered with respect to the long-term durability are related to the chemical resistance of the yarns, which may lead to degradation in their strength, as well as their interaction with the matrix, leading to changes in bonding with implications to the strength and toughness of the composite. It should be kept in mind that the cementitious matrix remains active even after reaching the maturity of 28 days and its microstructure, especially at the yarn–matrix interface, may change over time.

The evaluation of aging is based on the study of the strength of the yarns in simulated high-pH pore solutions and testing the tensile or flexural properties of the composite after natural or accelerated aging. Since some of the aging issues are associated with the physical and microstructural interactions of the bundled yarns and the cement matrix, several tests were developed to identify the micromechanics of the interaction in conjunction with microstructural characterization after accelerated or natural aging:

- Evaluation of pullout behavior using a pullout scheme that enables follow-up of the fracture of filaments during the test. A system of that kind was developed in RWTH Aachen (Banholzer, 2004; Banholzer and Brameshuber, 2004), based on simultaneous monitoring of load and displacement during pullout test along with filament breakage obtained by optical follow-up of light that is transmitted through the glass filaments and the darkening obtained when a filament breaks. This system was applied by Bentur et al. (2010) to study the aging behavior of pullout characteristics, as outlined in Chapter 4.
- Evaluation of the mechanical response of individual filaments in the strand by push-in tests and characterization of the changes due to aging. This test was developed by Zhu and Bartos (1997) and described in Figure 4.14.
- Special thin elements in which the bundled yarn is cast into a cement matrix. The loading scheme is such that the tensile force is carried by the reinforcement, applying a clamping scheme where the load is applied directly on the filament extending out of the matrix, as shown

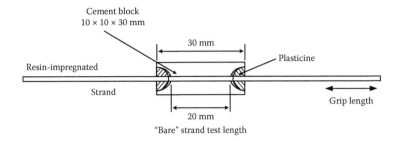

Cement block
10 × 10 × 30 mm

30 mm

Plasticine

Resin-impregnated

Strand

Grip length

20 mm
"Bare" strand test length

Figure 9.1 Schematic description of the SIC test. (From Litherland, K.L. et al., *Int. J. Cement Compos. Lightweight Concr.*, 6, 39, 1984.)

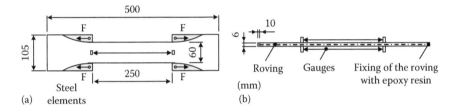

500

F   F

105   60

Steel   250
(a) elements   (mm)
Roving   Gauges   Fixing of the roving
          with epoxy resin
(b)

10
6

Figure 9.2 The TSP specimen for testing the strength of textile fabrics in cement: (a) top view and (b) cross section. (From Buttner, T. et al., Enhancement of the durability of alkali-resistant glass-rovings in concrete, in *International RILEM Conference on Material Science (MATSCI)*, Aachen, Germany, ICTRC, Vol. 1, 2010, pp. 333–342.)

in the strand-in-cement (SIC) test in Figure 9.1, developed in the United Kingdom, and the TSP test developed in Germany (Figure 9.2), where clamping is avoided.

• In situ loading in the SEM to observe the crack bridging of the yarn in notched specimens loaded in flexure (Bentur and Diamond, 1986; Trtik and Bartos, 1999).

It should be noted that the decline in mechanical properties upon aging was observed mainly in glass reinforcement but not in other types of materials such as carbon and polypropylene (Hannant, 1998; Mumenya et al., 2010). Thus, the discussion in this section will focus on the durability of TRC with glass yarns, highlighting the mechanisms that may lead to decline in properties as well as the means used to overcome this problem.

9.2.1 Chemical durability of the reinforcing yarns

The chemical durability of the yarns themselves is dependent on their resistance to the highly alkaline environment in the cement matrix, which is of concern mainly in glass, and to ultraviolet (UV) exposure, which may be an issue when polymeric materials are being used in the fabric. Most of the durability concerns of the fabric material have been with regard to the resistance of the glass, and therefore there is considerable bulk of knowledge and studies on this topic. Much less is available on polymeric materials, since their embedment in the matrix makes them largely immune to UV degradation.

9.2.1.1 Durability of glass reinforcement

The cement matrix is porous in nature and the pore fluid is a highly basic solution with pH values that can be as high as 13.5. This is largely due to the presence of Ca^{+2} ions associated with the $Ca(OH)_2$ hydrated product and the alkali ions, Na^+ and K^+, whose source is in soluble salts present in the Portland cement. A solution that is in contact only with $Ca(OH)_2$ will result in a pH of about 12.5 in equilibrium conditions. The very soluble alkalis, discharging Na^+ and K^+ into the pore solution, result in elevation of the pH level to about 13.5. Under these high-pH conditions, some yarn compositions, especially glass, are susceptible to a chemical attack by the solution that is highly concentrated with OH^- ions. In the case of glass, the result is a breakdown of the Si–O–Si network and the formation of defects on the surface associated with strength reduction (Figure 9.3).

This issue was a major concern in developing cement composites reinforced with chopped glass fibers (GRC) in the 1960s and 1970s. In order to overcome this problem, special glass compositions (designated alkali-resistant [AR]) (Majumdar and Nurse, 1974) were developed and used today to produce glass fabrics for TRC.

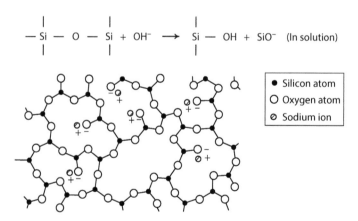

Figure 9.3 Schematic structure of the glass network and the degradation reaction in a solution of high pH. (From Hull, D., An Introduction to Composite Materials, Cambridge Solid State Science Series, Cambridge University Press, Cambridge, England, 1981.)

Table 9.1 Chemical composition of E- and AR glass types for polymer and cement reinforcement, respectively

	E-glass	AR glass
SiO_2	52.4%	71%
$K_2O + Na_2O$	0.8	11
B_2O_3	10.4	—
Al_2O_3	14.4	18
MgO	5.2	—
CaO	16.6	—
ZrO_2	—	16
Li_2O	—	1

Source: Majumdar, A.J. and Nurse, R.W., Glass fibre reinforced cement, Building Research Establishment Current Paper, CP79/74, Building Research Establishment, Watford, England, 1974.

The most common composition of the AR glass used for cement reinforcement is based on the incorporation of about 16% ZrO_2 into the Si–O–Si network, providing it with stabilization in the high-pH environment. The composition and properties of E-glass used for reinforcing polymer matrices and AR glass for cement reinforcement are presented in Tables 9.1 and 9.2. The mechanical properties of the two glass types are quite similar (Table 9.2).

The improved composition is observed by analyzing the chemical stability and changes in strength upon aging (Figures 9.4 and 9.5).

Earlier tests to assess the chemical durability of the glass fibers were based on extracting cement hydration solutions, or simulated solutions containing Ca^{+2} and alkali ions. However, in order to better simulate the actual environment around the reinforcing glass filaments, including influences associated with its special bundled structure, the unique testing of the SIC scheme was

Table 9.2 Mechanical properties of E- and AR glass types for polymer and cement reinforcement, respectively

	E-glass	AR glass
Density (kg/m³)	2,540	2,780
Tensile strength (MPa)	3,500	2,500
Modulus of elasticity (GPa)	72.5	70.0
Elongation at break (%)	4.8	3.6

Source: Majumdar, A.J. and Nurse, R.W., Glass fibre reinforced cement, Building Research Establishment Current Paper, CP79/74, Building Research Establishment, Watford, England, 1974.

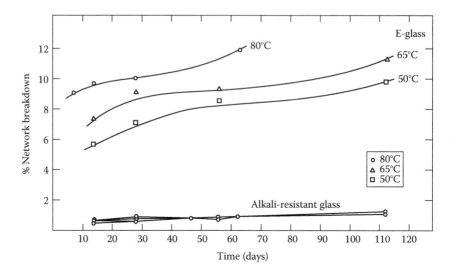

Figure 9.4 Effect of glass composition on the network breakdown in cement extract solutions. (From Larner, L.J. et al., *J. Non-Cryst. Solids*, 20, 43, 1976; Majumdar, A.J., Properties of GRC, in *Proceedings of the Symposium on 'Fibrous Concrete'*: *The Concrete Society*, London, U.K., 1980, pp. 48–68.)

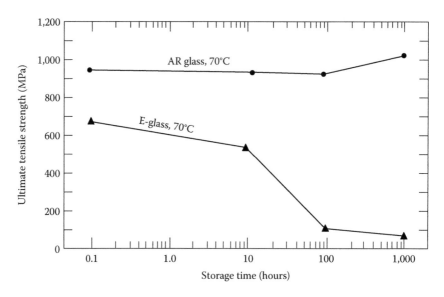

Figure 9.5 Effect of glass composition on tensile strength of fibers after exposure to Portland cement extract solution (pH = 13.4) at 70°C. (From Franke, L. and Overbeck, E., *Durab. Bldg. Mater.*, 4, 73, 1987.)

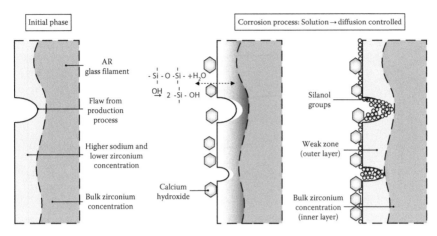

Figure 9.6 Formation of flaws in AR glass filaments and their growth in alkaline environment. (From Orlowsky, J. et al., *Mater. Struct.*, 38, 155, 2005.)

applied (Figure 9.1) and the results were compared with those of the composite parameters.

Several studies have been carried out to account for the improved stability of AR glass, some of them during the early stages of development of GRC composites, and some in more recent years, in view of the renewed interest that resulted from the development of TRC using glass fabrics. Several mechanisms were suggested to explain the improved durability performance of the glass when Z_rO_2 is present. The simplest one is that ZrO_2 stabilizes the network and, thus, even if there is leaching of Si, the network does not break down and an external layer that is rich in ZrO_2 remains and slows down any additional leaching process. Another mechanism suggested is the flaw formation that can occur in the zirconia-rich glass, but the growth of these flaws is arrested by the zirconia, as shown in Figure 9.6.

This mechanism of growth and arrest of flaws might explain some conflicting data in the literature, which support no strength loss in AR glass (Figure 9.5) in some cases and strength loss in other cases (Figure 9.7). This mechanism accounts for the observation that even if there is strength loss, it is stabilized at a level with sufficiently high residual strength.

9.2.1.2 Durability of polymeric reinforcement

The primary mechanism leading to degradation of polymeric reinforcement may be due to UV radiation. However, since the reinforcement is embedded in a cement matrix, it is quite likely that it is properly protected against such effects. This has been confirmed in a study by Hannant (1998) who investigated the durability of a cement composite reinforced with fibrillated

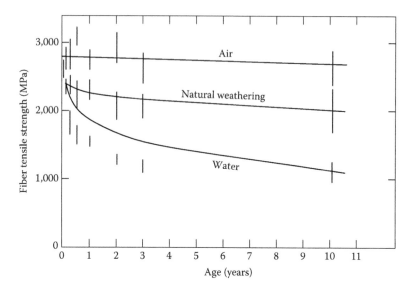

Figure 9.7 Strength of AR glass fibers removed from GRC composites after aging in different environments. (After West, J.M. and Majumdar, A.J., *J. Mater. Sci. Lett.*, 1, 214, 1982.)

network of polypropylene after 18 years of natural weathering. Two types of reinforcements were studied, one with thermal stabilizers added for processing purposes (film A) and one that included also a high UV stabilizer (film B). The strength of the two formulations after 18 years is presented in Figure 9.8, showing a mild reduction in strength of about 10% for film A and 20% for film B. It should be noted that the reduction occurred during

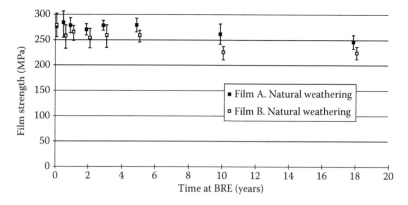

Figure 9.8 Effect of 18 years of natural weathering on the strength of the fibrillated polypropylene network reinforcement. (From Hannant, D.J., *Cement Concr. Res.*, 28(12), 1809, 1998.)

the first 10 years, and thereafter, up to 18 years, no additional change took place. It was observed that in these natural exposure conditions, the top surface of the cement paste in some of the samples had dissolved away locally, exposing the top one or two layers of the film to UV radiation. This was only a local effect and did not occur in the bulk where the reinforcement is completely surrounded by the cementitious matrix. Hannant (1998) commented that the performance of the film without UV stabilizer was even marginally better than that of the UV stabilized, suggesting that the protection effect of the matrix, even at cracks, is more than sufficient to provide UV protection.

9.2.1.3 Microstructural changes and aging

Earlier studies of the durability of GRC indicated that even when AR glass is being used, the properties of the composite undergo a decline when exposed to wet environment. The decline shows up as reduction in the tensile strength to a level close to first-crack strength (i.e., essentially the matrix strength) and loss of a major part of the postcracking load-bearing capacity, that is, embrittlement and loss of impact resistance, as demonstrated in Figure 9.9.

Extensive studies indicated that this reduction in properties was not associated with significant change in the strength of the AR glass reinforcement. This was confirmed by testing of the strength of filaments extracted from aged GRC composites as well as by observations of the surface of glass filaments in aged specimens that did not show any signs of damage or pits, which are characteristics of chemical degradation. Even if change in

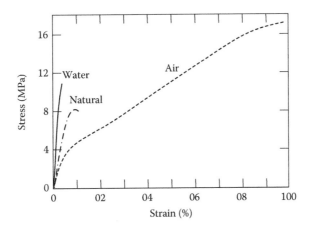

Figure 9.9 Effect of aging after 5 years on the stress–strain curves of AR-GRC composites. (From Singh, B. and Majumdar, A.J., J. Mater. Sci. Lett., 4, 967, 1985.)

reinforcement properties has taken place, it occurred at a time that is quite longer than the time when embrittlement occurred (Bentur, 1986, 1998).

Extensive microstructural studies have shown that during the time period at which embrittlement occurred, a major change took place at the reinforcement–matrix interface. This change is characteristic to the bundle reinforcement described in Chapter 4. In an unaged composite, the spaces between the core filaments are not readily filled with the particulate matrix, resulting in a reinforcement with a sleeve that is tightly bonded to the matrix and a core where the spaces are largely free. This is mainly the result of the fact that the spaces between the ~15 µm diameter filaments are too small to be penetrated by most of the cement grains whose median size is over 10 µm. The bundled reinforcement at this stage has some inherent flexibility, which shows up in the ability to undergo local bending across the crack surfaces, which are always somewhat shifted, and maintain a bridging effect in many of the filaments, without being fractured under the locally high deformation (Figure 4.9). The flexibility at this stage may show up by another mode, which is the telescopic failure (Figure 4.16), involving large slip deformations of the core filaments as long as they have some freedom to move one against the other, that is, they are largely free of the contact cross-links. The micromechanics of this system are discussed in Chapter 4.

Upon aging, the microstructure of the bundle changes drastically, with the core spaces being filled with hydration products. These are gradually deposited in the voids between the filaments when the composite is kept in a wet environment. The deposited hydration products can assume a variety of morphologies, depending on the composition of the matrix and the surface of the AR glass filaments.

The deposited hydration products tend to be rich in $Ca(OH)_2$, which is in the form of large crystals nucleating at the surfaces of the glass filaments (Figure 9.10a). The nature of these products can change and become more friable and porous, consisting of a greater proportion of CSH gel. This morphology develops when the matrix contains mineral admixtures that react with $Ca(OH)_2$ to form CSH (Figure 9.10b).

Modification of the size coating on the glass filaments was also found to have an influence on the nature of the deposited products, changing their tendency to be crystalline in nature to one that is more porous. This was clearly shown in reports on the second generation of AR glass reinforcement, CemFIL-2, which were identical in terms of the bulk glass composition, and the only change that was made was the nature of the size coating, which modifies the mode of nucleation of hydration products on the surface (Figure 9.10b) (Bentur, 1986).

The changes in the microstructure of the hydration products deposited between the filaments in the bundle upon aging correlated with the decline

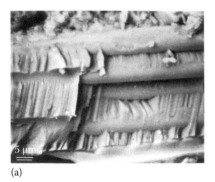

(a)   (b)

Figure 9.10 (a) The spaces between the filaments in aged GRC showing deposition of CH crystals in AR glass fire bundles and (b) relatively open space in similar bundles that were treated with microsilica. (From Bentur, A., Mechanisms of potential embrittlement and strength loss of glass fibre reinforced cement composites, in S. Diamond (ed.), *Proceedings: Durability of Glass Fibre Reinforced Concrete Symposium*, Prestressed Concrete Institute, Chicago, IL, 1986, pp. 109–123.)

in properties of GRC composites over time: deposition of more porous and friable products resulted in reduced rates of decline in mechanical properties (Figure 9.11).

These observations were made in the earlier days of the studies of the aging of GRC with chopped and randomly dispersed glass bundles. They have been confirmed in recent years in reports on the aging of TRC with AR glass fabrics.

In view of these relations between microstructure and aging, average values of the long-term performance of glass-reinforced cementitious matrices were compared to address the similarities between the older generation of GRC and the newer glass TRC.

In order to validate the observations of the relations between the deposition of hydration products in the filament bundles and aging effects, several novel methods were developed to facilitate quantification of the changes that take place in the AR glass bundle upon aging, as well as in situ characterization of the pullout of the bundle and its bridging across cracks. The results of characterization with such methods provide also a better understanding of the processes and enable to establish mechanisms and models to account for aging effects.

The push-in technique by Zhu and Bartos (1997) was described in Figure 4.14. It was applied for the study of an AR glass system with four matrices: control Portland cement (A), addition of 10% acrylic polymer (M1), replacement of the cement with 10% silica fume (M2), and 25% metakaolin (M3). Accelerated aging resulted in a considerable difference in behavior, with the metakaolin system preserving all the properties while the others losing a considerable part of them (Figure 9.12).

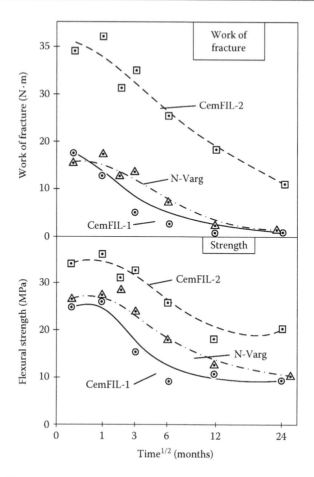

Figure 9.11 Changes in mechanical properties upon aging; systems with lower rates of decline are associated with more open hydration products deposition between the filaments (compare CemFIL-1 and CemFIL-2 in this figure with Figure 9.10). (From Bentur, A. et al., *J. Am. Ceram. Soc.*, 68, 203, 1985.)

The push-in test indicates that low push-in resistance was preserved in the core filaments only in the system with the metakaolin (Figure 9.13). Since this is also the system where mechanical properties were preserved upon aging (Figure 9.12), it clearly confirms the relation between aging resistance and preservation of the freedom of movement of the core filaments, either by elimination of growth of hydration products or having them in a friable nature.

A different type of method to follow the behavior of the individual filament in the strand was developed by Baholzer and Brameshuber (Banholzer, 2004; Banholzer and Brameshuber, 2004), namely, the FILT test (see Chapter 4).

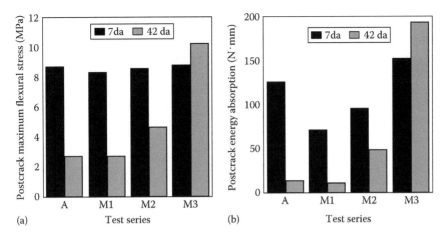

Figure 9.12 Effect of matrix composition on the postcracking strength (a) and toughness before and after aging (b) (A, control; M1, acrylic modification; M2, silica fume modification; M3, metakaolin modification; 7 da, accelerated aging for 7 days; 42 da, accelerated aging for 42 days). (From Zhu, W. and Bartos, P.J.M., *Cement Concr. Res.*, 27(11), 1701, 1997.)

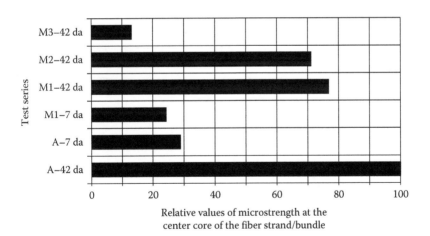

Figure 9.13 The effect of the matrix composition on the ratio between the push-in resistance of the core and sleeve filaments after accelerated aging for 7 and 42 days. A, control; M1, acrylic modification; M2, silica fume modification; M3, metakaolin modification; 7 da, accelerated aging for 7 days; 42 days, accelerated aging for 42 days. (From Zhu, W. and Bartos, P.J.M., *Cement Concr. Res.*, 27(11), 1701, 1997.)

It was used by these researchers to study the micromechanical behavior in TRC composites. Bentur et al. (2010) applied the FILT test for the study of the aging of systems in which the yarn was treated by impregnation with nanoparticles (microsilica, film- and non-film-forming polymers) to determine the breakage of individual filaments during the pullout and calculate at each stage the remaining active filaments. The pullout curves before and after aging are presented in Figure 9.14 for systems with microsilica particles, showing that all the systems exhibited curves with considerable

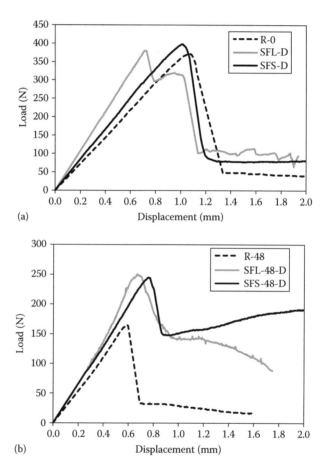

(a)

(b)

Figure 9.14 Pullout curves before (a) and after aging (b) of systems in which the bundle was impregnated with microsilica particles. (From Bentur, A. et al., Bonding and microstructure in textile reinforced concrete, in W. Brameshuber (ed.), *Proceedings of the International RILEM Conference on Materials Science: Textile Reinforced Concretes*, Aachen, Germany, RILEM Publications, Vol. I, 2010, pp. 23–33; Bentur, A. et al., *Cement Concr. Res.*, 47, 69, 2013; Yardimci, M.Y. et al., *Cement Concr. Compos.*, 33(1), 124, 2011.)

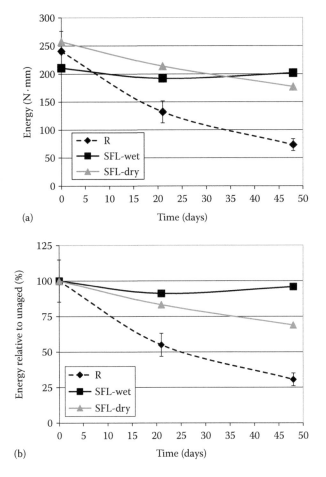

Figure 9.15 Effect of aging on (a) absolute and (b) relative pullout energy values for the systems treated with MS. (From Yardimci, M.Y. et al., *Cement Concr. Compos.*, 33(1), 124, 2011; Bentur, A. et al., *Cement Concr. Res.*, 47, 69, 2013.)

postcracking load-carrying capacity. After aging, the bond strength and the postpeak load-carrying capacity were reduced drastically in the reference system (marked R), while the ones that were treated with microsilica particles (marked SFL and SFS) retained their bond strength as well as considerable postpeak load-carrying capacity as exhibited by the energy absorbed during pullout (Figure 9.15).

This preservation of the bond behavior was consistent with the ability to preserve the telescopic mode of failure, as seen by the diagram of the active filaments in Figure 4.18, which presents the curve of the control and microsilica-treated bundle after aging. This difference in bond behavior after aging was consistent with the performance of the whole composite, showing preservation in properties after aging for the microsilica-treated system and embrittlement of the control. This will be discussed later on in this chapter.

The preservation of properties observed here occurred in systems in which telescopic mode of pullout could be maintained after aging, namely, that the core filaments could be engaged in slip after the peak load. This could take place if deposition of massive hydration products in the core bundle could be prevented, thus preserving the ability for slip of filaments in this zone, with the slip being frictional in nature. This is consistent with the low push-in resistance of the core filaments in systems where performance was maintained after aging, as seen in the results of Zhu and Bartos (1997) (Figures 9.12 and 9.13).

In situ SEM observations of notched specimens with aligned bundled glass filament reinforcement are consistent with the results and interpretations of the mode of failures of a bundle before and after aging (Bentur and Diamond, 1985; Trtik and Bartos, 1999).

Both studies have shown that the bridging of a crack in an unaged specimen involves considerable local flexure of the bridging bundle. This is the result of the unique characteristic of the crack microstructure: the crack surface is not in a single plane form, since some of its surfaces are inclined to the bundle (Figure 4.9). It can be seen that some of the filaments are broken while others undergo considerable deformation to bridge over the crack. Having locally high deformation while maintaining bridging with brittle glass filaments is made possible by mechanism of slippage of the core filament. They are free to move relative to one another in a telescopic mode, as described by Bartos (1987) (Figure 4.16).

In the case of aged specimens, the failure mode is brittle. Since the crack surfaces are in a planar form, perpendicular to the reinforcement, an aligned bundle with a fractured surface or pulled-out filaments with a very short pullout length are kept together (Trtik and Bartos, 1999).

9.2.2 Modifying microstructure to enhance durability performance

9.2.2.1 Fiber treatment

Improved performance of AR glass bundles can be obtained by two means of yarn treatment: applying a thin-sized surface coating or impregnation of the spaces in the bundle with polymer or nanoparticles.

Surface treatments can lead to a more stable surface that will suppress alkali attack, and can also change the nucleation and growth of the hydration products, to delay or prevent the formation of a dense interfacial matrix. The latter effect was apparently effective in the case

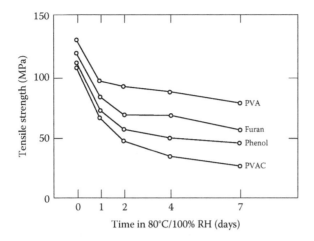

Figure 9.16 The effect of surface treatments of AR glass fibers on the strength retention of a strand in a Portland cement matrix. (From Hayashi, M. et al., Some ways to improve durability of GFRC, in S. Diamond (ed.), *Proceedings: Durability of Glass Fibre Reinforced Concrete Symposium*, Prestressed Concrete Institute, Chicago, IL, 1985, pp. 270–284.)

of the CemFIL-2 fibers with a modified size coating, as demonstrated in Figure 9.11. Bijen (1985) suggested that such effects can be brought about by changes in the zeta potential of the surface. He proposed that in CemFIL-2, the organic compounds in the coating may interfere with the precipitation of $Ca(OH)_2$. Additional results demonstrating the positive influence that can be brought about by treating the surfaces with polymers are shown in Figure 9.16.

Recent studies with AR glass fabrics confirmed these trends. Butler et al. (2009) studied the bonding of AR glass fabrics using two types of commercial sizing and showed considerable difference in the performance, which they accounted for by the bonding behavior upon aging in the cement matrix. With one of them (Vet3), the bond loss after aging in 40°C was 50% after 1 year, while with the other (Vet3), the loss was reduced to 20% (Figure 9.17). The difference between these two could be correlated with their influence on the microstructure of the hydration products deposited upon aging within the bundle. The improved performance was obtained when dense crystalline product formation was changed from a massive crystalline one to one that is less homogeneous and more coarsely structured and includes ettringite needles. The authors commented that it could not be determined whether damaging effects were caused primary by degradation of filaments or the changes in the interface morphology, although the latter seems more likely.

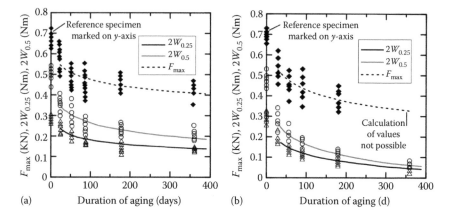

Figure 9.17 Maximum pullout force and pullout work at 0.25 and 0.55 mm crack openings as a function of duration of aging in 40°C wet environment of AR glass fabrics (product of Saint-Gobain) with two types of commercial size coating: (a) VET-03 and (b) VET-02. (From Butler, V. et al., *Cement Concr. Compos.*, 31, 221, 2009.)

Scheffler et al. (2009) and Gao et al. (2007) studied the improvement of the protective nature of the size coating achieved by incorporating in it small amounts of carbon nanotubes and nanoclays. The performance was evaluated by testing the strength in concentrated aqueous solutions (5 wt% NaOH, pH of 14) at 20°C and 40°C for periods of up to 30 days.

Nanoclay particles at 1 wt% were embedded in two types of styrene–butadiene dispersions, C1 (carboxylated and self-cross-linking, Tg 5°C, particle size 170 nm) and C2 (thermoplastic, Tg 8°C, particle size 130 nm). The C2 size coating was also prepared with addition of 3 wt% of carbon nanotubes. The coatings enhanced the retention of strength after aging from about 40% to about 75%, with the size coatings having the dispersed nanoparticles showing higher absolute strength values (Figure 9.18).

Impregnation by epoxy was reported by Xu et al. (2004) as a means for protection of the bundle. Their study dealt mainly with the influence of this treatment on the bond.

Buttner et al. (2010) evaluated the effect of the mode of epoxy impregnation on the durability performance. They compared epoxy that underwent cold hardening with a prepreg formulation that was cured after the impregnation by heat treatment at 120°C for 2 h. The performance was evaluated by means of testing in tension of reinforced TSP specimen. Accelerated aging of these specimens was carried out in 50°C water. The loss of strength during the accelerated aging period of the control and the epoxy-impregnated fabrics are shown in Figure 9.19. The control lost about 60% of its strength and the prepreg epoxy-impregnated

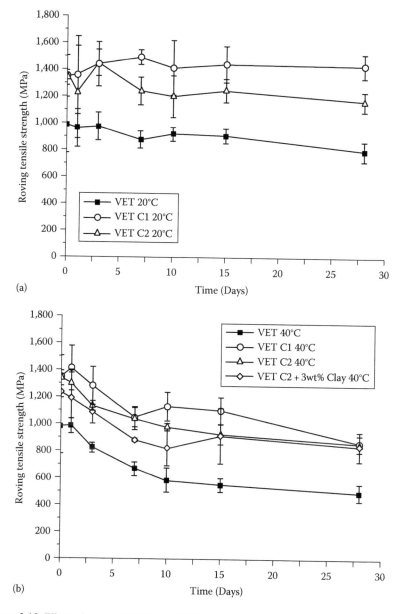

(a)

(b)

Figure 9.18 Effect of aging at 20°C and 40°C on the strength of AR glass bundled roving with C1, C2, and C3+carbon nanotube size coatings. (From Scheffler, S. et al., *Compos. Sci. Technol.*, 69, 531, 2009.)

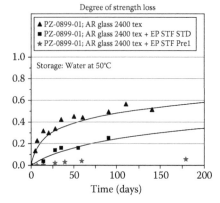

Degree of strength loss

▲ PZ-0899-01; AR glass 2400 tex
■ PZ-0899-01; AR glass 2400 tex + EP STF STD
★ PZ-0899-01; AR glass 2400 tex + EP STF Pre1

Storage: Water at 50°C

	Reference tensile strength (N/mm^2)[a]
AR glass 2400 tex	690
AR glass 2400 tex EP STF STD	1,540
AR glass 2400 tex EP STF Pre1	1,440

[a]Determined with the TSP test.

Figure 9.19 Effect of aging in 50°C water on the strength of control and impregnated AR glass fabrics evaluated by the TSP test. (From Buttner, T. et al., Enhancement of the durability of alkali-resistant glass-rovings in concrete, in *International RILEM Conference on Material Science (MATSCI)*, Aachen, Germany, ICTRC, Vol. 1, 2010, pp. 333–342.)

reinforcement retained almost all of its strength; the cold hardening one demonstrated an intermediate behavior of about 30% loss.

A different approach of impregnation is based on the deposition of nanoparticles within the spaces in the filaments before the production of the composite. This is done by treatment of the fabric with a slurry containing the particles that are being absorbed in the vacant spaces, followed by removal of the excess slurry and incorporation into the cementitious matrix (Bentur, 1989; Bentur and Diamond, 1987a). This approach was first applied by using microsilica slurry, and the performance of the cementitious composites was evaluated with and without modification of the matrix with microsilica (Figure 9.20).

It can be seen that the impregnation of the bundled yarns with microsilica resulted in a marked improvement even when the matrix was not modified: the loss on aging was reduced from about 90% to 40%. Additional modification of the matrix enhanced the performance further to reduce the toughness loss upon aging to about 20% only. On the other hand, modification of the matrix only with microsilica had only a marginal improvement, leading to 80% reduction in toughness compared to the 90% reduction in plain cement matrix. SEM observation indicated that the impregnated yarn could maintain the telescopic mode of failure after aging.

These results indicate that the nanoparticles can be deposited in between the filaments in the yarn (Figure 9.20a) where they are most effective in preserving the ductile telescopic failure mode. This is achieved by preventing deposition of hydration products in between the core filaments in the bundle, or changing their nature from crystalline to more porous CSH.

(a)

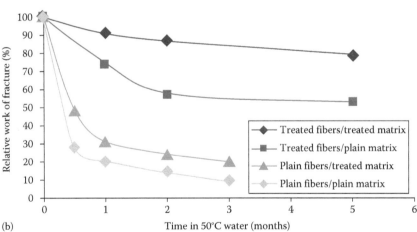

(b)

Figure 9.20 The effect of impregnation of AR glass bundles with nanosized microsilica (~100 nm) on the aging performance in different matrices: (a) the impregnation with the nanosized microsilica and (b) effect of matrix composition with microsilica replacement (plain vs. treated) and treatment of the yarns with the microsilica slurry (plain vs. treated). (From Bentur, A., *J. Mater. Civil Eng.*, 1, 167, 1989.)

The microscopical observations suggest also another effect, that is, the formation of a dense layer of hydration products around the perimeter of the bundle, due to the pozzolanic reaction of the microsilica, slowing down or preventing any ingress of ions into the bundle, to preserve the nature of its core. If one considers alkaline chemical degradation, it is expected that the massive presence of the microsilica within the bundle will reduce the pH locally, which may have a favorable effect on preventing or slowing down chemical degradation.

The concept of impregnating with nanoparticles was further studied by considering microsilica particles of different sizes as well as polymer particles (film- and non-film-forming ~100 nm particles) and evaluating the aging performance of the composite and the pullout behavior of treated and nontreated yarns (Bentur et al., 2010, 2013; Cohen and Peled, 2010; Yardimci et al., 2011). The stress–strain curves before and after aging are shown in Figure 9.21, demonstrating that the microsilica-treated composites lost some strength and ductility upon aging but remained highly ductile and strong; their properties after aging were similar to those of the control TRC composite before aging.

Observations of the differences in the trends induced by the various nanoparticles can bring about some insights into the mechanisms involved. The treatment with the non-film-forming polymer (Figure 9.21) did not yield improved aging performance, suggesting that the positive influence of the microsilica is associated also with its composition. The film-forming polymer resulted in low strength but extremely ductile TRC composite, which retained all its tensile properties upon aging (Figure 9.21). This behavior might be attributed to the complete engulfing of the bundle with the film, providing a completely different mode of stress transfer and isolation from changes within the bundle that might be caused by deposition of hydration products. This may account for the reduced bond, the high ductility, and its preservation upon aging.

The aging trends of the composite correlated quite well with the bonding performance. The low tensile strength of the film-forming polymer was consistent with low bond but with marked postpeak performance and its preservation upon aging. The systems that preserved ductility upon aging showed a pullout curve, which maintained after aging a considerable postpeak load-bearing capacity. There was similarity between the trends of the effects of aging on composite properties and bond characteristics (Figure 9.22).

A similar approach to enhance durability was reported by Buttner et al. (2010) using a different technique in which Portland cement (with particles smaller than 6 μm) and microsilica particles were dispersed in a solvent prepared with polyvinyl alcohol (PVA) or polyvinyl acetate (PVAc). Three such dispersions were made, with cement:polymer and microsilica:polymer ratio of 80:20 and a ternary dispersion of cement:microsilica:polymer ratio

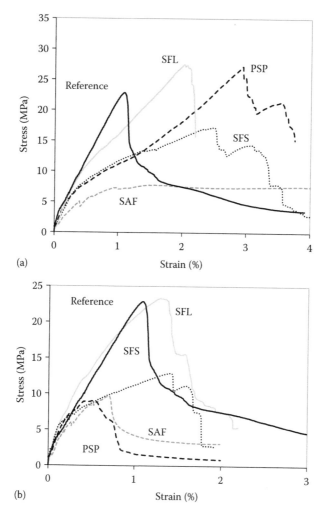

Figure 9.21 Effect of treatment of the yarns in the fabric on the tensile behavior of TRC before (a) and after aging (b); Reference, no treatment; SFL, treatment with 100 μm microsilica; SFS, treatment with 50 μm treatment; PSP, treatment with non-film-forming polymer; SAF, treatment with film-forming polymer. (From Cohen, Z. and Peled, A., *Cement Concr. Res.*, 40(10), 1495, 2010.)

of 40:40:80. The impregnation was achieved by pultrusion of the yarns through the dispersion. The performance was determined by the testing of TSP specimens (Figure 9.2). Accelerated aging took place in wet environment at 50°C. The impregnation with the cement-only slurry did not result in improvement in aging, while the dispersions that contained microsilica showed considerable improvement, with the composition having microsilica only preventing reduction in tensile strength (Figure 9.23).

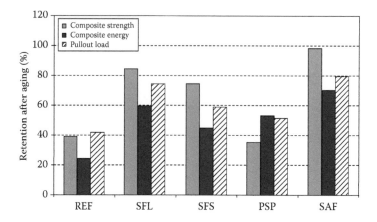

Figure 9.22 Effect of the treatment of the yarns in the TRC on the aging retention of properties after accelerated aging. (From Bentur, A. et al., Bonding and microstructure in textile reinforced concrete, in W. Brameshuber (ed.), *Proceedings of the International RILEM Conference on Materials Science: Textile Reinforced Concretes*, Aachen, Germany, RILEM Publications, Vol. 1, 2010, pp. 23–33; Bentur, A. et al., *Cement Concrete Res.*, 47, 69, 2013; Cohen, Z. and Peled, A., *Cement Concr. Res.*, 40(10), 1495, 2010.)

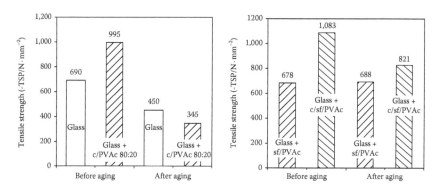

Figure 9.23 Effect of treatment of the yarns on the preservation of strength after accelerated aging. (From Buttner, T. et al., Enhancement of the durability of alkali-resistant glass-rovings in concrete, in *International RILEM Conference on Material Science (MATSCI)*, Aachen, Germany, ICTRC, Vol. 1, 2010, pp. 333–342.)

9.2.2.2 Matrix composition

Modification of the cementitious matrix was one of the strategies taken at the earliest stages of the development of GRC to control the aging effects by changing the nature of the pore solution (i.e., reducing pH) as well as the microstructure at the interfaces and within the bundle. The most notable are the modification of Portland cement with mineral additives such as fly ash, blast furnace slag, silica fume, and metakaolin, which react with the $Ca(OH)_2$ and have an overall influence on reducing the alkalinity of the pore solution. A more extreme strategy was to use non-Portland cement matrices, such as high alumina cement and supersulfated cement, or cement modified with polymer latex (Majumdar, 1980).

These results show that all these strategies have a positive influence on the property retention after aging. Of particular interest is the influence of mineral admixtures added to the cementitious matrix, as a replacement of part of the Portland cement, since this is the more practical route to be taken. Modifications with low or moderate reactivity pozzolans such as fly ash had only a modest influence on reducing the loss of properties on aging, as seen in Figure 9.24 and as reported extensively in numerous publications (e.g., Leonard and Bentur, 1984; Proctor et al., 1982; Singh and Majumdar, 1981, 1985; Singh et al., 1984).

These additives did not prevent a loss in strength and toughness, but they did slow down the rate of loss. Singh and Majumdar (1981) reported that the best results, in terms of strength and toughness retention, were obtained with a 40% fly ash content. They pointed out, however, that these high-level additions were accompanied by some reduction in the initial properties. The extent of the improvement in durability varied with different fly ashes and natural pozzolans (Leonard and Bentur, 1984; Singh and Majumdar, 1981). Leonard and Bentur (1984) reported that such differences could not be correlated with the pozzolanic activity, but rather with the effect of the fly ash on the interfacial microstructure developed.

In view of this limited improvement, special attention has been given to the use of more reactive mineral additives such as blast furnace slag (Bentur, 1986, 1989; Bentur and Diamond, 1985, 1987; Bergstrom and Gram, 1984; Brandt and Glinicki, 2003; Hayashi et al., 1985; Marikunte et al., 1997; Peled et al., 2005; Purnell et al., 1999, 2000; Rajczyk et al., 1997; Yilmaz and Glasser, 1991; Zhu and Bartos, 1997).

The replacement of part of the Portland cement with 10%–25% silica fume brought about a modest slowing down of the degradation process (Bentur, 1989; Bentur and Diamond, 1985, 1987a; Hayashi et al., 1985; Rajczyk et al., 1997; Singh and Majumdar, 1985) as demonstrated in Table 9.3.

A combination of blast furnace slag and dimension-stabilizing admixture was found to be much more effective (Peled et al., 2005), and its influence was attributed to the elimination of CH and its deposition between the filaments in the strand. Metakaolin was reported to be extremely effective in drastically reducing the rates of loss (Bentur, 1989; Marikunte et al., 1997; Purnell et al., 1999, 2000; Zhu and Bartos, 1997) as seen for example in Table 9.3. The effectiveness of metakaolin could be explained by its influence on the changes in the bonding between the filaments, especially as shown from the data of Zhu and Bartos (1997), based on the push-in characterization of the filaments in the bundle after aging: the bonding of

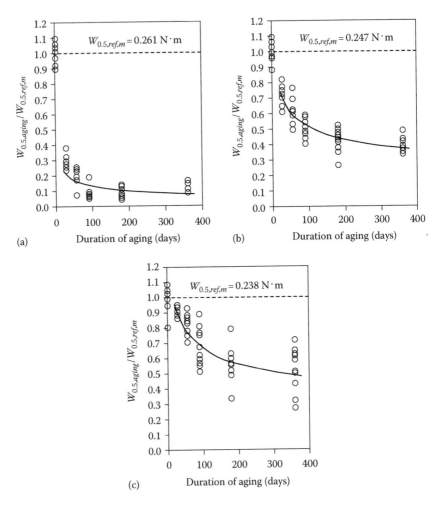

Figure 9.24 Effect of aging on the pullout work up to 0.5 mm crack opening of AR glass fabrics in a CEMI matrix, control (a), CEMI with fly ash and microsilica replacement (b), and CEMIII with fly ash and microsilica replacement (c). (From Butler, M. et al., *Mater. Struct.*, 43, 1351, 2010.)

the inner, core filaments, increased and became as high as that of the outer, sleeve filaments, in the case of the Portland cement and silica fume matrices, but remained less than 20% of the bond of the outer filaments in the case of the metakaolin matrix (Figure 9.13). This may reflect differences in the nature of the hydration products and their deposition within the filaments.

The greater effectiveness of applying the more reactive mineral admixture when used in combination with CEMIII was recently confirmed in

Table 9.3 Effect of 25% replacement of cement with silica fume and metakaolin on the toughness retention of AR glass fiber composites after 50°C accelerated aging

Matrix composition	Toughness retention after accelerated aging at 50°C (%)	
	28 days	84 days
Portland cement	37.0	12.9
25% silica fume replacement	43.6	29.9
25% metakaolin replacement	78.7	71.0

Source: Adapted from Marikunte, S. et al., *J. Adv. Cement Based Mater.*, 5, 100, 1997.
Note: The values in the table are relative to unaged composite.

studies of AR glass TRC composites (Butler et al., 2009, 2010). Butler et al. studied the effect of a matrix with CEMI cement, and its partial replacement with fly ash and microsilica (CEMI/FA/MS), and matrices with CEMIII cement with partial replacement with fly ash and microsilica (CEMIII/FA/MS), or metakaolin instead of the microsilica (CEMIII/FA/MK). The reduced alkalinity of the matrices was estimated by their pH level, measured in the pore solution or by pore expression from the hardened matrix. The highest pH of about 12.7 in suspension was obtained with the control, CEMI cement, and the lowest ones, of about 11.8–12.0, with the CEMIII compositions. The CEMI with fly ash and microsilica replacement was in between, at about 12.4. These are values in suspension after aging for 360 days at 40°C.

The composite with the CEMI (control) matrix lost all its postcracking load-bearing capacity after 28 days of accelerated aging at 40°C, while the matrices with mineral admixture modification retained their ductility and tensile strength and even improved it after 360 days of accelerated aging.

These aging trends of the composite correlated quite well with the influence of aging on bond behavior (crack-bridging force and work). Bond values were reduced upon aging, to nonmeasurable values for the control, and by about 40% for the matrices with fly ash and microsilica replacements. These influences could be observed for the postpeak pullout behavior, as quantified by the pullout work up to 0.5 mm crack opening (Figure 9.24).

The differences in aging behavior were also reflected in the microstructure of the hydration products deposited within the filaments in the bundle after aging: In the control CEMI matrix composite, massive deposits of CH crystals could be observed upon aging. The $Ca(OH)_2$ crystals as well as the brittle shells around the filaments tended to shatter already at small displacement and tended to wedge themselves unfavorably and exert notching at the

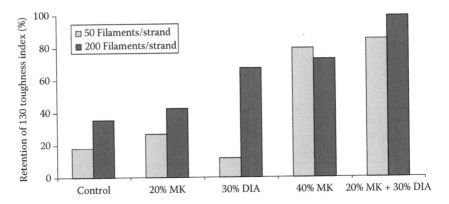

Figure 9.25 Effect of bundle size (number of filaments) and mineral admixture (metakaolin, MK; Diatomite, DIA) on the retention of the I-30 toughness index after accelerated aging at 50°C for 84 days. (Plotted from Brandt, A.M. and Glinicki, M.A., Effects of pozzolanic additives on long-term flexural toughness of HPGFRC, in A.E. Naaman and H.W. Reinhardt (eds.), *Fourth International Workshop on High Performance Fibre Reinforced Cement Composites (HPFRCC 4)*, Ann Arbor, MI, RILEM Publications, Bagneux, France, 2003, pp. 399–408.)

filaments as well as lateral punctual pressure (Butler et al., 2009). In the modified matrix and in the CEMIII cement, mainly CSH was deposited around the filaments, and the microstructure was more open, hardly without a massive crystalline CH. It was noted that in all of the compositions of the matrices, there were hardly any defects seen on the surface of the filaments that might represent chemical attack. However, such defects were not ruled out since they may not be observable in the ESEM (Butler et al., 2009, 2010, 2011).

The combined influence of yarn size (i.e., number of filaments per yarn) and matrix modification, used as a means for improved durability, was reported by Brandt and Glinicki (2003), applying metakaolin and diatomite. An increase in the size of the bundle resulted in an enhanced toughness retention (Figure 9.25) in the systems in which CH was formed (control and systems with lower content of replacements, 20%–30%). The influence of the bundle size could be accounted for by the bonding and flexural mechanisms outlined in Section 9.3: with a larger number of filaments, the relative influence of the $Ca(OH)_2$, which preferentially deposits at the external filaments, becomes smaller. With larger replacement contents (40%–50%), where the $Ca(OH)_2$ is completely eliminated, the toughness retention is much higher, and the system is practically insensitive to the bundle size. This behavior can be interpreted in terms of the elimination of the mechanism of deposition of crystalline material and other hydration products that may bind the filaments in the bundle.

9.3 Aging mechanisms

Review of the aging trends and their modifications by matrix and yarn treatments can serve as a basis for resolving the mechanisms that may lead to the decline in the properties of composites with cementitious matrix and glass yarn reinforcement.

It is well established that the aging of glass-reinforced cementitious composites cannot be simply attributed to alkali attack of the glass fibers by the highly alkaline cementitious matrix (Bentur, 1998; Bentur and Mindess, 2007). Two mechanisms must be considered to account for the reduction in strength and loss in toughness of this composite:

1. Chemical attack of the glass fibers—chemical attack mechanism
2. Growth of hydration products between the glass filaments—microstructural mechanism

There is some controversy regarding the relative importance of these two mechanisms. Resolving these aspects of the aging process is not merely of academic interest; it is essential to clarify this question if composites of improved long-term performance are to be developed and used.

9.3.1 Chemical attack mechanism

The nature of the chemical attack, its causes, and the methods used to alleviate its adverse effect by changing the glass composition have been discussed in this chapter and more in-depth treatment is provided in Bentur and Mindess (2007). The extent of this problem can be evaluated most readily by determining the change in the strength of glass bundles exposed to an alkaline environment, either in a solution or in a cementitious matrix. Even with commercial AR glass fibers, a reduction in strength of glass filaments removed from GRC aged in a wet environment could be observed (Figure 9.7), as well as in SIC tests (Litherland et al., 1981; West and Majumdar, 1982) and TSP tests (Scheffler et al., 2009). Thus, even the commercial AR glass fibers are not completely immune to chemical attack. However, data such as those shown in Figure 9.7 indicate that the reduction in strength takes place mainly during the first 2–3 years, and then the strength becomes stable, for at least up to 10 years. Also, the residual strength of the fibers is quite high, being 1,000 MPa in the data in Figure 9.7 of West and Majumdar (1982), more than 500 MPa in the SIC tests reported by Proctor et al. (1982), and 500 MPa reported by Scheffler et al. (2009). This is sufficiently high to provide useful reinforcing effects.

9.3.2 Microstructural mechanisms

The marked microstructural changes that can take place at the fiber–matrix interface as the composite ages in a humid environment are the result of the special bundled structure of the reinforcing strand, where hydration products gradually grow into the spaces between the filaments, which are largely empty at early ages of few months. This has been described and modeled in terms of sleeve and core parts of the bundle (Chapter 4). In Portland cement matrix, the products that deposit upon aging are usually denser and more crystalline, mainly CH. They become more porous if mineral additives are available in the matrix or are being impregnated into the bundle during the production of the composite (Bentur and Diamond, 1987a; Jaras and Litherland, 1975; Mills, 1981; Stucke and Majumdar, 1976). It has been suggested that the densification of the matrix microstructure at the glass interface can lead to embrittlement (Bentur et al., 1985; Leonard and Bentur, 1984; Stucke and Majumdar, 1976). This can be explained in terms of two mechanisms, or a combination of these:

1. Effective increase in the bonding within the filaments in the bundle, leading to loss in flexibility of this reinforcing unit, reducing its ability to undergo ductile telescopic pullout, or to accommodate locally large deformations in flexure (Bentur, 1998; Katz and Bentur, 1996; Stucke and Majumdar, 1976)
2. Development of stress concentrations resulting in notching or surface flaw enlargement, which lead to static fatigue effects (Purnell et al., 1999, 2000, 2001; Yilmaz and Glasser, 1991)

9.3.2.1 Internal bonding in the bundle and loss of its flexibility

Investigators have shown that the interfacial bond strength of glass filaments surrounded completely by cement matrix ranges from about 1 MPa (Oakley and Proctor, 1975) to 5 MPa (Zhu and Bartos, 1997), which causes the pullout to change from slipping of fibers at the low range of bond to fracture at the high bond levels. Results of the effect of aging on pullout tests indicated an increase in the bond strength with aging when single filaments were studied (De Vekey and Majumdar, 1970), but reduction when the filaments are in the form of a bundle. This reduction takes place in conjunction with deposition of hydration products in the core filaments as clearly observed in SEM and quantified by push-in characterization by Zhu and Bartos (1997). When these effects take place, the mode of pullout failure changes from telescopic failure to a brittle failure as shown by FILT test of aging systems. This is consistent with the report by Mobasher and Li (1995) showing that the changes in the pullout of glass fibers upon aging could be accounted for by an increase in bond as well as the stiffness of the interfacial matrix. Using a fracture mechanics model, they calculated the increases in adhesional resistance, frictional resistance, and stiffness at the interface from 0.62 N/mm, 0.59 N/mm, and

0.04 mm^{-1} to 1.23 N/mm, 0.88 N/mm, and 0.15 mm^{-1}, respectively (for 3 days of accelerated aging). These changes could account for a change in the pullout curve, from one that exhibits elastic–quasiplastic behavior to one that shows a sharp decline after the peak.

This change in the nature of the reinforcement and its ability to engage in considerable slip may correlate with the embrittlement of the composite, which is the most significant mark of aging. If ideally aligned reinforcement is considered, then the reduction can be accounted for by the loss in the slip capacity as seen in the results of pullout tests. When this capacity is maintained, the composite can retain its toughness. An alternative explanation is that even in aligned composite under tension, the interaction of the crack and the bundle involves some local bending, as observed by in situ SEM observations (Bentur and Diamond, 1984). Bentur and Diamond (1986) have shown by in situ testing in an SEM that such local flexure can occur even if the glass fiber strand is oriented perpendicular to the crack. The interaction of the crack with the glass fiber strand is associated with some shift in the crack path and the development of local bending (Figure 4.9). In order to bridge across the crack, the glass filaments must be able to accommodate large flexural deformations. If the spaces between the filaments are filled with hydration products, the strand becomes rigid since the freedom of movement of one filament relative to the other is lost or reduced; as a result, it cannot accommodate the large local flexural deformations, and brittle failure will occur. Stucke and Majumdar (1976) showed by a simple calculation that the densification of the matrix microstructure around the glass filaments in aged GRC could lead to local flexural stresses in the glass filament that may exceed its tensile strength. Thus, fiber fracture and embrittlement can occur even if the glass filament does not lose strength due to a chemical attack.

One may expect that if this type of mechanism is indeed effective, it might show up in other materials that are used as reinforcement. Hannant (1998) studied the aging of fibrillated networks of polypropylene that were used to produce thin-sheet cement composites and evaluated the change in properties of the reinforcement itself and the properties of the composite. The reinforcement maintained its strength over the 18-year period studied with only a mild decline of 10%–20%, which seemed to be leveled off. In the composite itself, considerable densifying of the interfacial region could be seen, but there was no decline in the properties of the composite, and the bond strength remained constant at about 0.4 MPa in spite of the microstructural change. Hannant (1998) suggested, based on observation of the structure of the reinforcement, that the bond may be controlled by shearing within the film, and this would be unaffected by the microstructural changes. This is at the base of the sleeve–core behavior modeled by Ohno and Hannant (1994) (for more details, see Chapter 4).

9.3.2.2 Flaw enlargement and notching

The internal bonding and loss of flexibility mechanisms that have been applied to account for reduction in mechanical properties due to microstructural changes have been challenged by Purnell and coworkers (Purnell et al., 1999, 2000, 2001), based on the observation of aging of GRC with different types of matrices: Portland cement with 20% metakaolin and calcium sulfoaluminate cement.

Aging curves of strength changes at 60°C of GRC with different matrices indicated that in modified matrices (Portland cement with 20% metakaolin and calcium sulfoaluminate cement), a two-stage process is obtained, with strength increasing at the beginning and reducing later on. This behavior cannot fit the linear relation between strength and inverse of log time, which was used by Litherland et al. (1981) to develop plots of the rate of strength loss in terms of relative time displacement (i.e., the ratio between the time in actual conditions relative to the time in standard lab conditions required to lead to similar strength loss) versus $1/T$, to construct Arrhenius-type relations (Figure 9.26).

This relation did not hold true for matrices with modified composition, containing 20% of metakaolin or calcium sulfoaluminate and anhydrite (shrinkage compensating cement) (Figure 9.27).

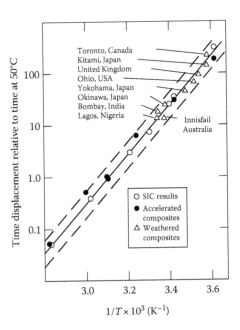

Figure 9.26 Normalized Arrhenius plots of strength retention (SIC test and composite flexural strength) for specimens in accelerated and natural weathering. (From Litherland, K.L. et al., Cement Concr. Res., 11, 455, 1981.)

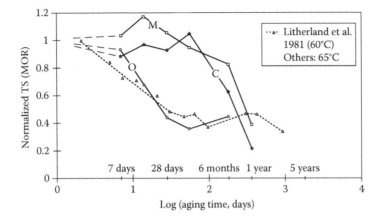

Figure 9.27 Arrhenius plots for aging of GRC in terms of normalized tensile and flexural strength of composites (strength after aging relative strength before aging) versus log of aging time for composites with Portland cement matrix (O), modified matrices containing 20% MK (M), shrinkage compensating cement, containing calcium sulfoaluminate and anhydrite at 8:3:1 ratio (C); comparison is made in the figure with Arrhenius plots reported by Litherland et al. (1981), for Portland cement matrix. (From Purnell, P. et al., *Compos. Appl. Sci. Manuf.*, 30, 1073, 1999.)

The plot for the Portland cement matrix curve (O) seems to follow the Arrhenius linear relation, but not the two other matrices, showing two slopes in the plots, suggesting a two-stage process. Purnell et al. (1999) concluded therefore that the effect of temperature is not just the acceleration of one type of process.

Based on these observations, Purnell et al. (1999) suggested a model that is based on the enlargement of preexisting flaws on the surface of the filaments, which takes place by a stress corrosion or static fatigue mechanism, which could better simulate the experimental results (Figure 9.28). They reported that this conclusion is supported by SEM observations that in some of the modified matrices, there was growth of hydration products, and yet, the aging in mechanical properties took a different course than that of a Portland cement matrix (Purnell et al., 2000).

Several subfailure driving stresses for such processes were considered: (1) thermal stresses developed at the higher temperature aging, or during temperature cycling in service, due to mismatch in the thermal coefficient of expansion of the filaments and the matrix; (2) precipitation of CH and its nucleation at preexisting flaws; and (3) preferential leaching of components from the glass surface. Based on the relation between the rate of growth of a critical flaw, which is dependent on the induced stress, OH^- concentration, and the temperature, and invoking a fracture mechanics approach, they determined the aging curves in terms of a characteristic parameter for each matrix, k, which was

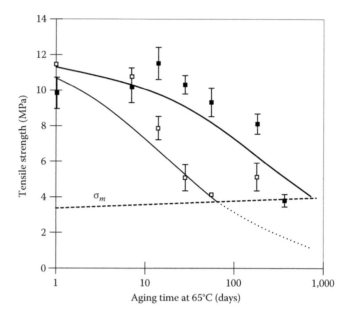

Figure 9.28 Modeling the degradation of aged GRC made of modified matrices by a model based on flaw enlargement. Continuous lines, derived from model; bold, matrix M; fine, matrix O; points, experimental data; closed, matrix M; open, matrix O; σ_m, matrix strength (BOP). (From Purnell, P. et al., *Compos. Appl. Sci. Manuf.*, 30, 1073, 1999.)

obtained by curve fitting. The *k* parameter was shown to fit Arrhenius-type equations for aging at different temperatures, for the various matrices studied.

The notching of the glass filaments as an aging mechanism was also suggested by Yilmaz and Glasser (1991), based on the observation of growth of CH crystals at the glass surface, which seem to be inducing notches into it. This may be consistent with the observations by Purnell et al. (2000).

9.3.2.3 Combined mechanisms

The two mechanisms identified earlier may be acting simultaneously, or in sequence, and therefore it is difficult perhaps to distinguish between them, since they are all associated with deposition of hydration products between the filaments in the strand.

Leonard and Bentur (1984) observed that when the hydration products grown are porous in nature (as is the case in modified Portland cement matrix), the rate of degradation in mechanical properties is reduced. This was explained in terms of lower level of bond increase induced by such products, relative to that obtained when $Ca(OH)_2$ is deposited, which is characterized by a dense crystalline structure and intimate contact with the filaments. However, one can also make the argument that such porous hydration products are not inducing stress concentrations and notches, consistent with the model of Purnell et al. (2001). Therefore, one cannot rule out that the increase in bond and notching/crack growth mechanisms are occurring simultaneously, with both being dependent on the nature of hydration products deposition. Van Itterbeeck et al. (2009) and Purnell et al. (2008) presented a unified model that is based on the growth of flaws whether induced by chemical attack or growth of hydration products. The model was calibrated against experimental results.

In an extreme case, when the spaces between the filaments were filled with minute silica fume particles that did not react with the surrounding matrix (Bentur, 1989), no aging occurred, although the spaces were filled with relatively dense material. This is an indication that indeed, it is not just filling of the space between the filaments that leads to aging, but rather the nature of the material deposited, and its interaction with the glass. A weak interaction will not lead to aging, and this can be predicted in terms of the various mechanisms suggested, whether bonding or crack growth.

In conjunction with the two-stage process observed by Purnell et al. (1999) in the modified matrices, one should note that this has been quantified by considering the bonding and flexural mechanism simultaneously: the bond mechanism leading to an increase in strength at the early stage of aging and the flexural mechanism becoming more effective later on, leading to strength reduction. Such an approach was used to model the aging of carbon fibers, and the same concept might be applied to glass fibers, to explain an increase in strength upon aging, followed by a decrease in extremely dense matrices (Katz and Bentur, 1996).

9.3.3 Effectiveness of various aging mechanisms in controlling long-term performance

The question of the relative importance of the two major aging mechanisms described earlier is not clear, since both processes, fiber degradation and microstructural changes, occur at the same time, and it is difficult to devise a critical experiment to distinguish between the two. However, a critical analysis of the data published in the literature enables us to reach some conclusions at least regarding the qualitative trends of the effects of the two mechanisms.

In the case of GRC with AR glass fibers, one would expect that if chemical attack was a major factor in controlling the long-term performance, the durability would be sensitive to the pH of the matrix. Proctor et al. (1982) reported that there is such a correlation (Table 9.2), while Leonard and Bentur (1984) showed that differences in durability characteristics could not be correlated with the pH of the matrix (Table 9.4). This is consistent also with observations by Purnell et al. (1999) that when using modified matrices

Table 9.4 Effect of cement slurry pH on strength retention of CemFIL-1 fiber

Cement	pH	SIC strength (MPa)[a]
RHPC	13.0	500
RHPC[b] + 20% Danish diatomite	12.6	700
Supersulfated cement	12.0	800
RHPC[b] + 40% precipitated silica	11.0	950

Source: Proctor, B.A. et al., *Composites*, 13, 173, 1982.

[a] Strand-in-cement strength after accelerated aging for 2 months in 50°C water; similar trends were observed for the performance of the GRC composite.
[b] Rapid hardening Portland cement.

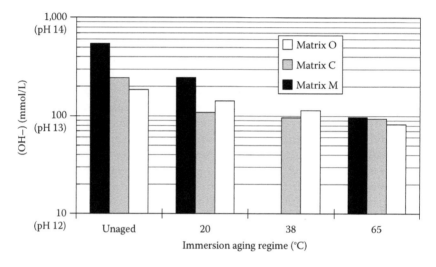

Figure 9.29 Comparison of pore solution pH of the GRC composite with Portland cement matrix (O) and the modified matrices (C and M), unaged and after aging for 6 months. (From Purnell, P. et al., *Compos. Appl. Sci. Manuf.*, 30, 1073, 1999.)

where the difference in aging characteristics was drastic, the difference in the pH was significant only during the first period of aging (Figure 9.29).

When considering such apparent contradictions, it should be borne in mind that the changes in pH were obtained by variations in the composition of the matrix, which were accompanied also by changes in the microstructure. Therefore, one cannot rule out the possibility that the improved aging performance of the lower pH matrices in Table 9.4 is the result of differences in the microstructure of the matrix around the glass filaments. The differences in the performance of the GRC composites whose results are shown in Table 9.5 were explained on the basis of differences in microstructure

Table 9.5 Effect of matrix pH on retention of flexural strength and toughness of GFRC composite with AR glass fiber

Cement	pH	Retention of mechanical property (% of unaged composite)	
		Flexural strength	Toughness
Portland cement	13.02	44	5
Portland cement + 35% fly ash B	12.91	48	29
Portland cement + 35% fly ash A	12.93	92	76

Source: Leonard, S. and Bentur, A., Cement Concr. Res., 14, 717, 1984.

(Leonard and Bentur, 1984). The marked improvement in the durability of composites made using the Japanese low-alkali CGC cement was attributed to microstructural changes at the interface rather than to its low alkalinity (Hayashi et al., 1985; Tanaka and Uchida, 1985).

It is evident that some loss in strength of the AR glass fibers in aged GRC is taking place (West and Majumdar, 1982) (Figure 9.7). It is interesting to note that the curve appears to be leveling off, which is similar to the trend observed for the loss of the flexural strength of the composite. On the basis of extensive analysis, Litherland et al. (1981) demonstrated that there is a quantitative relation between the strength retention of the glass fibers (estimated by the SIC test) and the flexural strength retention of the GRC composite. This may indicate that chemical degradation is a dominant factor controlling the durability of GRC.

However, if the toughness of the composite is considered, the GRC becomes brittle at about the point at which the decline in its modulus of rupture has leveled off (Bentur et al., 1985; BRE, 1976). These trends are demonstrated in Figure 9.30, which show that the composite loses practically all of its toughness, but only part of its flexural strength. Thus, although a correlation can be established between the flexural strength of the composite and the strength of the glass fibers (Litherland et al., 1981), no such correlation is evident between the glass fiber strength and the toughness of the composite. In spite of the fact that the glass fibers retain a considerable portion of their strength even after prolonged aging (Figure 9.7), the composite becomes brittle and loses more than 95% of its original toughness. The embrittlement observed at this stage can be accounted for by microstructural changes. In the specific example in Figure 9.30, at the time the composite became brittle (1/2 year in water at 20°C), the filaments were engulfed with massive CH. In several other studies (Bentur et al., 1985; Hayashi et al., 1985; Leonard and Bentur, 1984), it was also shown that improvement in the durability of GRC composites made with different types of AR glass fibers or blended cement matrices could be correlated with microstructural changes, but not with chemical attack on the glass fiber.

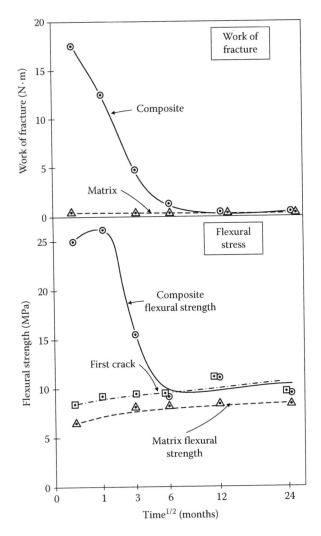

Figure 9.30 Effect of aging in water at 20°C on the flexural strength and work of fracture (area under the load–deflection curve) of GRC composite with AR glass fiber. (From Bentur, A. et al., *J. Am. Ceram. Soc.*, 68, 203, 1985.)

Examination of the glass surface of filaments removed from composites after 1/2 year of aging in water at 20°C, when some of them became brittle (CemFIL-1), while others remained ductile (CemFIL-2), showed in all of them hardly any visible surface damage (Bentur et al., 1985). In a few filaments, some roughening of the surface was seen, but such roughening was shown to result only in a limited strength loss of the fibers. On the other hand, in the composites that became brittle after 1/2 year in water at 20°C, dense CH had engulfed the filaments and the failure mode was brittle (Bentur et al., 1985). In the other composites that at this stage of aging were still maintaining most of their toughness, hardly any hydration products were deposited around the filaments and the failure mode was ductile.

It should be noted that after prolonged aging (5 months at 50°C water), signs of more severe chemical attack could be observed in all of the GRC composites (Bentur, 1986). However, the embrittlement of the composites occurred at a much earlier stage, when no such damage was seen, but when the matrix around the glass filaments became dense.

This discussion suggests that the major cause of the embrittlement of GRC reinforced with AR glass fibers is the microstructural changes resulting in deposition of $Ca(OH)_2$ around the glass filaments. At the stage when the composite becomes brittle (~5–30 years of natural weathering or ~1 year in water at 20°C), chemical attack of the glass fibers is very mild. There is evidence that prolonged exposure to a wet environment may result in a marked increase in the damage to the glass fibers due to the chemical attack. However, the embrittlement occurs at an earlier stage, when the chemical attack is very mild, if it occurs at all.

In a study of the effect of non-Portland cement matrices on the durability of E- and AR glass fibers (Bentur et al., 1997), it was shown that in the case of E-glass, a markedly improved performance was obtained when the alkalinity of the matrix was sufficiently low to delay or prevent alkali attack and the spaces between the filaments were not filled with hydration products, or if filled with hydration products, they tended to be very porous and friable. In the case of AR glass fibers, the condition for marked durability improvement was associated mainly with the latter condition, either (1) prevention of filling of the spaces between the filaments or (2) deposition of hydration products that are porous and friable.

Table 9.6 Effect of aging mechanisms on reduction in mechanical properties of GRC

Type of fiber	Aging period	Effect of aging mechanism on reduction in mechanical properties	
		Chemical degradation of fibers	Growth of dense hydration products
E-glass	Short (<1 year)	Very effective	Mildly effective
AR glass	Short (<1 year)	Not effective	Not effective
	Medium (5–40 years)	Mildly effective	Very effective
	Long (30–50 years)	Effective	Very effective

Source: Bentur, A. and Mindess, S., *Fibre Reinforced Cementitious Composites*, 2007.

Based on the studies reviewed earlier, the effectiveness of the various mechanisms involved in loss in strength and toughness is summarized in Table 9.6 (Bentur and Mindess, 2007). The table provides a qualitative assessment of the influence of these processes, for cement composite with E- and AR glass fibers, at different periods of natural aging. These aging periods should be viewed as rough estimates, giving an order of magnitude only. The filling of spaces referred to in this table is for dense hydration products.

9.4 Long-term performance of TRC components

TRC composites are applied in structures in a variety of ways, ranging from stand-alone thin-sheet components, through elements that are incorporated in reinforced concrete structures such as in permanent formwork, to materials that are used for repair of deteriorated structures. In all of these applications, the TRC is expected to deliver protection, namely, isolate the structure from the environment by preventing penetration of fluids and deleterious materials, and achieving all of these when it is used as a thin-sheet component. This is of particular importance when used as a repair material and combined in a reinforced concrete element where it is expected to act as an external skin to facilitate protection of the reinforcing steel while keeping the cover quite small in depth.

The characteristics of TRC make it potentially suitable to meet these challenges: its matrix is usually made of relatively low w/c ratio, 0.40–0.45, and in many instances it contains mineral additives that are known to improve durability performance by reducing the penetrability into the cementitious material. On top of it, the composite is strain hardening in nature, which implies that during service its cracking will be controlled to be within a small width, much smaller than those occurring in conventional reinforced concrete. These characteristics are common to most of the currently available strain-hardening cement composites (SHCC), whether made of fabric reinforcement or discrete short fibers, and thus these aspects of performance of TRC can be highlighted by considering also the know-how available in strain-hardening composites in general.

Several overviews of SHCC have been reported in recent years (Ahmed and Mihashi, 2007; Mechtcherine, 2012; van Zijl et al., 2012). In these publications, special emphasis was given to the resistance of these materials to penetration, considering especially the impact of the reduced cracking on penetration of water, air, and gas, as well as chloride ions. Within this context, special attention was given to the ability of cracks to heal, which is more likely to be feasible in systems where crack width is small. These aspects are highlighted in the following section.

9.4.1 Penetration of fluids

The penetration of fluids has been evaluated with respect to the cracking of the TRC composite, considering the special advantages that can be imparted due to the superb crack control in this composite. In conventional reinforced concrete, cracking is often the weak link in assuring reduced penetration. The concrete matrix can be made extremely impermeable by the use of low water/binder ratio mixes, but this is of limited advantage if cracks cannot be controlled. This is well reflected in standards that call upon restriction of crack width to the range of 0.1–0.4 mm (Ahmed and Mihashi, 2007).

The influence of crack width on the transport properties is presented in Figure 9.31, which is based on compilation of various references.

It should be noted that when the crack width is 0.1 mm or bigger (Figure 9.31), the penetration increases by more than an order of magnitude relative to the matrix. However, the codes and standards recommend 0.1 mm width only in extreme exposure conditions. In TRC composites, the crack width can be controlled to be even at smaller values, demonstrating the marked improved performance that can be achieved in this composite.

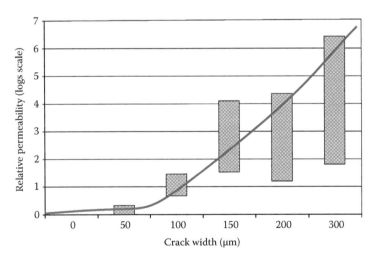

Figure 9.31 Effect of crack width on relative permeability of water flow, based on compilation from the literature. (From Bentur, A. and Mitchel, D., *Cement Concr. Res.*, 38, 259, 2008; Compiled from Schiessl, P., Cracking of concrete and durability of concrete structures, in *AFREM-CCE*, St Remy les Chevreuse, France, 1988; Aldea, C.M. et al., *Mater. Struct.*, 32(219), 370, 1999; Reinhardt, H.W. and Jooss, M., *Cement Concr. Res.*, 32, 981, 2002; Schiessl, P. and Raupach, M., *ACI Mater. J.*, 94(1), 56, 1997; Burlion, N. et al., *Cement Concr. Res.*, 33, 679, 2003.)

The strong dependency on crack width can be accounted for by model calculation of flow through cracks, which is theoretically a function of the cube of the crack width:

$$q_o = \frac{gIlw^3}{12v} \qquad (9.1)$$

where

q_o is the rate of water flow through an idealized smooth crack, m²/s
g is the gravity acceleration
I is the pressure gradient, h/d, where h is the height of the fluid column on the inlet side and d is the crack length in the flow direction (i.e., the thickness of the specimen in a permeability test)
l is the crack length at right angle to the flow direction
w is the crack width
v is the kinematic viscosity, m²/s

However, for cementitious systems, especially at the low width values, the observed flow is much smaller than predicted by the theoretical equation, and this has been explained in terms of the microstructure of the crack, which is not the form of ideal separation by two parallel planes, but rather a tortuous structure, with zones that might be smaller than the imprint seen on the surface of the material. Therefore, it is common to make a correction to the equation using a flow coefficient ξ, which is between 0 and 1 and accounts for the tortuosity of the crack:

$$q_o = \xi \frac{gIlw^3}{12v} \qquad (9.2)$$

Mihashi et al. (2003) have shown that for steel fibers, this coefficient could be extremely low. It could be expected that the tortuosity would be bigger for smaller crack width, implying that flow would be inversely proportional to the power of 4 of the crack width. However, Mechtcherine and Lieboldt (2011) could not resolve a clear-cut dependence of this coefficient with the imposed strain level but confirmed that for TRC composites with crack width in the range of up to 125 μm, the coefficient is extremely low, that is, in the range of 0.28×10^{-3} to 0.75×10^{-3}.

The evaluation of the effect of crack width in SHCC in general and TRC in particular has been carried out by studying the flow of fluids through composites that were precracked, either by tensile loading or by restrained shrinkage. Some characteristic values reported by Lieboldt and Mechtcherine (2010) for tensile loading of glass TRC are 13–65 μm for 0.2% strain and 46–125 μm for 0.4% strain. These values depend on the nature of the yarn. Yarns of a smaller number of filaments (1,600 vs. 3,200) give a composite

Table 9.7 Average crack width obtained in restrained shrinkage test of TRC composites in which the material is placed on a rack with roughened surface geometry

	Plain matrix	PE	Steel	Coated carbon	AR glass	Noncoated carbon	PVA bundle	PVA mesh
Average crack width (µm)	175.5 (33)	112.6 (31)	49.3 (35)	38.4 (23)	66.6 (28)	35.5 (24)	8.8 (4)	14.6 (4)

Source: Pourasee, A. et al., *J. Mater. Civil Eng.*, 23(8), 1227, 2011.

Note: The numbers in parenthesis denote the coefficient of variation in %.

with smaller values in the range mentioned. Pourasee et al. (2011) generated cracking in laboratory tests using restrained shrinkage conditions, by having the TRC produced on a rough steel rack. They observed also cracks with width smaller than 100 µm (Table 9.7), which is much smaller than that obtained for plain matrix at 175 µm (Table 9.7).

The effect of the nature of the yarn in the fabric and the strain in the TRC composite on oxygen and water permeability was reported by Lieboldt and Mechtcherine (2010) and Mechtcherine and Lieboldt (2011). The permeability tests were carried out while the specimens were under load. The variables were the yarn material (carbon and AR glass), the bundled nature of the AR glass yarn, and a polymer coating—aqueous styrene–butadiene dispersion at a level of 10% weight. The effect of these parameters at different strain levels is presented in Figures 9.32 and 9.33 for oxygen and water permeability, respectively.

The permeability to both fluids shows a nonlinear marked increase, which starts at about 0.2% strain. It can be seen that yarns with lower tex value increase the permeability. These trends can be explained in terms of increase in crack width induced by these changes: coating reduces the bond, while the decrease in tex reduces the reinforcement content (the same number of fabric layers were used in the study, and therefore lower tex implies smaller reinforcement content). Compiling the data and presenting all of it in terms of oxygen permeability versus crack width suggest that a single curve can account for the observations in the different systems. This implies that crack width is the most important parameter in controlling permeability. The shape of the curve is very similar to that drawn on the basis of compilation of data of other cementitious systems (Figure 9.31), and both indicate that a marked rise in permeability starts to take place when the crack width exceeds to the 50–100 µm range.

Pourasee et al. (2011) studied the ingress of water through cracks in loaded TRC made from multifilament yarn fabrics of carbon, AR glass, PVA, and PE. They characterized the movement path of the water by imaging it using X-ray radiography. They observed that water penetrates through the cracks

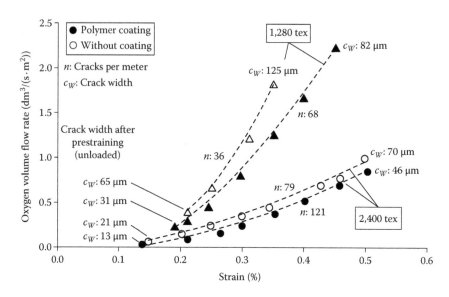

Figure 9.32 Oxygen permeability of TRC with different types of AR glass fabric reinforcement under different levels of tensile loading. (From Lieboldt, M. and Mechtcherine, V., Transport of liquids and gases through textile reinforced concrete, in *International RILEM Conference on Materials Science (MATSCI)*, Aachen, Germany, ICTRC, Vol. 1, 2010, pp. 343–351; Mechtcherine, V. and Lieboldt, M., *Cement Concr. Compos.*, 33, 725, 2011.)

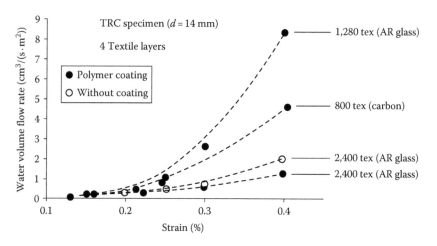

Figure 9.33 Water permeability of TRC with different types of AR glass fabric reinforcement under different levels of tensile loading. (From Lieboldt, M. and Mechtcherine, V., Transport of liquids and gases through textile reinforced concrete, in *International RILEM Conference on Materials Science (MATSCI)*, Aachen, Germany, ICTRC, Vol. 1, 2010, pp. 343–351; Mechtcherine, V. and Lieboldt, M., *Cement Concr. Compos.*, 33, 725, 2011.)

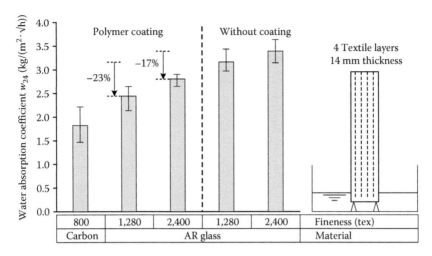

Figure 9.34 Capillary absorption of TRC through the yarns showing the effect of yarn fineness and coating. (From Lieboldt, M. and Mechtcherine, V., Transport of liquids and gases through textile reinforced concrete, in *International RILEM Conference on Materials Science (MATSCI)*, Aachen, Germany, ICTRC, Vol. 1, 2010, pp. 343–351; Mechtcherine, V. and Lieboldt, M., *Cement Concr. Compos.*, 33, 725, 2011.)

and thereafter follows a path within the yarn through the open spaces in the bundled filaments. Coating of the yarns slowed down this process and full coating prevented this movement. These observations are consistent with the reports by Lieboldt and Mechtcherine (2010) and Mechtcherine and Lieboldt (2011) showing a marked capillary water absorption through the bundled yarn when immersing the composite in water with the yarns perpendicular to the water surface (Figure 9.34). Increasing the yarn size (higher tex) increases the absorption (more spaces for penetration), while coating reduces it (spaces eliminated or reduced).

The reported trends of penetration through the spaces in the multifilament yarns raise the issue whether this mechanism of penetration should be taken into account in addition to penetration through the matrix and cracks. The current discussion suggests that the crack width is the overriding parameter. However, there is room to further validate that this is indeed the case.

9.4.2 Penetration of chlorides

Several studies have compared the performance of strain-hardening cement composites and fiber-reinforced concretes with respect to resistance to chloride penetration. Special attention was given to evaluation with respect to the performance of normal- and high-strength concrete to address the

Table 9.8 Effect of deflection on crack width and effective chloride diffusion for conventional concrete and SHCC of the ECC type

Type of composite	Beam deformation (mm)	Average crack width (μm)	Effective diffusion coefficient (m²/s 10⁻¹²)
Mortar	0.00	0	10.58
	0.50	~50	33.28
	0.70	~150	35.54
	0.80	~300	126.53
	0.83	~400	205.76
ECC	0.00	0	6.75
	0.50	~0	8.10
	1.00	~50	27.99
	1.50	~50	37.50
	2.00	~50	54.22

Source: Sahmaran, M. et al., *ACI Mater. J.*, 104(6), 303, 2007.

influence of cracks. Although these were not done directly on TRC, one would expect that the properties of SHCC with respect to cracking would be similar, whether TRC or some other SHCC. Sahmaran et al. (2007) carried out ponding and immersion tests in chloride solution evaluating the performance of conventional concrete and SHCC (ECC) beams that were loaded in flexure, to determine chloride transport in the cracked material. The crack width–deflection relations of the two materials can be evaluated from the data in Table 9.8, which shows the beam deflection of conventional mortar and ECC, along with the crack width and the resulting effective diffusion coefficient. For the same deflection, the diffusion is much higher in the conventional beam and increases with deflection, while the diffusion in the SHCC is kept low and increases very mildly with deflection. This difference in behavior can be clearly correlated with the cracking behavior: the SHCC can undergo deflection without increasing the crack width, as expected in a composite that undergoes multiple cracking. Keeping the crack width as low as 50 μm enables to keep the diffusion low while the beam is undergoing considerable deflection.

9.4.3 Crack healing

Studies dealing with penetration through cracks have resolved that when the crack width is sufficiently small, usually less than 100 μm, crack healing can take place (Bentur and Mindess, 2007; Jia et al., 2010; Kunieda et al., 2012; Lieboldt and Mechtcherine, 2010; Mechtcherine and Lieboldt, 2011; Mihashi and Nishiwaki, 2012; Nishiwaki et al., 2012; Reinhardt and Jooss, 2003; Yang et al., 2009; Yu et al., 2010). The cracks have their own microstructure, which is tortuous and not necessarily parallel walled. The tortuous nature shows up in width that is changing along the crack, being different than the imprint observed on the surface. Also, the crack surfaces that have some contacts with each other are able to transfer load by interlocking mechanism. Such effects have shown up also in fracture mechanics testing where stress transfer across cracks was seen to be taking place, that is, a tension-softening behavior.

These effects can take place when the cracks are sufficiently small, apparently less than about 100 µm. Since there are always remnants of unhydrated grains, continued hydration may supply sufficient hydration products to deposit in these narrow spaces to heal the crack. The fact that the mechanics of TRC leads to crack control by the multiple-cracking process to width less than 100 µm makes this class of composites particularly effective for generating crack healing. This also shows up in the very low tortuosity coefficient, being less than 1×10^{-3}.

Lieboldt and Mechtcherine (2010) and Mechtcherine and Lieboldt (2011) explored the nature of self-healing in TRC composites. The effectiveness of healing could be readily quantified and confirmed by measuring the water penetration as a function of time, as seen in Figure 9.35. The decline is attributed to crack healing, and this could be confirmed by SEM observation showing the closure of 20 µm cracks after about 1 month, with the closure being facilitated by deposits of calcium carbonates. The slower rate of self-healing shown in Figure 9.35 takes place in composites where the crack

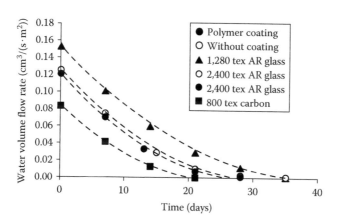

Figure 9.35 Time-dependent reduction in the permeability to water of cracked TRC. (From Lieboldt, M. and Mechtcherine, V., Transport of liquids and gases through textile reinforced concrete, in *International RILEM Conference on Materials Science (MATSCI)*, Aachen, Germany, ICTRC, Vol. 1, 2010, pp. 343–351; Mechtcherine, V. and Lieboldt, M., *Cement Concr. Compos.*, 33, 725, 2011.)

width was initially bigger, that is, composites with polymer coating of yarns having lower reinforcement content (smaller tex values).

These trends and observations are very similar to those reported for other types of strain-hardening cement composites produced with short fibers, such as ECC, where the cracks are usually 100 μm and less.

Yu et al. (2010) and Yang et al. (2009) reported the formation of calcium carbonate in the cracks, which accounted for self-healing. In the study of this class of strain-hardening composites, the environment for self-healing was simulated by wetting and drying cycles of cracked composites, and self-healing was quantified in terms of testing the resonance frequency and its restoration to levels before cracking as well as water permeability. Tests were carried out in a system where a single crack was formed and studied as well as in the composite itself.

Single crack testing results are presented in Figure 9.36. The effect of cracking on reduction in the resonant frequency before conditioning is compared with the resonant frequency after wet–dry cycling. It can be clearly seen that if the crack width is kept below 50 μm, full recovery of the

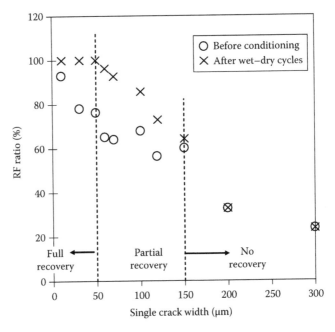

Figure 9.36 Effect of crack width on the effectiveness of self-healing as estimated by frequency resonant measurements. (From Yang, Y. et al., *Cement Concr. Res.*, 39, 382, 2009.)

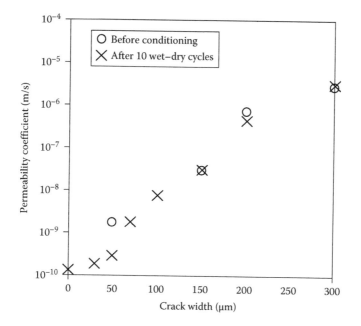

Figure 9.37 The effect of crack width on the effectiveness of self-healing as estimated by water permeability measurements. (After Yang, Y. et al., *Cement Concr. Res.*, 39, 382, 2009.)

resonant frequency can take place. In the cracking range between 50 and 150 μm only partial recovery can occur, while above 150 μm no recovery can take place.

Similar trends of the effectiveness of crack healing could be obtained by assessing the permeability to water (Figure 9.37). Full recovery of the permeability characteristics could be obtained when the crack width was 50 μm. At 150 μm crack, no recovery was observed.

The required number of cycles to achieve full recovery by self-healing was about four, as seen in Figure 9.38, for specimens loaded initially at strain levels of up to 3.0%. The maximum crack widths corresponding to the different strain levels were 50, 70, 60, 60, and 90 μm for 0.3%, 0.50%, 1.0%, 2.0%, 3.0% strain, respectively. It can be seen that when the maximum crack width is 90 μm (3.0% strain), full recovery does not take place.

The trends reported for the strain-hardening composites are similar to ones reported for cracked cementitious matrices, showing that above 100 μm cracking complete self-healing is unlikely to take place, as demonstrated, for example, in Figure 9.39.

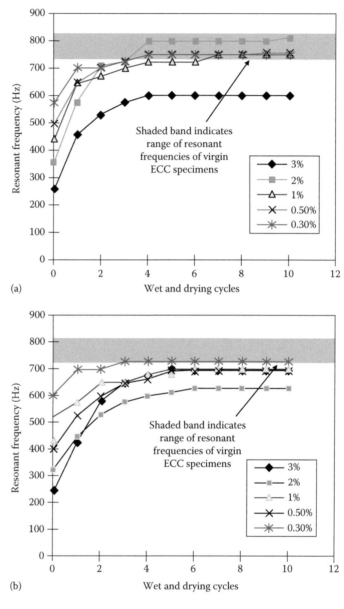

Figure 9.38 Recovery of resonant frequency as a function of the number of wet/dry cycles: (a) water/air cycles and (b) water/hot air cycles. (After Yang, Y. et al., *Cement Concr. Res.*, 39, 382, 2009.)

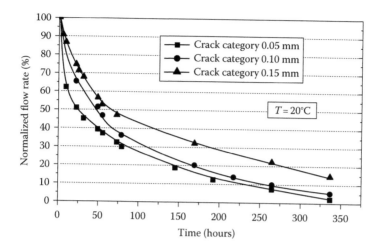

Figure 9.39 Decrease in the normalized flow rates due to self-healing of concrete specimens loaded to crack width of 50, 100, and 150 μm. (From Reinhardt, H.W. and Jooss, M., *Cement Concr. Res.*, 32, 981, 2002.)

References

Ahmed, S. F. U. and Mihashi, H., A review on durability properties of strain hardening fibre reinforced cementitious composites, *Cement and Concrete Composites*, 29, 265–376, 2007.

Aldea, C. M., Shah, S. P., and Karr, A., Permeability of cracked concrete, *Materials and Structures*, 32(5), 370–376, 1999.

Banholzer, B., Bond of a multi-filament yarn embedded in a cementitious matrix, Doctoral dissertation, RWTH Aachen, Aachen, Germany, 2004.

Banholzer, B. and Brameshuber, W., Tailoring of AR-glass filament/cement based matrix bond—Analytical and experimental techniques, in *Sixth RILEM Symposium on Fibre-Reinforced Concrete (FRC)—BEFIB*, Varenna, Italy, September 20–22, 2004, pp. 1443–1452.

Bartos, P. J. M., Telescopic failure in a bundled fibre reinforcement, in *From Materials Science to Construction Materials: Proceedings of the RILEM Symposium*, Paris, France, Chapman & Hall, London, U.K., 1987, pp. 539–546.

Bentur, A., Mechanisms of potential embrittlement and strength loss of glass fibre reinforced cement composites, in S. Diamond (ed.), *Proceedings: Durability of Glass Fibre Reinforced Concrete Symposium*, Prestressed Concrete Institute, Chicago, IL, 1986, pp. 109–123.

Bentur, A., Silica fume treatments as means for improving durability of glass fibre reinforced cements, *Journal of Materials in Civil Engineering*, 1, 167–183, 1989.

Bentur, A., Durability of fibre reinforced cementitious composites, in J. P. Skalny and S. Mindess (eds.), *Materials Science in Concrete—V*, The American Ceramic Society, Westerville, OH, 1998, pp. 513–536.

Bentur, A., Ben Bassat, M., and Schneider, D., Durability of glass fibre reinforced cements with different alkali resistant glass fibres, *Journal of the American Ceramic Society*, 68, 203–208, 1985.

Bentur, A. and Diamond, S., Fracture of glass fibre reinforced cement, *Cement and Concrete Research*, 14, 31–42, 1984.

Bentur, A. and Diamond, S., Effects of direct incorporation of microsilica into GFRC composites on retention of mechanical properties after aging, in S. Diamond (ed.), *Proceedings: Durability of Glass Fibre Reinforced Concrete Symposium*, Prestressed Concrete Institute, Chicago, IL, 1985, pp. 337–351.

Bentur, A. and Diamond, S., Effect of aging of glass fibre reinforced cement on the response of an advancing crack on intersecting a glass fibre strand, *International Journal of Cement Composites and Lightweight Concrete*, 8(4), 213–222, 1986.

Bentur, A. and Diamond, S., Direct incorporation of silica fume into strands as a means for developing GFRC composites of improved durability, *International Journal of Cement Composites and Lightweight Concrete*, 9, 127–136, 1987a.

Bentur, A. and Diamond, S., Aging and microstructure of glass fibre cement composites reinforced with different types of glass fibres, *Durability of Building Materials*, 4, 201–226, 1987b.

Bentur, A., Kovler, K., and Odler, I., Durability of some glass fibre reinforced cementitious composites, in K. C. G. Ong, J. M. Lau, and P. Paramasivam (eds.), *Fifth International Conference on Structural Failure, Durability and Retrofitting*, Singapore, 1997, pp. 190–199.

Bentur, A. and Mindess, S., *Fibre Reinforced Cementitious Composites*, Taylor and Francis, London and New York, 2007.

Bentur, A. and Mitchel, D., Materials performance lessons, *Cement and Concrete Research*, 38, 259–272, 2008.

Bentur, A., Tirosh, R., Yardimci, M., Puterman, M., and Peled, A., Bonding and microstructure in textile reinforced concrete, in W. Brameshuber (ed.), *Textile Reinforced Concretes: Proceedings of the International RILEM Conference on Materials Science*, Aachen, Germany, RILEM Publications, Vol. 1, 2010, pp. 23–33.

Bentur, A., Yardimci, M., and Tirosh, R., Preservation of telescopic bonding upon aging of bundled glass filaments by treatments with nano-particles, *Cement and Concrete Research*, 47, 69–77, 2013.

Bergstrom, A. S. G. and Gram, H. E., Durability of alkali-sensitive fibres in concrete, *International Journal of Cement Composites and Lightweight Concrete*, 6, 75–80, 1984.

Bijen, J., A survey of new developments in glass composition, coatings and matrices to extend service lifetime of GFRC, in S. Diamond (ed.), *Proceedings: Durability of Glass Fibre Reinforced Concrete Symposium*, Prestressed Concrete Institute, Chicago, IL, 1985, pp. 251–269.

Brandt, A. M. and Glinicki, M. A., Effects of pozzolanic additives on long-term flexural toughness of HPGFRC, in A. E. Naaman and H. W. Reinhardt (eds.), *Fourth International Workshop on High Performance Fibre Reinforced Cement Composites (HPFRCC 4)*, Ann Arbor, MI, RILEM Publications, Bagneux, France, 2003, pp. 399–408.

BRE, A study of the properties of Cem-FIL/OPC composites, Building Research Establishment Current Paper CP 38/76, Building Research Establishment, Watford, England, 1976.

Burlion, N., Skoczylas, F., and Dubois, T., Induced anisotropic permeability due to drying of concrete, *Cement and Concrete Research*, 33, 679–687, 2003.

Butler, M., Hempel, S., and Mechtcherine, V., Modeling of ageing effects on crack-bridging behavior of AR-glass multifilament yarns embedded in cement-based matrix, *Cement and Concrete Research*, 41, 403–411, 2011.

Butler, M., Mectcherine, V., and Hempel, S., Experimental investigations on the durability of fibre-matrix interfaces in textile reinforced concrete, *Cement and Concrete Composites*, 31, 221–231, 2009.

Butler, M., Mechtcherine, V., and Hempel, S., Durability of textile reinforced concrete made with AR glass fibre: Effect of the matrix composition, *Materials and Structures*, 43, 1351–1368, 2010.

Buttner, T., Orlowsy, J., Raupach, M., Hojczyk, M., and Weichold, O., Enhancement of the durability of alkali-resistant glass-rovings in concrete, in *International RILEM Conference on Material Science (MATSCI)*, Aachen, Germany, ICTRC, Vol. 1, 2010, pp. 333–342.

Cohen, Z. and Peled, A., Controlled telescopic reinforcement system of fabric-cement composites: Durability concerns, *Cement and Concrete Research*, 40(10), 1495–1506, 2010.

De Vekey, R. C. and Majumdar, A. J., Interfacial bond strength of glass fibre reinforced cement composites, *Journal of Materials Science*, 5, 183–185, 1970.

Franke, L. and Overbeck, E., Loss in strength and damage to glass fibres in alkaline solutions and cement extracts, *Durability of Building Materials*, 4, 73–79, 1987.

Gao, S. L., Mader, E., and Plonka, R., Nanostructures coatings of glass fibres: Improvement of alkali resistance and mechanical properties, *Acta Materialia*, 55, 1043–1052, 2007.

Hannant, D. J., Durability of polypropylene fibres in Portland cement-based composites: Eighteen years of data, *Cement and Concrete Research*, 28(12), 1809–1817, 1998.

Hayashi, M., Sato, S., and Fujii, H., Some ways to improve durability of GFRC, in S. Diamond (ed.), *Proceedings: Durability of Glass Fibre Reinforced Concrete Symposium*, Prestressed Concrete Institute, Chicago, IL, 1985, pp. 270–284.

Hull, D., *An Introduction to Composite Materials*, Cambridge Solid State Science Series, Cambridge University Press, Cambridge, England, 1981.

Jaras, A. C. and Litherland, A. C., Microstructural features in glass fibre reinforced cement composites, in A. Neville (ed.), *Fibre Reinforced Cement and Concrete: Proceedings of the RILEM Symposium*, The Construction Press, Lancaster, England, 1975, pp. 327–334.

Jia, H. Y., Chen, W., Yu, M. X., and Hua, Y. E., The microstructure of self-healed PVA ECC under wet and dry cycles, *Materials Research*, 13(2), 225–231, 2010.

Katz, A. and Bentur, A., Mechanisms and processes leading to changes in time in the properties of carbon fibre reinforced cement, *Advanced Cement Based Materials*, 3(1), 1–13, 1996.

Kunieda, M., Choonghyun, K., Ueda, N., and Nakamura, H., Recovery of protective performance of cracked ultra high performance-strain hardening cementitious composites (UHP_SHCC) due to autogenous healing, *Journal of Advanced Concrete Technology*, 10, 313–322, 2012.

Larner, L. J., Speakman, K., and Majumdar, A. J., Chemical interactions between glass fibres and cement, *Journal of Non-Crystalline Solids*, 20, 43–74, 1976.

Leonard, S. and Bentur, A., Improvement of the durability of glass fibre reinforced cement using blended cement matrix, *Cement and Concrete Research*, 14, 717–728, 1984.

Lieboldt, M. and Mechtcherine, V., Transport of liquids and gases through textile reinforced concrete, in *International RILEM Conference on Materials Science (MATSCI)*, Aachen, Germany, ICTRC, Vol. 1, 2010, 343–351.

Litherland, K. L., Maguire, P., and Proctor, B. A., A test method for the strength of glass fibres in cement, *International Journal of Cement Composites and Lightweight Concrete*, 6, 39–45, 1984.

Litherland, K. L., Oakley, D. R., and Proctor, B. A., The use of accelerated aging procedures to predict the long term strength of GRC composites, *Cement and Concrete Research*, 11, 455–466, 1981.

Majumdar, A. J., Properties of GRC, in *Proceedings of the Symposium on 'Fibrous Concrete'*, The Concrete Society, London, U.K., 1980, pp. 48–68.

Majumdar, A. J. and Nurse, R. W., Glass fibre reinforced cement, Building Research Establishment Current Paper, CP79/74, Building Research Establishment, Watford, England, 1974.

Marikunte, S., Aldea, C., and Shah, S. P., Durability of glass fibre cement composites: Effect of silica fume and metakaolin, *Journal of Advanced Cement Based Materials*, 5, 100–108, 1997.

Mechtcherine, V., Towards a durability framework for structural elements and structures made of or strengthened with high-performance fibre-reinforced composites, *Construction and Building Materials*, 31, 94–104, 2012.

Mechtcherine, V. and Lieboldt, M., Permeation of water and gases through cracked textile reinforced concrete, *Cement and Concrete Composites*, 33, 725–734, 2011.

Mihashi, H. and Nishiwaki, T., Development of self-healing and self-repairing concrete—State-of-the-art report, *Journal of Advanced Concrete Technology*, 10, 170–184, 2012.

Mihashi, H., Nishiwaki, T., and de Leite, J. P. B., Effectiveness of crack control on durability of HFRCC, in A. Naaman and H. W. Reinhardt (eds.), *Fourth International Workshop on High Performance Fibre Reinforced Cement Composites (HPFRCC 4)*, Ann Arbor, MI, RILEM Publications, Bagneux, France, 2003, pp. 437–450.

Mills, R. H., Preferential precipitation of calcium hydroxide on alkali resistant glass fibres, *Cement and Concrete Research*, 11, 689–697, 1981.

Mobasher, B. and Li, C. Y., Modeling of stiffness degradation of the interfacial transition zone during fibre debonding, *Journal of Composite Engineering*, 5, 1349, 1995.

Mumenya, S. W., Tait, R. B., and Alexander, M. G., Mechanical behavior of textile concrete under accelerated ageing conditions, *Cement and Concrete Composites*, 32, 580–588, 2010.

Nishiwaki, T., Koda, M., Yamada, M., Mihashi, H., and Kikuta, T., Experimental study on self-healing capacity of FRCC using different types of synthetic fibres, *Journal of Advanced Concrete Technology*, 10, 195–206, 2012.

Oakely, D. R. and Proctor, B. A., Tensile stress-strain behaviour of glass fibre reinforced cement composites, in A. Neville (ed.), *Fibre Reinforced Cement and Concrete: Proceedings of the RILEM Symposium*, The Construction Press, Lancaster, England, 1975, pp. 347–359.

Ohno, S. and Hannant, D. J., Modeling the stress-strain response of continuous fibre reinforced cement composites, *ACI Materials Journal*, 91(3), 306–312, 1994.

Orlowsky, J., Raupach, M., Cuypers, H., and Wastiels, J., Durability modelling of glass fibre reinforcement in cementitious environment, *Materials and Structures*, 38, 155–162, 2005.

Peled, A., Jones, J., and Shah, S. P., Effect of matrix modification on durability of glass fibre reinforced cement composite, *Materials and Structures*, 38, 163–171, 2005.

Pourasee, A., Peled, A., and Weiss, J., Fluid transport in cracked fabric-reinforced cement based composites, *Journal of Materials in Civil Engineering*, 23(8), 1227–1238, 2011.

Proctor, B. A., Oakley, D. R., and Litherland, K. L., Development in the assessment and performance of GRC over 10 years, *Composites*, 13, 173–179, 1982.

Purnell, P., Buchanan, A. J., Short, N. R., Page, C. L., and Majumdar, A. R., Determination of bond strength in glass fibre reinforced cement using petrography and image analysis, *Journal of Materials Science*, 35, 4653–4659, 2000.

Purnell, P., Cain, J., van Itterbeeck, P., and Lesko, J., Service life modeling of fibre composites: A unified approach, *Composite Science and Technology*, 68, 3330–3336, 2008.

Purnell, P., Short, N. R., and Page, C. L., A static fatigue model for the durability of glass fibre reinforced cement, *Journal of Materials Science*, 36, 5385–5390, 2001.

Purnell, P., Short, N. R., Page, C. L., Majumdar, A. J., and Walton, P. L., Accelerated ageing characteristics of glass-fibre reinforced cement made with new cementitious matrices, *Composites Part A: Applied Science and Manufacturing*, 30, 1073–1080, 1999.

Rajczyk, K., Giergiczny, E., and Glinicki, M. A., The influence of pozzolanic materials on the durability of glass fibre reinforced cement composites, in A. M. Brandt, V. C. Li, and I. H. Marshal (eds.), *Proceedings of the International Symposium on Brittle Cement Composites 5*, Warsaw, Poland, BIGRAF and Woodhead Publishers, Warsaw, Poland, 1997, pp. 103–112.

Reinhardt, H. W. and Jooss, M., Permeability and self-healing of cracked concrete as a function of temperature and crack width, *Cement and Concrete Research*, 32, 981–985, 2002.

Sahmaran, M., Li, M., and Li, V. C., Transport properties of engineered cementitious composites under chloride exposure, *ACI Materials Journal*, 104(6), 303–310, 2007.

Scheffler, S., Gao, S.L., Plonka, R., Mader, E., Hempel, S., Butler, M., and Mechtcherine, V., Interphase modification of alkali-resistant glass fibres and carbon fibres for textile reinforced concrete I: Fibre properties and durability, *Composites Science and Technology*, 69, 531–538, 2009.

Schiessl, P., *Cracking of Concrete and Durability of Concrete Structures*, AFREM-CCE, St Remy les Chevreuse, France, 1988.

Schiessl, P. and Raupach, M., Laboratory studies and calculations on the influence of crack width on chloride-induced corrosion of steel in concrete, *ACI Materials Journal*, 94(1), 56–62, 1997.

Singh, B. and Majumdar, A. J., Properties of GRC containing inorganic fillers, *International Journal of Cement Composites and Lightweight Concrete*, 3, 93–102, 1981.

Singh, B. and Majumdar, A. J., The effect of fibre length and content on the durability of glass reinforced cement—Ten year results, *Journal of Materials Science Letters*, 4, 967–971, 1985.

Singh, B., Majumdar, A. J., and Ali, M. A., Properties of GRC containing PFA, *International Journal of Cement Composites and Lightweight Concrete*, 6, 65–74, 1984.

Stucke, M. S. and Majumdar, A. J., Microstructure of glass fibre reinforced cement composites, *Journal of Materials Science*, 11, 1019–1030, 1976.

Tanaka, M. and Uchida, L., Durability of GFRC with calcium silicate: $C_4A_3\bar{S} - C\bar{S}$ -slag type low alkaline cement, in S. Diamond (ed.), *Proceedings: Durability of Glass Fibre Reinforced Concrete Symposium*, Prestressed Concrete Institute, Chicago, IL, 1985, pp. 305–314.

Trtik, P. and Bartos, P. J. M., Assessment of glass fibre reinforced cement by in-situ SEM bending test, *Materials and Structures*, 32, 140–143, 1999.

Van Itterbeeck, P., Purnell, P., Cuypers, H., and Wastiels, J., Study of strength durability models for GRC: Theoretical overview, *Composites Part A*, 40, 2020–2030, 2009.

van Zijl, G. P. A., Wittman, F. H., Oh, B. H., Kabele, P., Toledo Filho, R. D., Fairbarin, E. M. R., Slowik, V. et al., Durability of strain hardening cement based composites (SHCC), *Materials and Structures*, 45, 1447–1463, 2012.

West, J. M. and Majumdar, A. J., Strength of glass fibres in cement environments, *Journal of Materials Science Letters*, 1, 214–216, 1982.

Xu, S., Kruger, M., Reinhardt, H. W., and Ozbolt, J., Bond characteristics of carbon, alkali resistant glass and aramid textiles in mortar, *Journal of Materials in Civil Engineering*, 16(4), 356–364, 2004.

Yang, Y., Lepech, M. D., Yang, E. H., and Li, V. C., Autogenous healing of engineered cementitious composites under wet-dry cycles, *Cement and Concrete Research*, 39, 382–390, 2009.

Yardimci, M. Y., Tirosh, R., Larianovsky, P., Puterman, M., and Bentur, A., Improving the bond characteristics of AR-glass strands by microstructure modification technique, *Cement and Concrete Composites*, 33(1), 124–130, 2011.

Yilmaz, V. T. and Glasser, F. P., Reaction of alkali-resistant glass fibres with cement, Part 2: Durability in cement matrices conditioned with silica fume, *Glass Technology*, 32, 138–147, 1991.

Yu, J. H., Chen, W., Yu, M. X., and Hua, Y. E., The microstructure of self-healed PVA ECC under wet and dry cycles, *Materials Research*, 13(2), 225–231, 2010.

Zhu, W. and Bartos, P. J. M., Assessment of interfacial microstructure and bond properties in aged GRC using a novel microindentation method, *Cement and Concrete Research*, 27(11), 1701–1711, 1997.

Chapter 10

Repair and retrofit with TRC

The current practice of repair and retrofit with textiles made up of high-quality yarns such as glass and carbon is based on bonding or wrapping the relevant component with the textile and impregnating it with polymer using hand lay-up technology. A similar technology can be employed using a cementitious matrix, which includes some inherent benefits such as greater compatibility with the concrete substrate. Several studies have been carried out to determine the extent of enhancement in performance that can be achieved with such systems. The improved performance includes protection against environmental effects, which have led to corrosion processes, as well as mechanical strengthening to compensate for the loss of resistance due to the deterioration, or enhancement of mechanical performance in nondeteriorated structures where retrofit is required to meet additional loading requirements or new standards for resistance, such as earthquake effects. Retrofitting with textile-reinforced concrete (TRC) has been evaluated also for use in masonry structures (e.g., Garmendia et al., 2014; Papanicolaou et al., 2008).

Protection against environmental effects was discussed in Chapter 9, and the issues covered there are relevant also to repair applications. This chapter will focus on mechanical strengthening. The strengthening mechanisms can relate to shear, tension/flexural behavior, and compression. The application of TRC is achieved by the layers of impregnated textile at the bottom or sides of concrete components, as well as confinement, which can be obtained in wrapping of a whole component such as a column. Strengthening can be mobilized to enhance static resistance as well as dynamic loading.

The literature covering repair with textile-reinforced cementitious matrix uses the terms textile-reinforced mortar (TRM) along with textile-reinforced concrete. Since the term TRC refers to fine-grained concrete, the terminology in this chapter will be TRC for both.

10.1 Shear strengthening

The studies of shear strengthening provide insight into the efficiency of the TRC itself, in terms of its compositional parameters and number of layers, compare its performance with that of fiber-reinforced polymer (FRP), and consider the mode of bonding, interfacial bonding as well as anchoring. In addition, the geometry of the TRC reinforcement is addressed, U shaped around the bottom and side portions of a beam, as well as wrapping around a whole component, either orthogonal or spiral.

Shear strengthening of T-shaped beams was studied by Bruckner et al. (2006, 2008) using AR glass multiaxial textile. TRC was applied by wrapping symmetrically around the sides and the bottom of the beam to obtain a U-shaped external reinforcement. The combination of interfacial bonding of the TRC and mechanical anchoring was evaluated (Figure 10.1).

The effect of the number of layers and anchoring was evaluated by loading tests of full-size beams. The beams were conventionally reinforced to a level that provided higher flexural resistance so that the failure would be by shear. Results are presented in Table 10.1 and Figure 10.2, showing that increasing the number of layers of TRC did not provide enhanced resistance, since failure was by debonding. To avoid such failure and mobilize additional TRC layers for reinforcement, it was necessary to provide mechanical bonding by anchorage. Yet, even with mechanical anchoring, the overall strengthening effect did not exceed 16%.

U-shaped and side reinforcement shear strengthening was also studied by Larbi et al. (2010) (Figure 10.3). Several variables of the reinforcement configuration and the composition of the reinforcing layer were evaluated (Figure 10.3): cementitious matrix with glass textile, phosphate cement with glass textile, ultrahigh-performance composite with short metallic fibers, and carbon-fiber-reinforced polymer (CFRP). Bonding was carried out with epoxy resin cured at room temperature.

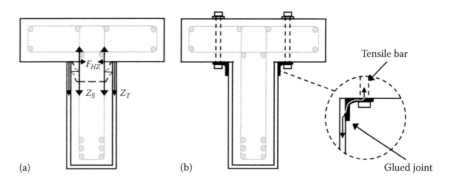

Figure 10.1 Bonding of the TRC without (a) and with (b) mechanical anchoring. (After Bruckner, A. et al., *Mater. Struct.*, 41, 407, 2008.)

Table 10.1 Strengthening effects of TRC, ultimate failure load divided by the average failure load of the reference beam

	Two layers without anchorage	Four layers without anchorage	Six layers without anchorage	Two layers with anchorage	Four layers with anchorage	Six layers with anchorage
Strengthening effect	1.01	1.01	1.07	1.06	1.09	1.16
Failure mode	Fracture	Debonding	Debonding	Fracture	Fracture	Fracture

Source: Bruckner, A. et al., *Mater. Struct.*, 41, 407, 2008.

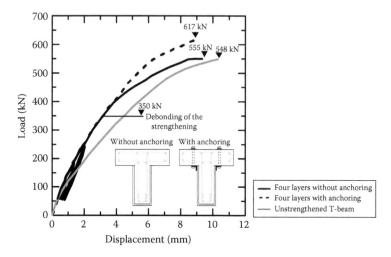

Figure 10.2 Load–displacement curves of strengthened T-beams. (After Bruckner, A. et al., *Mater. Struct.*, 41, 407, 2008.)

The performance was evaluated by means of three-point bending, and the results are shown in Figure 10.4 and Table 10.2.

When considering the U-shaped reinforcement, the TRC composite is as reliable as the carbon–polymer composite. Three different types of failure modes were observed:

1. Shear failure with considerable cracking at an angle of 45° for the reference unstrengthened beam, beam reinforced externally with bonded CFRP and mortar–glass composite
2. Debonding at the reinforcement interface, between the concrete and the phosphate cement–glass composite
3. Flexural failure in the system reinforced with ultrahigh-performance matrix (UHPM)-short metallic fibers

The shear strengthening achieved by the U-shaped composites seemed to be relatively small with regard to the load level. Much more dramatic improvements could be achieved by wrapping of the whole component, that is, jacketing. Jacketing is perhaps more complex to apply on-site, yet it is potentially more efficient as demonstrated by Triantafillou and Papanicolaou (2006). They evaluated shear strengthening by jacketing, either conventional or spirally applied (Figure 10.5). They investigated the effect of three parameters: textile reinforced with cementitious matrix (TRC) versus polymer matrix composite, conventional wrapping versus spiral, and the bonding agent being either epoxy resin or polymer-modified cement. The strengthened

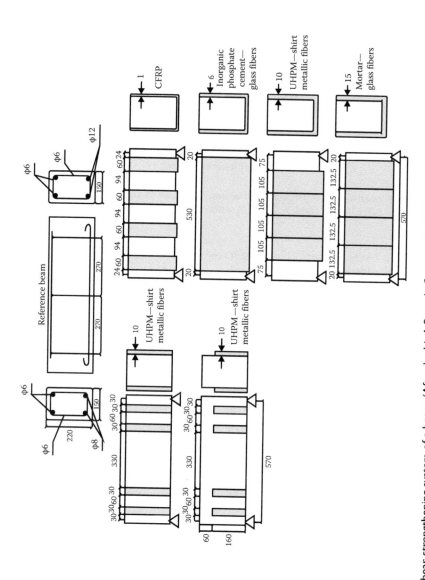

Figure 10.3 Shear strengthening system of a beam. (After Larbi, A.S. et al., *Construct. Build. Mater.*, 24, 1928, 2010.)

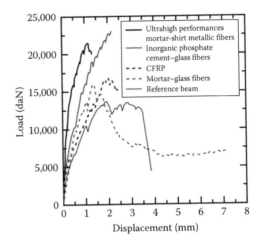

Figure 10.4 Load–displacement curves of beams reinforced to achieve higher shear resistance. (After Larbi, A.S. et al., *Construct. Build. Mater.*, 24, 1928, 2010.)

Table 10.2 Loading results of beams reinforced with TRC to achieve higher shear resistance

Reinforcement scheme	Kind of reinforcement	Ultimate load (N)	Displacement at ultimate load (mm)	Ultimate displacement
U shaped	Reference beam	138	2.2	3.7
	CFRP	169	2.3	3.7
	Phosphate cement–glass	233	2.1	2.1
	UHPM-short metallic fibers	216	1.4	1.4
	Mortar–glass	162	1.3	7
Side reinforcement	Reference beam	142	9.2	9.2
	UHPM-short metallic fibers	164	7	7
	UHPM-short metallic fibers	138	2.2	

Source: Larbi, A.S. et al., *Construct. Build. Mater.*, 24, 1928, 2010.

components were reinforced concrete beams with 150 × 300 mm cross sections, which were deficient in shear reinforcement.

The strengthening effects achieved with different systems are presented in Figure 10.6 in terms of load–deflection curves: control (C), the single- and two-layer TRC (M1 and M2, respectively), the single- and two-layer

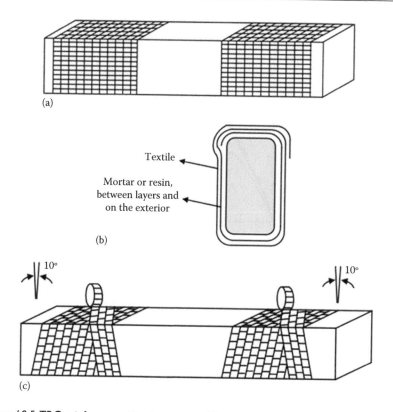

Figure 10.5 TRC reinforcement in shear spans: (a) conventional jacket, (b) layers of textile or resin, and (c) spirally applied strips. (After Triantafillou, T.C. and Papanicolaou, C.G., *Mater. Struct.*, 39, 93, 2006.)

resin–matrix composite (R1 and R2, respectively), the spirally applied two-layer TRC (M2-s), and the control without external reinforcement (C). Specimens C, M2, R2, and M2-s were loaded monotonically, while M1 and R1 were subjected to cyclic loading. It was concluded that the TRC wrapping of the beams was effective in increasing shear resistance significantly, with two layers being sufficient to prevent sudden shear failure. One layer was effective only when the bonding was with the epoxy agent (R1 vs. M1).

10.2 Flexural strengthening

The efficiency of flexural reinforcement by TRC was found to be quite high, as demonstrated in several studies. A range of parameters influencing the efficiency were evaluated, which included the number of layers, the nature of the reinforcing textile in the TRC, and the modification of the

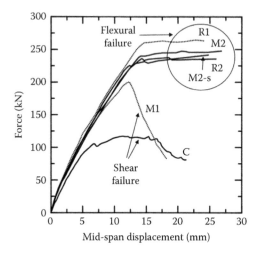

Figure 10.6 Force–midspan displacement curves for all the beams tested. Dashed lines indicate cyclic loading (M1 and R1) while full lines are static loading. (After Triantafillou, T.C. and Papanicolaou, C.G., *Mater. Struct.*, 39, 93, 2006.)

cementitious matrix in the TRC. Usually, the reinforcement is achieved by bonding of a strip of the composite at the bottom of the beam. However, also flexural reinforcement by wrapping was evaluated under static and dynamic conditions.

The effect of the number of layers of TRC bonded to the bottom of a beam was studied experimentally and analytically by Schladitz et al. (2012). The reinforcing system consisted of polymer-coated carbon textile with a clearance of 10.8 mm in the longitudinal direction and 18 mm in the lateral direction. The TRC reinforcement was 6.5 m long and bonded to the concrete beam surface, as shown in Figure 10.7.

The concrete surface was prepared for bonding by sandblasting, followed by placement of successive layers of 3 mm thick fine-grained concrete and

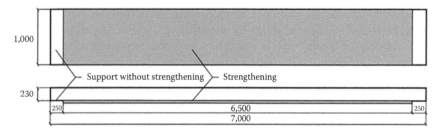

Figure 10.7 Arrangement of TRC reinforcement of the beam, bottom and side views. (After Schladitz, F. et al., *Eng. Struct.*, 40, 317, 2012.)

textile reinforcement. This buildup of the reinforcement was continued until the required layers of reinforcement were achieved, with the final layer being that of 3 mm fine-grained concrete.

The average strength of the carbon rovings was 1,200 MPa, and the strain to failure was 1.2%. The cementitious matrix was a fine-grained concrete with 1 mm maximum grain size and average compressive and flexural strength of 89 and 5.7 MPa, respectively.

The externally reinforced beams were loaded in flexure and load–deflection curves were obtained. The effect of the number of layers on the flexural behavior and ultimate failure moment are presented in Figure 10.8 and Table 10.3, demonstrating a significant reinforcement effect that increases monotonically with the number of layers of textile reinforcement. Schladitz et al. (2012) presented an analytical treatment and developed a model that provided predictions that are consistent with the data, as shown in Figure 10.8.

The effect of the reinforcing textile in the TRC on its efficiency for flexural strengthening was evaluated by Bruckner et al. (2006) using a range of strengthening layers. The strengthening effects are presented in Figure 10.9. They were quite considerable, resulting in an increase in the load-carrying capacity as well as serviceability. The displacement of the strengthened slabs, related to the service load, is smaller than the displacement of the nonstrengthened beam, because of the fine crack patterns induced by the TRC reinforcement.

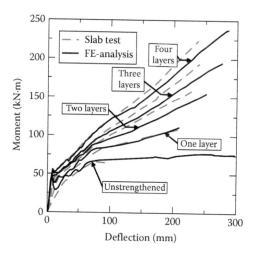

Figure 10.8 Moment–deflection relations for TRC reinforcement with 1,200 MPa strength and 1.2% ultimate strain, comparing experimental results and model simulation. (After Schladitz, F. et al., *Eng. Struct.*, 40, 317, 2012.)

Table 10.3 Calculated failure moments of beams reinforced with different types of TRC

Test specimen	Calculated failure moment (kN·m)
1—Unstrengthened slab	62
2—Single layer of reinforcement	101
3—Two layers of reinforcement	144
4—Three layers of reinforcement	188
5—Four layers of reinforcement	232

Source: Schladitz, F. et al., *Eng. Struct.*, 40, 317, 2012.

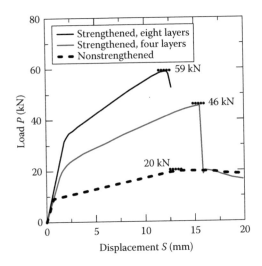

Figure 10.9 Load–displacement curves of beams reinforced with TRC to achieve enhanced bending performance. (After Bruckner, A. et al., *Mater. Struct.*, 39, 741, 2006.)

The influence of matrix and textile treatment in TRC on flexural strengthening was reported by Yin et al. (2014). The textile consisted of carbon yarns with E-glass yarns used to position them in place. The 10 mm mesh size textile was impregnated with polymer, and sand mixture to improve its bonding to matrix. The properties of the textile are provided in Table 10.4. The cementitious matrix of the TRC consisted of Portland cement, fly ash, silica fume, and silica sand with a maximum size of 1.2 mm, a water/binder ratio of 0.388, and a compressive strength of 55 MPa. The concrete beam was 120 mm in width and 210 mm in height, with the TRC placed at the bottom 120 mm width of the beam. Loading of the beam was by means of four-point scheme with 1,800 mm span and 600 mm constant bending moment zone.

Table 10.4 Properties of the textiles used for the TRC reinforcement

Yarn type	Carbon	E-glass	Impregnated carbon yarn
Number of filaments per yarn	12,000	4,000	12,000
Tensile strength (MPa)	4,660	3,200	4,100
Modulus of elasticity (GPa)	231	65	180
Ultimate strain (%)	2.0	4.5	2.3
Yarn tex (g/km)	801	600	—
Density (g/cm³)	1.78	2.58	—

Source: Yin, S. et al., *ASCE J. Mater. Civil Eng.*, 26, 320, 2014.

The results of the loading tests are summarized in Table 10.5, providing the yield load P_y, the ultimate load P_u, and the crack width at the yielding stage w_y. Load–maximum crack width curves demonstrating the effect of the number of textile layers are shown in Figure 10.10. The influence of the surface treatment on the curves for a given number of textile layers of reinforcement (not shown here) was small. These results indicate that the TRC could provide a reinforcing effect that increases monotonically with the number of layers, yet the increase was not dramatic—less than 30%. However, the TRC reinforcement had a dramatic influence on the reduction in the crack width.

A completely different geometry for flexural strengthening was applied by Tsesarsky et al. (2013, 2015), by wrapping concrete components and evaluating their performance in static and dynamic (impact) flexural loadings. They investigated the effect of the reinforcing yarn in the TRC (carbon, glass, and low-modulus polyethylene [PE]) and the modification of the cementitious matrix with film-forming polymer, chopped polypropylene (PP) fibers, fly ash, and silica fume. The properties of the fabric and stand-alone TRC are provided in Table 10.6 and Figure 10.11.

The flexural performance of the wrapped concrete was significantly enhanced by the TRC reinforcement (Figure 10.12). The nature of strengthening observed in Figure 10.12 reflects to a large extent the properties (strength and stiffness) of the reinforcing textile, carbon > glass > PE.

The effect of matrix modification was quite significant in glass TRC but had only a very small influence in PE TRC (Figure 10.13). The greater influence in the glass TRC could be accounted for by the special effects of the modifications on the bonding with the glass rovings, where the more efficient modifications enabled better bonding with the inner filaments in the roving, due to penetration of the modifying agents or particles into the roving.

Under dynamic loading, the performance of the wrapped concrete was enhanced with the TRC reinforcement, and the extent of reinforcement

Table 10.5 Values of load, crack width, deflection, and ductility of TRC-reinforced beams

Specimen	Textile surface treatment + number of layers	Yield load (Py)		Ultimate load (Pu)		Crack width at yielding (wy)	
		kN	% increase over control	kN	% increase over control	mm	% decrease relative to control
Beam 1	0	48.6	0	58.6	0	0.8	0
Beam 2	0	48.9		59.0		1.2	
Beam 3	Epoxy+1	53.8	10	64.5	10	0.43	57
Beam 4	Epoxy+1	54.0	11	65.6	12	0.42	58
Beam 5	Epoxy+2	59.5	22	74.7	27	0.32	68
Beam 6	Epoxy+3	59.0	21	85.0	45	0.26	74
Beam 7	Epoxy+fine sand+1	55.0	13	63.0	7	0.22	78
Beam 8	Epoxy+fine sand+1	56.0	15	65.8	12	0.24	76
Beam 9	Epoxy+fine sand+2	58.6	20	75.9	29	0.26	74
Beam 10	Epoxy+fine sand+2	57.0	17	75.0	28	0.30	70
Beam 11	Epoxy+fine sand+2	58.0	19	75.3	28	0.28	72
Beam 12	Epoxy+coarse sand+1	55.0	13	62.0	5	0.32	68
Beam 13	Epoxy+coarse sand+1	55.0	13	63.8	9	0.28	72
Beam 14	Epoxy+coarse sand+1	55.1	13	63.6	8	0.22	78
Beam 15	Epoxy+coarse sand+2	58.9	21	71.8	22	0.24	76

Source: Yin, S. et al., ASCE J. Mater. Civil Eng., 26, 320, 2014.

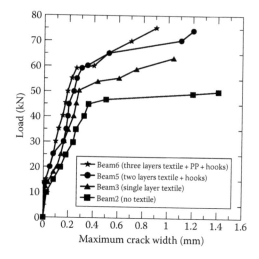

Figure 10.10 Experimental curves of load versus maximum crack width at the bottom surface of a loaded beam showing the effect of the number of textile layers. (After Yin, S. et al., *ASCE J. Mater. Civil Eng.*, 26, 320, 2014.)

Table 10.6 Properties of the textile and the pultruded TRC

Fabric	Yarn diameter (mm)	Yarn density (mm)	TRC thickness (mm)	Yarn tensile strength (MPa)	TRC tensile strength (MPa)	Volume fraction of fabric (%)
PE	0.5	0.37	9	240	1.4	3
Glass	0.3	0.28	8	1,372	4.5	1
Carbon	1.15	0.20	11	2,200	22	8

Source: Tsesarsky, M. et al., *Construct. Build. Mater.*, 44, 514, 2013.

reflected, like in static loading, the properties of the TRC, with the carbon TRC giving the highest reinforcement and the PE the lowest one (Figure 10.14).

However, the relative increase in dynamic loading was smaller than that obtained in static loading. It was suggested that this apparent lack of proportional improvement implies that the performance of the reinforced concrete under impact loading is controlled by the ductile behavior of the TRC composite and not by the concrete core. The concrete core (referring to the reinforced concrete) exhibits enhanced strength under dynamic loading, which is characteristic to brittle materials but not ductile ones.

Aging effects of glass TRC were also recorded, as seen by reduction impulse load ratio over time (Figure 10.15).

The performance of the strengthening with TRC should also be assessed from the point of view of enhanced durability performance. Leiboldt and

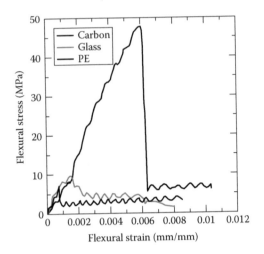

Figure 10.11 Flexural stress–strain curves for the carbon, glass, and PE TRC. (After Tsesarsky, M. et al., *Construct. Build. Mater.*, 44, 514, 2013.)

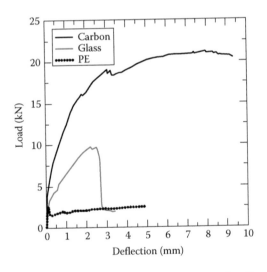

Figure 10.12 Representative load–displacement curves for static bending tests of PE, glass, and carbon TRC. (After Tsesarsky, M. et al., *Construct. Build. Mater.*, 44, 514, 2013.)

Mechtcherine (2013) investigated this aspect by evaluating the capillary transport of water through TRC repaired and strengthened cracked structures. The TRC layer was extremely effective in reducing the water ingress even when the TRC layer was cracked. This could be attributed to the much smaller cracks that develop in TRC, ~20 μm, compared to more than

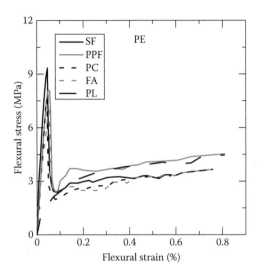

Figure 10.13 Representative stress–strain curves for TRC in static flexural tests. SF, silica fume; PL, polymer; PPF, polypropylene fibers; FA, fly ash; PC, plain cement. (After Tsesarsky, M. et al., *Mater. Struct.*, 48, 471, 2015.)

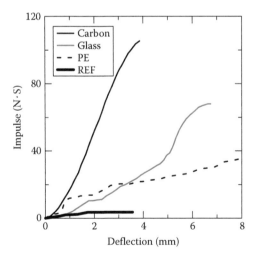

Figure 10.14 Representative impulse–deflection curves for PE, glass, and carbon TRC strengthening under impact loading for a pendulum drop height of 60 mm. (After Tsesarsky, M. et al., *Construct. Build. Mater.*, 44, 514, 2013.)

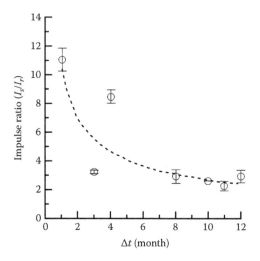

Figure 10.15 Impulse ratio (I_s and I_r are impulse of sample and reference, respectively) as a function of age for fully wrapped glass TRC strengthened elements. (After Tsesarsky, M. et al., *Mater. Struct.*, 48, 471, 2015.)

100 μm in the concrete alone. Some advantage was found in AR glass TRC compared to carbon TRC, but this could be attributed to the higher fineness of the AR glass TRC (2,400 tex in the AR glass TRC compared to 800 tex in the carbon TRC), which resulted in smaller cracks in the AR glass TRC (15 vs. 20 μm). This highlights the significance of the geometry of the reinforcement for achieving enhanced performance.

10.3 Compression strengthening in columns and column–beam joints

Reinforcement of columns to achieve enhanced compression performance is most readily achieved by wrapping the component with TRC, and the jacket achieved provides enhancement of properties through confinement. The efficiency of such reinforcement was evaluated for static compressive loading, for enhancement of buckling resistance, and for strengthening to dynamic response of column–beam joints. Since the reinforcing effect has to do with confinement mechanisms, attention was also given to the shape of the reinforced column, rectangular versus circular, and to geometries that are in between.

The effect of different types of textiles in TRC and a comparison with the efficiency of FRP reinforcement were reported by Peled (2007). Cylindrical specimens, 150 mm in diameter, were damaged by loading them to peak load and thereafter descending up to 90% of the maximum load, and unloaded. The damaged specimens were wrapped with TRC or FRP and loaded again in compression to determine the load–deformation behavior. The TRC systems used for the repair consisted of four layers of textiles impregnated with 0.40 w/c ratio pastes, with the textiles being woven PE and knitted PP, glass, and Kevlar. The properties of the fabrics were characterized in terms of strength and modulus of elasticity (Table 10.7). The fabric stiffness was quantified by the ASTM D1388 test, which relates to the ease in its application of the TRC by wrapping (Figure 10.16).

The strengthening effect of TRC is presented in Figure 10.17 and Table 10.8, showing that the TRC restores the compressive strength and provides additional benefits through enhancing the ductility by the confinement effects.

The influences observed can be correlated with the strength and modulus of elasticity of the reinforcement (Figure 10.18), with glass and Kevlar providing higher strength ratios, compared to PE and PP (Table 10.8).

The situation is quite different with regard to restoration of the compressive modulus of elasticity, which remains lower than the undamaged concrete. In this case, the retention of the modulus of elasticity seems to be better in the PE and PP in spite of their lower modulus compared to glass and Kevlar. In this case, a correlation could be found between the retention of the modulus and the fabric stiffness (Figure 10.19).

It is noted that the repair efficiencies, as estimated from the strength and modulus efficiency ratios, are better in TRC than in FRP (Table 10.8). This may have to do with the higher modulus of TRC, which is induced by the cementitious matrix, which is more rigid than a polymer matrix.

The influence of the number of layers and a comparison of TRC with FRP on column strengthening were studied by Triantafillou et al. (2006) and Triantafillou (2012), in axially loaded concretes, with and without conventional reinforcement. The concretes were wrapped with the composite until the desired number of layers (2–6) was obtained. The matrix in the TRC was polymer-modified mortar, and for the FRP, epoxy resin was used. The stress–strain curves of the plain concrete with the external reinforcement are presented in Figure 10.20 and for the reinforced concrete in Figure 10.21.

These results indicate that TRC reinforcement can provide a significant increase in compressive strength and deformation capacity, which increases with the number of reinforcing layers. Although the efficiency of the TRC is less than that of the FRP, the failure mode of the TRC jacket is more gradual and less abrupt, due to the slowly progressive fracture of the individual filaments.

The reports reviewed clearly indicate that confinement is an important mechanism, and if so, the geometry of the column should have a significant

Table 10.7 Properties and geometry characteristics of yarns and fabrics used for TRC reinforcement by wrapping

Yarn type	Yarn nature	Number of filaments	Bundle diameter (mm)	Fabric type	Fabric density (yarns/cm)	Tensile strength (MPa)	Modulus of elasticity (GPa)
PVA	Bundle	200	0.80	Woven	622	920	36
PE	Monofilament	1	0.25	Short-weft warp knitted	—	260	1.76
PP	Bundle	100	0.40	Weft insertion warp knitted	9	500	6.9
Kevlar	Bundle	325	0.25	Weft insertion warp knitted	16	2,300	44
Carbon	Bundle	24,000	—	Weft insertion warp knitted	4	2,200	240
Glass	Bundle (coated)	—	0.80	Bonded	4	1,360	78

Source: Peled, A., ASCE J. Compos. Construct., 11, 514, 2007.

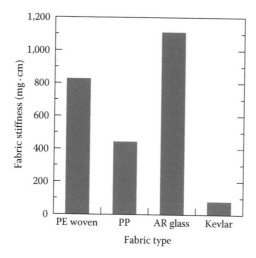

Figure 10.16 The stiffness properties of the fabrics used for TRC for wrapping of cylindrical specimens. (After Peled, A., *ASCE J. Compos. Construct.*, 11, 514, 2007.)

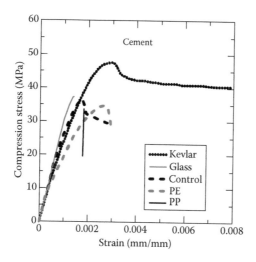

Figure 10.17 Compression response of TRC repaired cylinders with different types of TRC. (After Peled, A., *ASCE J. Compos. Construct.*, 11, 514, 2007.)

influence on the TRC reinforcing efficiency. Ortlepp et al. (2009) studied this influence by evaluating the reinforcement efficiency around columns of different geometries (Figure 10.22).

The TRC reinforcement consisted of a fine-grained cementitious matrix with a maximum aggregate size of 1 mm and with 80 MPa compressive

Table 10.8 Compression properties of TRC repaired concrete specimens

Matrix type	Fabric type (series number)	Strength, relative to control	Modulus of elasticity, relative to control	Strain at peak load, relative to control
Cement (TRC)	Woven (1)	1.01	0.78	1.26
	PP (2)	1.03	0.75	1.19
	Glass (3)	1.18	0.80	0.93
	Kevlar (4)	1.24	0.55	1.59
Epoxy (FRP)	PP	0.93	0.66	1.70
	Glass	1.06	0.49	0.64

Source: Peled, A., *ASCE J. Compos. Construct.*, 11, 514, 2007.

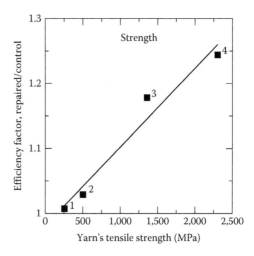

Figure 10.18 Efficiency factor of the repaired cylinders as a function of the yarn tensile strength; the series number presented in Table 10.8. (After Peled, A., *ASCE J. Compos. Construct.*, 11, 514, 2007.)

strength, reinforced with different types of textiles and different number of layers, as outlined in Table 10.9.

The combined effect of the nature of the TRC and the geometry of the column on the relative reinforcement enhancement is presented in Figure 10.23. It can be seen that the load increase is small for the fine-grained matrix, regardless of the geometry of the column. For rectangular-shaped column, the reinforcement is small, regardless of the reinforcement in the TRC. However, as the shape becomes more circular, the reinforcing efficiency of the TRC increases significantly, with the increase being higher for the stiffer TRC. These trends indicate that the load fraction carried by

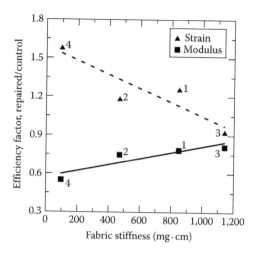

Figure 10.19 The influence of fabric stiffness on the modulus and strain efficiency factors of the TRC repaired cylinders; the series number presented in Table 10.8. (After Peled, A., *ASCE J. Compos. Construct.*, 11, 514, 2007.)

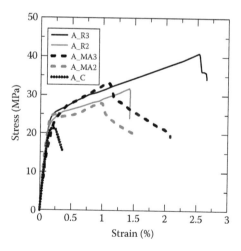

Figure 10.20 Stress–strain curves in compression of the plain concretes, control (A_C0), external TRC reinforced with two and three layers (A_MA2 and A_MA3, respectively), and external FRP reinforced with two and three layers (A_R2 and A_R3, respectively). (After Triantafillou, T., Innovative textile-based cementitious composites for retrofitting of concrete structures, Chapter 13, in M.N. Fardis (ed.), *Innovative Materials and Techniques in Concrete Construction: ASCE Workshop*, Springer, 2012, pp. 209–223.)

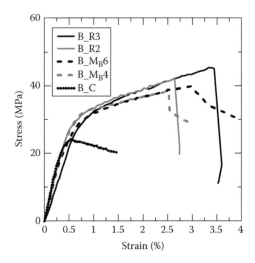

Figure 10.21 Stress–strain curves in compression for the conventionally reinforced control (B_C), externally reinforced TRC with four and six layers (B_MB4 and B_MB6, respectively), and FRP reinforced with two and three layers (B_R2 and B_R3, respectively). (After Triantafillou, T., Innovative textile-based cementitious composites for retrofitting of concrete structures, Chapter 13, in M.N. Fardis (ed.), *Innovative Materials and Techniques in Concrete Construction: ASCE Workshop*, Springer, 2012, pp. 209–223.)

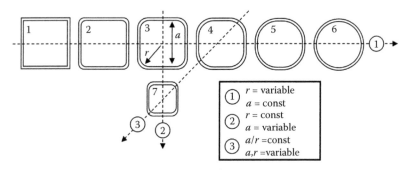

Figure 10.22 The seven different geometries of columns evaluated with regard to the efficiency of TRC reinforcement. (After Ortlepp, R. et al., *Adv. Mater. Sci. Eng.*, 29, 5, 2009.)

the normal forces within the fine-grained concrete coat is small and that the enhancement achieved by the TRC composite is largely due to confinement of the core concrete.

The use of TRC for enhancement of the buckling resistance of columns was reported by Bournas and Triantafillou (2011). They investigated the

Table 10.9 The compositional variables of the TRC composites applied for reinforcing columns of different shapes

Series number	Reinforcing textile	Fineness (tex)	Yarn distance (mm)	Yarn cross section (mm²/m)	Layers	Stiffness (MN/m)
1	Control	—	—	—	—	—
2	Matrix only	—	—	—	0	—
3	AR glass	1,200	7.2	61.1	6	27.3
4	Carbon	800	7.2	62.5	2	27.9
5	Carbon	3,500	10.8	179.6	3	109.0

Source: Ortlepp, R. et al., *Adv. Mater. Sci. Eng.*, 29, 5, 2009.

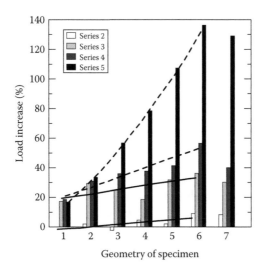

Figure 10.23 The influence of the column geometry (marked 1–7; see Figure 10.22) and the nature of the reinforcing layer (marked series 2–4; see Table 10.9) on the relative load increase of the reinforced column. (After Ortlepp, R. et al., *Adv. Mater. Sci. Eng.*, 29, 5, 2009.)

enhancement of buckling resistance by confinement of reinforced columns with TRC as well as other composites: CFRP jacket, four-layer carbon TRC of equal stiffness and strength, and four-layer glass TRC of lower stiffness and strength. The buckling performance was studied under simulated seismic excitation and compared with nonretrofitted RC column. Buckling was initiated in all the columns immediately after the yielding in compression due to the nearly zero axial resistance (stiffness) of the steel rebars. In the nonretrofitted concrete, the bar buckling accelerated the cover separation and spalling. Although buckling was not prevented in the retrofitted columns, the confinement, CFRP as well as TRC, provides a significant delay, ranging from three to seven loading cycles. Also, the initiation of the bar buckling was not accompanied by lateral strength degradation. The strengthening composites provided a means for redistribution of the compressive over-load to the core. The TRC reinforcement was able to provide higher local deformations compared to the CFRP, and this was attributed to its ability to deform outward without early fiber fracture. Overall, the composite confinement enabled a substantial deformation capacity from the initiation of the bar buckling until recording of the conventional failure. This reserve increased with the stiffness of the TRC confinement.

The use of TRC and FRP for strengthening of RC beam–column joints was studied by Al-Salloum et al. (2011). They evaluated the performance of FRP (glass reinforced [GFPR] and carbon reinforced [CFRP]) and TRC in the case of seismically deficient joints. The TRC was carbon textile with polymer-modified cementitious matrix. The design approach for the external reinforcement was based on providing composite reinforcement to replace the missing joint shear reinforcement and the inadequately detailed steel bars. Reinforcement consisted of bonding of a single layer of CFRP or double layer of GFRP. Double layer of TRC was estimated to provide equivalent reinforcement. The bonding to the beam–column joint was achieved by a combination of interfacial bonding and mechanical bolt anchoring, as seen in Figure 10.24. To facilitate bonding to the TRC, the surface of concrete, was conditioned by sandblasting. the TRC was hand laid and wrapped around the concrete, with the cementitious matrix being laid in 2 mm thick layers, in between the carbon textile reinforcement. The CFRC and GFRP were epoxy bonded to the concrete surface.

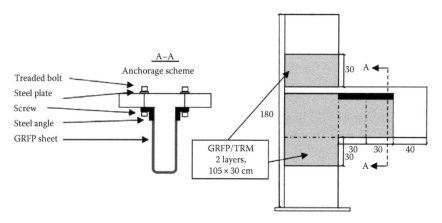

Figure 10.24 Schematic representation of GFRP and TRC reinforcement applied to the as-built joint. (After Al-Salloum, Y. et al., *ASCE J. Compos. Construct.,* 15, 920, 2011.)

Seismic loading was simulated by a horizontal loading scheme, and the loading cycles were controlled by the peak displacement until failure. It was concluded that the TRC could effectively improve the shear strength and ductility of the joint to a level that is comparable to the FRP systems, if a sufficient number of layers are being used. Caution should be taken before increasing the number of layers, to assure that the increased stiffness does not adversely affect the load sharing between the members and does not result in early debonding if not prevented by the mechanical anchoring.

10.4 Bonding

The bonding of the TRC reinforcement to the concrete surface is of prime importance in controlling the reinforcing efficiency. In the previous sections, the effect of various treatments on the flexural, compressive, and shear reinforcement efficiencies was highlighted, considering interfacial bonding and mechanical anchoring.

Bruckner et al. (2006) discussed the influence of bond properties. In order to transfer load between the old concrete and the reinforcing TRC layer, a minimum bond length is needed. They presented the relations between the bond length and the bond force for TRC reinforcement with different numbers of layers (Figure 10.25). For the flexural strengthening shown in Figure 10.9, a bond length of approximately 50–60 mm is required for six-layer TRC. For four-layer TRC, the bond length needed is less than 40 mm because the ultimate tensile force of the strengthening layer is smaller.

Ortlepp et al. (2006) analyzed the bonding mechanisms with the objective of developing new approaches to their modeling. They considered

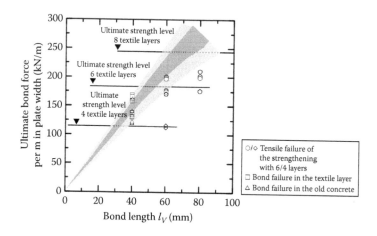

Figure 10.25 Ultimate bond force per meter width as a function of bond length. (After Bruckner, A. et al., *Mater. Struct.*, 39, 741, 2006.)

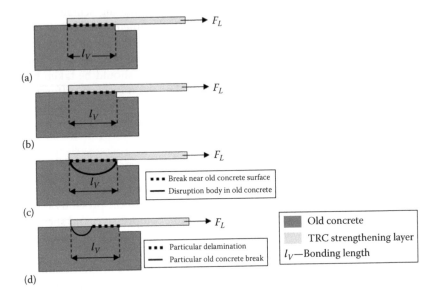

Figure 10.26 Schematic description of modes of bond failures: (a) delamination of cementitious matrix in the textile layer, (b) failure in the new–old concrete joint, (c) failure in the old concrete, and (d) mix of different types of failure. (After Ortlepp, R. et al., *Cement Concr. Compos.*, 28, 589, 2006.)

four modes of bond failure: delamination in the textile layer, failure in the bond joint, failure in the old concrete, and mix of different types of modes (Figure 10.26).

In the modeling of the bond, they took into account the development of cracking (Figure 10.27) and considered the contribution of the cracked and uncracked zones (Figure 10.28).

The analytical treatment of bond proposed by Ortlepp et al. (2006) is based on consideration of the contribution to the bond from the cracked

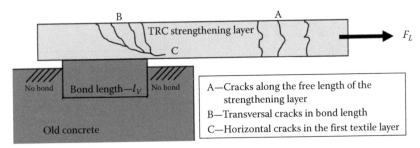

Figure 10.27 Schematic illustration of the cracking in bonding. (After Ortlepp, R. et al., *Cement Concr. Compos.*, 28, 589, 2006.)

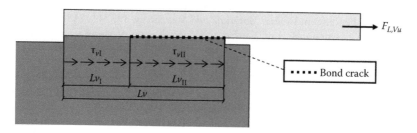

Figure 10.28 Symbols for modeling the contribution of the cracked and uncracked zones to bonding. (After Ortlepp, R. et al., *Cement Concr. Compos.*, 28, 589, 2006.)

and uncracked portions, $l_{V,II}$ and $l_{V,I}$, respectively (Figure 10.28), to the overall ultimate bond force $F_{L,Vu}$:

$$F_{L,Vu} = b_L x \left(l_{V,I} x \tau_{Vu,I} + l_{V,II} x \tau_{V,II} \right) \tag{10.1}$$

where
   $F_{L,Vu}$ is the ultimate bond force
   b_L is the width of the strengthening layer
   $l_{V,I}$, $l_{V,II}$ are the length of the uncracked and cracked range of the bond length, respectively
   $\tau_{Vu,I}$ is the ultimate bond stress (shear strength) in $l_{V,I}$
   $\tau_{V,II}$ is the bond stress (shear stress) in $l_{V,II}$

The failure along the bond crack takes place in the layer of the textile reinforcement. In this case, the shear bonding force is transferred only by the fine-grained concrete between the yarns by crack friction. Thus, the total bond stress, $\tau_{V,II}$, along the layer is a function of the ratio of the fine-grained concrete matrix:

$$\tau_{V,II} = \tau_{V,II,fc} k_{A,eff} \tag{10.2}$$

In the uncracked region, there is a need to consider failure in possibly two different layers. The sudden failure of the remainder part can continue in the textile layer or in the old concrete substrate.

References

Al-Salloum, Y., Siddiqui, N. A., Elsanadedy, H. M., Abadel, A. A., and Aquel, M. A., Textile reinforced mortar versus FRP as strengthening material for seismically deficient RC beam-column joints, *ASCE Journal of Composites for Construction*, 15, 920–933, 2011.

Bournas, D. A. and Triantafillou, T. C., Bar buckling in RC columns confined with composite materials, *ASCE Journal of Composites for Construction*, 15, 393–403, 2011.

Bruckner, A., Ortlepp, R., and Curbach, M., Textile reinforced concrete for strengthening in bending and shear, *Materials and Structures*, 39, 741–748, 2006.

Bruckner, A., Ortlepp, R., and Curbach, M., Anchoring of shear strengthening for T-beams made of textile reinforced concrete (TRC), *Materials and Structures*, 41, 407–418, 2008.

Garmendia, L., Marcos, I., Garbin, E., and Valluzi, M. R., Strengthening of masonry arches with textile reinforced mortar: Experimental behavior and analytical approaches, *Materials and Structures*, 47, 2067–2080, 2014.

Larbi, A. S., Contamine, R., Ferrier, E., and Hamelin, P., Shear strengthening of RC beams with textile reinforced concrete (TRC) plate, *Construction and Building Materials*, 24, 1928–1936, 2010.

Leiboldt, M. and Mechtcherine, V., Capillary transport of water though textile-reinforced concrete applied in repairing and/or strengthening cracked TRC, *Cement and Concrete Research*, 52(1), 53–62, 2013.

Ortlepp, R., Hampel, U., and Churbach, M., A new approach for evaluating bond capacity of TRC strengthening, *Cement and Concrete Composites*, 28, 589–597, 2006.

Ortlepp, R., Lorenz, A., and Curbach, M., Column strengthening with TRC: Influence of the column geometry onto the confinement effect, *Advances in Materials Science and Engineering*, 29, 5, 2009.

Papanicolaou, C. G., Triantafillou, T. C., Papathansiou, M., and Karlos, K., Textile reinforced mortar (TRM) versus FRP as strengthening material of URM walls: Out-of-plane cyclic loading, *Materials and Structures*, 41, 143–157, 2008.

Peled, A., Confinement of damaged and nondamaged structural concrete with FRP and TRC sleeves, *ASCE Journal of Composites for Construction*, 11, 514–522, 2007.

Schladitz, F., Frenzel, M., Ehlig, D., and Churbach, M., Bending load capacity of reinforced concrete slabs strengthened with textile reinforced concrete, *Engineering Structures*, 40, 317–326, 2012.

Triantafillou, T., Innovative textile-based cementitious composites for retrofitting of concrete structures, Chapter 13, in M. N. Fardis (ed.), *Innovative Materials and Techniques in Concrete Construction: ASCE Workshop*, Springer, Dordrecht, Heidelberg, London, New York, 2012, pp. 209–223.

Triantafillou, T. C. and Papanicolaou, C. G., Shear strengthening of reinforced concrete members with textile reinforced mortar (TRM) jackets, *Materials and Structures*, 39, 93–103, 2006.

Triantafillou, T. C., Papanicolaou, C. G., Zissimopoulos, P., and Laoudrekis, T., Concrete confinement with textile-reinforced mortar jackets, *ACI Structural Journal*, 103, 28–37, 2006.

Tsesarsky, M., Peled, A., Katz, A., and Anteby, I., Strengthening concrete elements by confinement within textile reinforced concrete (TRC) shells—Static and impact properties, *Construction and Building Materials*, 44, 514–523, 2013.

Tsesarsky, M., Katz, A., Peled, A., and Sadot, O., Textile reinforced concrete (TRC) shells for strengthening and retrofitting of concrete elements: Influence of admixtures, *Materials and Structures*, 48, 471–484, 2015.

Yin, S., Xu, S., and lv, H., Flexural behavior of reinforced concrete beams with TRC tension zone cover, *ASCE Journal Materials in Civil Engineering*, 26, 320–330, 2014.

Chapter 11

Innovative applications of textile reinforced concrete (TRC) for sustainability and efficiency

11.1 Potential for TRC integration in novel construction

The cost-effectiveness of using cement-based materials for residential buildings is becoming increasingly apparent. Next to steel, cement-based composites have demonstrated the highest strength-to-weight ratio of any building material. They are noncombustible and will not contribute fuel to the spread of a fire. Cement composites will not rot, warp, split, crack, or creep and are 100% recyclable. Furthermore, they are not vulnerable to termites, fungi, or organisms. The dimensional stability sensitivity is mostly during the early hydration stages and can be accommodated easily in design. Furthermore, the expansion or contraction with moisture movement can be optimized and designed into the final product. The connection-based design will be an integral part and will be appropriate so that full moment rotation characteristics are incorporated in the design of panels. The consistent material quality of cement products developed in a precast operation allows them to be produced in strict accordance with national standards. Finally, there is the opportunity to train the labor pool with the production of precast-cement-based products.

By integrating the reinforcement within the material design, one can creatively use standard and nonconventional shapes to obtain economical designs. The use of textile-reinforcement systems expedites manufacturing and construction processes that utilize lightweight sections with materials at a fraction of the cost of steel and significantly more robust than wood. By saving materials costs, reducing deadweight, and streamlining will result in significant cost benefits. From a design perspective, an elastoplastic design approach results in reductions in material weight and improvements in ductility. Improved tensile and flexural strengths and shrinkage crack control result in a reduced number of surface cracks, narrower crack widths, and extended service life. Potential applications include highly ductile members, cladding panels, cast-in-place forms, roofing and exterior wall elements, fatigue- and impact-resistant infrastructure, industrial components, and transportation structures (Mobasher, 2011; Zhu et al., 2009).

From a sustainability and durability perspective, designs of elements for a variety of structures using a tensile serviceability approach address a large class of structural components that are not just limited to beams, panels, retaining walls, elevated slabs, structural vaults, buried structures, pipes and culverts, and environmental structures. This is especially important as we move toward sustainable construction materials. Therefore, the areas of utilization are well defined, and potential applications are rather enormous. The modeling approaches proposed for TRC are applicable at a multiscale level for flexural members, impact, and repair and retrofit, which allows the computation of load–deformation response of nonlinear materials (Mobasher, 2003, 2011; Mobasher and Barsby, 2012; Mobasher and Ferraris, 2004).

Several case study examples of the potential for innovation with textile-reinforced concrete (TRC) will be demonstrated in this chapter (Mobasher and Li, 1996; Mobasher et al., 1990, 1997, 2006a,b). Textile–cement composites encompass all of these characteristics, and at the same time, they are of high-performance strain hardening, demonstrating high tensile strength as well as superb ductility. Their production process is very flexible. The free-form nature of shape and reverse curvature in the construction can be used in order to generate panels and shapes that benefit from added structural stiffness offered by the three-dimensional (3D) shape. These can be functional such as airfoils for small power generation wind and water blades. Figures 11.1a and b show single or doubly curved elements with increased stiffness for use as wind turbine elements, or panels for outdoor structural shade elements. Free-form panels can also be used for large-scale roof elements (see Figure 11.1c). The size of elements may range from small to large of the order of meters.

The promising combination of fabric reinforcement in cement composite products and the pultrusion process is expected to lead to an effective novel technique to produce a new class of high-performance fabric–cement composite materials. Such a breakthrough may open the way for a multitude of new products and applications such as honeycomb structural members, as shown in Figure 11.2.

11.2 Structural shapes using TRC materials

A main innovative area is in the development of structural shapes from TRC materials. Such an effort is novel and unique. It is expected that the structural shapes such as angles, C-channels, hat sections, and closed tubes can be manufactured and used in the construction of wall and panel elements. Such a technology will be a direct competitor for wood-based construction elements such as 2 × 4 and 2 × 6 as well as light-gage steel members that are increasingly used in construction, paneling, and framing applications.

(a)   (b)

(c)

Figure 11.1 Development of single or doubly curved elements with increased stiffness for use as (a) wind turbine elements, (b) panels for outdoor structural shade elements, or (c) temporary or permanent shelter elements using partially curved integrated elements.

Figure 11.2 Development of honeycomb structural elements for axial and shear resistance as well as free-form interlacing structural triangular elements.

Computer-controlled pultrusion equipment can be used in conjunction with pultrusion dies to make structural shapes such as C-channel and hat- and T-sections of lengths of up to 2 m. The long-term objective is to develop sustainable, lightweight, ductile, and optimized structural shapes and document their mechanical response using the available structural mechanics tools such as finite strip method. These are shown in Figure 11.3 and promise a host of new construction products for competing and replacing wood- and steel-based elements.

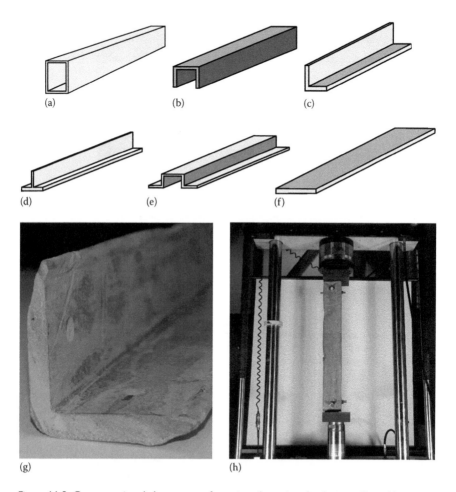

Figure 11.3 Cross-sectional shapes manufacturing plan using the Arizona State University cement composite pultrusion equipment: (a) closed cell, (b) C-channel, (c) equal leg angle, (d) T-section, (e) hat section, (f) flat board, (g) structural L3 × 3 × 0.5 in., 4 ft long section with four layers of ARG textile (1.0%), and (h) experimental setup large-scale testing of structural shapes in tension and compression.

11.3 Modular and panelized cementitious construction systems

Figure 11.4 represents innovative approaches of using TRC to make honeycomb elements as well as 3D integrated triangular brick elements. These types of components can be used for panel-making or truss-making operations. Panelized concrete construction is beneficial for the builder as it provides a light framing material, with less of a need to cull or sort straight walls, square corners, less scrap and waste (20% for lumber), and price stability (compared to the price volatility of steel). Benefits to the consumer include higher strength, safer structures, less maintenance, slower aging, fire safety, and stronger connections (e.g., screwed vs. nailed). From a design perspective, the fabrication of the truss and wall systems can be easily automated. Unlike wood, changes to truss or wall can be made in the field to correct fabrication errors or design errors. Figure 11.4 also shows directionally oriented elements for shading or exterior façade applications in buildings.

11.4 Cast-in-place modular homes

Cast-in-place construction using fully integrated molds is attractive for the development of energy-efficient and sustainable homes. This technique is applicable to many regions of the world. Multicomponent wall construction uses a system of fully integrated molds that are set up and cast in place in a single operation completing the interior and exterior walls and roof, as shown in Figure 11.5. By using a unique system of metallic forms and accessories that are assembled into a complete mold, interior and exterior walls, roofs, and stairs are constructed in one cast with a minimum number of workers. An integrated mold system requires a series of interacting

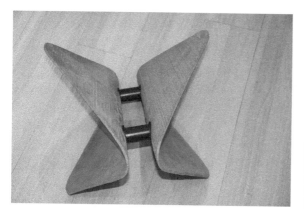

Figure 11.4 Development of directionally oriented elements for shading or exterior façade applications in buildings.

Figure 11.5 Schematic details of cast-in-place modular homes.

and integrating metal and composite forms that are applicable to a variety of floor plans and applications. A unique set of forms allows interior and exterior walls and roof to be constructed in one cast with minimum labor. With a life expectancy exceeding 50 years, a potential for additions to the original structure exists.

The process of setup of molds, concrete casting, setting, and removal of molds can take place over a course of 2 days, thus allowing for a quick turnaround time and an efficient operational procedure. Homes can be efficiently designed since the wall system is integrated in the structural design; however, the main intention is to create a sufficiently effective thermal barrier that reduces the need for air-conditioning. Mechanical and thermal performance of monolithic concrete wall systems using TRC materials as permanent or reusable form work is therefore an important aspect of design (Mobasher et al., 2011).

Using integrated metallic and TRC forms in the construction of concrete homes as shown in Figure 11.6, a single mold can be reused several times, or alternatively, the molds are used as stay-in-place forms. Currently, more than 80,000 homes are built annually in Central and South America, a figure that is expected to double in the next 5 years. This construction technique is

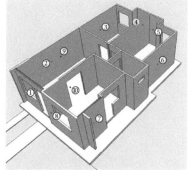

Figure 11.6 The finished cast-in-place home and the model of the residential structure as a cast-in-place home.

applicable to many regions of the world and is particularly attractive where steel and wood are scarce. Interacting and integrating forms that are applicable to a variety of floor plans and applications can be designed to yield efficient systems if the interaction of materials, structures, and environment is addressed.

11.5 Sandwich composites with TRC skin-aerated fiber reinforced concrete (FRC)

In several areas of application, the strength of the thin TRC panels can be utilized in the context of a sandwich element. Sandwich construction provides for the highest strength-to-weight ratio and stiffness-to-weight ratio defined in terms of specific strength and specific stiffness. The skin is primarily involved in strength, stiffness, and load-carrying capacity, whereas the core carries the shear stresses and provides a thermal barrier and thermal mass. The use of the tensile strength and stiffness of TRC when employed as the skin element significantly improves the characteristics of the overall system. Other components of the system such as the core can be made with different materials such as expanded polystyrene (EPS), or aerated concrete products.

A sandwich construction can be optimized for both structural and thermal characteristics, especially in areas where both parameters are design considerations. The majority of energy consumption of buildings is due to heating and cooling; hence, thermal characteristics of building materials affect the energy performance. Thermal conductivity and heat capacity describe the ability of heat to flow, or be retained in a section in the presence of a differential temperature (Budaiwi et al., 2002).

The use of TRC as a component of a sandwich skin panel in conjunction with fiber-reinforced aerated concrete (FRAC) has potential for wall and roof elements since the sandwich construction allows for longer spans, ductile composites, thermally efficient elements, and lighter-weight structural components.

Cellular structures exhibit a considerable amount of residual strength after reaching the peak strength (Gibson and Ashby, 1997). While autoclaving provides higher strength because of the homogenization of the reaction products and reduces shrinkage potential when compared to moist curing, its cost may be prohibitive. Elimination of autoclaving process lowers the strength and introduces inhomogeneity when compared with autoclaved aerated concrete (AAC); however, short fibers have a direct effect on the initiation, growth, and bridging of the cracks caused during the plastic stage or later on due to mechanical forces, shrinkage, or heating–cooling cycles. The advantages include: thermal mass, airtightness, whole wall coverage, reduction in HVAC costs and energy consumption. Furthermore, fire resistance is improved such that an 8 in. bearing wall provides a 4 h UL fire rating. Comparative evaluation of the embodied energy data indicates that lightweight concrete could result in 70% and 40% energy reduction, when compared with concrete and brick, respectively.

Aerated concrete can serve the purpose of a high-efficiency and thermally efficient construction material. Reduction of the heat loss in buildings decreases energy consumption and reduces the cost of heating and cooling (ACI 122R, 1996). Lightweight FRAC are manufactured using Portland cement, fly ash, quicklime, gypsum, water, and aluminum powder (or paste) (ACI 523.2R, 1996). Compared to AAC, the high temperature and pressure of the autoclaving process are eliminated and curing is performed at room temperature. The air-pore structure in aerated concrete is in the range of 0.1–1 mm in diameter with approximately 80% of the volume of the hardened material made up of pores with a general ratio of 2.5:1.0 of air pores to micropore, as shown in Figure 11.7 (Holt and Raivio, 2005; Neithalath and Ramamurthy, 2000). The effect of fiber content on the mechanical characteristics of AFRC mixtures with a range of volume fractions has been compared to AAC materials, as shown in Figure 11.8 (Bonakdar et al., 2013). Experimental results indicate that while AFRC may exhibit lower average compressive strength and higher variability than AAC,

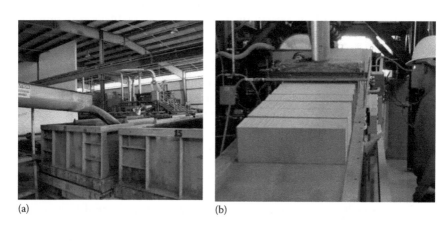

(a)   (b)

(c)   (d)   (e)

Figure 11.7 Manufacturing of aerated concrete panels as a core material (a) fresh slurry poured into large (8 × 1.2 × 0.6 m) steel molds, (b) large loaves (8 × 1.2 × 0.6 m) cut using wheel blades into blocks, (c) schematics of the proposed textile-reinforced aerated concrete sandwich composite system, (d) finished sandwich panel, and (e) beam test coupon harvested from panel.

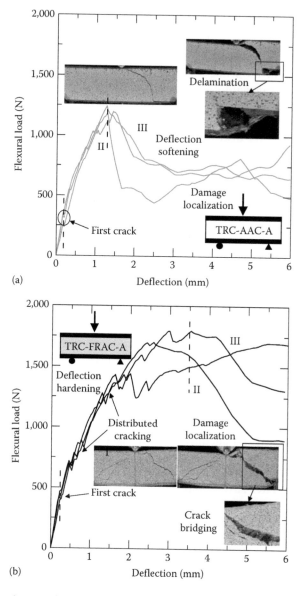

Figure 11.8 Replicates of sandwich TRC and aerated concrete (TRC–AAC and TRC–FRAC) samples tested under static loading conditions. (a) Sandwich ductile TRC skin and brittle AAC core and (b) sandwich ductile TRC skin and ductile FRAC core. (From Dey, V. et al., *Mater. Des.*, 86, 187, 2015.)

its thermal properties are as comparable, while its flexural toughness is as much as two orders of magnitude higher than AAC. This added ductility due to the bridging action of the fibers allows for the design of the material based on an elastic–quasiplastic approach, as well as a great core material for sandwich composites made with TRC (Dey et al., 2014).

Procedures for modeling thermal analysis based on steady-state or transient conditions depend on two main parameters of thermal conductivity and heat capacity and are used to relate the thermal loading to the temperature distribution through the wall system. The steady-state approach is used for the temperature distribution profile and requires thermal conductivity measurements. Once a steady-state condition is reached, the difference in temperature between the hot and cold surfaces is used in the calculation of the thermal conductivity. The transient analysis determines the temperature distribution and other heat flow parameters that vary over a period of time. The transient analysis is used to address the response of 2D and 3D composite wall systems extended to simulate the results of experimental data obtained of three single room recording chambers that were constructed with different wall systems (Mobasher et al., 2011).

Typical load–deflection trends and damage mechanisms of TRC–AAC and TRC–FRAC are shown in Figure 11.8 (Dey et al., 2015). The load–deformation behavior shows three dominant zones. Zone I is defined as the initial linear-elastic behavior up to roughly 25% of the peak load and is terminated by the initiation of the first crack, designated as the limit of proportionality. In zone II, increasing deformation produces multiple cracks in the tension face and the core element resulting in deflection hardening. This feature is more dominant for the TRC–FRAC composites (see Figure 11.8b). The nonlinearity and stiffness degradation continues gradually until the ultimate load capacity. In the post peak range, or zone III, load capacity and stiffness reduce drastically, and ductility is exhausted due to localization after saturation of cracking leading to widening of a major crack (Silva et al., 2011). Delamination of the skin–core layer is also observed in this zone, as evident especially for TRC–AAC composites (see Figure 11.8a). Fiber reinforcement in the FRAC promotes crack-bridging mechanism (see Figure 11.8b), resulting in higher residual strength and enhanced energy absorption, especially under quasistatic loading (Bonakdar et al., 2013). Considering the mechanical characteristics, the two ductile systems are characterized as ductile core–ductile skin for TRC–FRAC and brittle core–ductile skin for TRC–AAC.

11.6 Computational tools for design of TRC components: Case study

Computational mechanics generate opportunities for better modeling and design of infrastructure systems. Implementation of new material models in finite element and also other solution methodologies allows for a more realistic analysis in order to close the gap between materials, properties, analysis, modeling, and design (Mobasher et al., 1999). Various research groups are involved in developing these methods and associated design guides (di Prisco et al., 2004; RILEM TC162-TDF, 2000, 2003; Soranakom and Mobasher, 2009; Vandewalle, 2004).

Tensile response has a dominant effect on the performance of TRC materials, and potential for the use of textiles in improving ductility in tensile regions can be realized. The proposed closed-form equations for generating moment–curvature response from material tests can be used in conjunction with crack localization rules to predict flexural response of a beam, or a panel (Soranakom and Mobasher, 2008, 2009).

Several types of panel tests (ASTM C1550-02, 2002; Bernard, 2000; EFNARC, 1996) have been used to simulate the stress state in plate-type specimen and evaluate the flexural capacity of ductile panels. Test results produce a load–deflection response that is used to calculate the energy absorption as a material index. As the size, thickness, and fixity of the panel change, load, deflection, and energy absorption also change (Soranakom and Mobasher, 2007). A theoretical model is used with a constitutive model to correlate the experimental results obtained using different boundary conditions. An inverse analysis of load–deflection response of these materials is described along with the results of back-calculation of stress–strain response for various TRC samples. The experimental results of the beam flexural tests can be utilized in back-calculation of constitutive stress–strain response. It is shown that flexural strength test methods may overestimate the nominal strength in tension significantly (Soranakom and Mobasher, 2010).

Closed-form relationships that can be used to compute the load–deflection results of a nonlinear TRC material are used to explain the differences between the tensile and flexural strengths for both strain-hardening (Soranakom and Mobasher, 2007; Soranakom et al., 2006) and strain-softening type composites (Mobasher et al., 2014; Soranakom and Mobasher, 2007a,b). Figure 11.9 shows the general application of a structural panel designed for water distribution system. Numerical simulation results for a rectangular panel (h = 70 in., L = 45 in., and t = 4 in.) with rebar and a post-crack residual tensile strength of μ = 0.8 using a medium mesh are presented in Soranakom and Mobasher (2007). The deflection distribution normal to the plane can be computed, and the von Mises stress distribution is used as a serviceability criterion.

Finite element software, such as ABAQUS (Hibbitt and Sorensen, 2004), are also used to create virtual testing platforms and conduct inverse analysis to find the material parameters from the load–deflection response and ductility characteristics of a structure. This process is necessary since, in the absence of experimental load–deflection results, no other procedure exists to correlate the material properties with the structural ductility requirements. Back-calculation procedures also serve to reduce the experimental data using a model to formulate the crack patterns observed in tests (Figure 11.10).

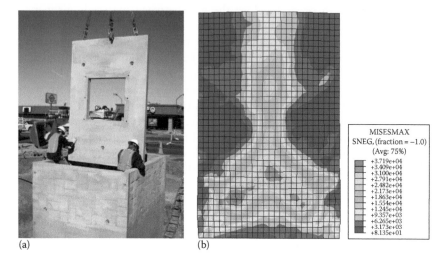

(a)   (b)

Figure 11.9 (a) Wall panels used for water distribution box structures and (b) numerical simulation results for a rectangular panel (*h* = 70 in., *L* = 45 in., and *Thk.* = 4 in.) with rebar, 323 psi (2.2 MPa), postcrack tensile strength (approximate μ = 0.8) using medium mesh: showing von Mises stress distribution.

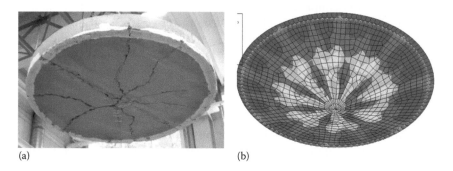

(a)   (b)

Figure 11.10 (a) Experimental failure yield lines and (b) numerical simulation of yield lines based on finite element modeling of a round panel continuously supported round panel.

11.7 Natural fiber systems

The enhanced strength and ductility of natural fiber TRC systems is primarily governed by the action of fibers that bridge the cracks to transfer the loads, allowing a distributed microcrack system to develop. Natural fiber systems such as sisal are directly applicable as reinforcing materials for cementitious systems, show very good engineering properties, have low environmental impact, and cost less than high-performance fibers.

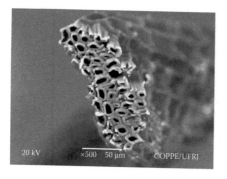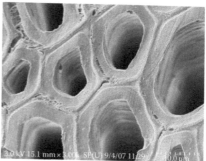

Figure 11.11 The sisal fiber microstructure composed of several fiber cells linked by the middle lamellae, which consist of lignin and hemicellulose.

Average annual production of sisal fiber is in the range of 153,000 tons (1996–2000, Prosea 2003). The sisal fiber microstructure offers great potential for natural TRC composites. The microstructure of sisal fiber showing the long hollow fibers and the cell walls that are connected by means of lignin is shown in Figure 11.11. To improve the long-term durability, as much as 50% of cement was replaced by calcined clays (Toledo Filho et al., 2009). This matrix gives the composite enhanced durability against fiber degradation and also provides adequate rheology for the fiber volume fractions proposed in these systems. The multiple cracking behavior achieved by this composite is governed by interfacial bond characteristics between fiber and matrix.

Cement composites reinforced with continuous aligned sisal fibers demonstrate a tension-hardening with multiple cracking behavior (Silva et al., 2009, 2010a), high tolerance to fatigue loading (Silva et al., 2010b), and high energy absorption capacity under dynamic loading. This type of composite system is reinforced with up to five layers of continuous fibers, resulting in a total volume fraction of 10%.

References

ACI 122R, Guide to thermal properties of concrete and masonry systems, American Concrete Institute, Farmington Hills, MI, 1996.

ACI 523.2R, Guide for precast cellular concrete floor, roof, and wall units, American Concrete Institute, Farmington Hills, MI, 1996.

ASTM C1550-02, Standard test method for flexural toughness of fiber reinforced concrete (using centrally loaded round panel), West Conshohocken, PA, 2002.

Banthia, N., Yan, C., and Sakai, K., Impact resistance of fiber reinforced concrete at subnormal temperatures, *Cement and Concrete Composites*, 20, 393–404, 1998.

Bentur, A. and Mindess, S., *Fiber Reinforced Cementitious Composites*, Elsevier, London, U.K., 1990.

Bernard, E. S., Behaviour of round steel fiber reinforced concrete panels under point loads, *Materials and Structures*, 33, 181–188, 2000.

Bonakdar, A., Babbitt, F., and Mobasher, B., Physical and mechanical characterization of fiber-reinforced aerated concrete (FRAC), *Cement and Concrete Composites*, 38, 82–91, 2013.

Budaiwi, I., Abdou, A., and Al-Homoud, M., Variations of thermal conductivity of insulation materials under different operating temperatures: Impact on envelope-induced cooling load, *Journal of Archaeological Engineering*, 8(4), 125–132, 2002.

Dey, V., Bonakdar, A., and Mobasher, B., Low-velocity flexural impact response of fiber-reinforced aerated concrete, *Cement and Concrete Composites*, 49, 100–110, May 2014.

Dey, V., Zani, G., Colombo, M., Di Prisco, M., and Mobasher, B., Flexural impact response of textile-reinforced aerated concrete sandwich panels, *Materials & Design*, 86, 187–197, 2015.

di Prisco, M., Failla, C., Plizzari, G. A., and Toniolo, G., Italian guidelines on SFRC, in M. di Prisco and G. A. Plizzari (eds.), *Proceeding of the International Workshop on Advances in Fiber Reinforced Concrete: From Theory to Practice*, Bergamo, Italy, September 24–25, 2004, pp. 39–72.

European Specification for Sprayed Concrete (EFNARC), European federation of national associations of specialist contractors and material suppliers for the construction industry (EFNARC), Hampshire, U.K., 1996.

Gibson, L. J. and Ashby, M. F., *Cellular Solids, Structure and Properties*, Cambridge University Press, Cambridge, U.K., 1997.

Hibbitt, K. and Sorensen, P., *ABAQUS, Analysis User's Manual Version 6.5*, Vols. II–V, ABAQUS, Pawtucket, RI, 2004.

Holt, E. and Raivio, P., Use of gasification residues in aerated autoclaved concrete, *Cement and Concrete Research*, 35, 796–802, 2005.

Mobasher, B., Micromechanical modeling of filament wound cement-based composites, *ASCE, Journal of Engineering Mechanics*, 129(4), 373–382, 2003.

Mobasher, B., *Mechanics of Fibre and Textile Reinforced Cement Composites*, CRC Press, Boca Raton, FL, 2011, pp. 480.

Mobasher, B. and Barsby, C., Flexural design of strain hardening cement composites, in J. Barros (ed.), *Proceedings pro088: Eighth RILEM International Symposium on Fiber Reinforced Concrete: Challenges and Opportunities (BEFIB 2012)*, Guimarães, Portugal, 2012, pp. 906–917.

Mobasher, B., Chen, S. Y., Young, C., and Rajan, S. D., Cost-based design of residential steel roof systems: A case study, *Structural Engineering and Mechanics*, 8(2), 165–180, 1999.

Mobasher, B., Dey, V., Cohen, Z., and Peled, A., Correlation of constitutive response of hybrid textile reinforced concrete from tensile and flexural tests, *Cement and Concrete Composites*, 53, 148–161, 2014.

Mobasher, B. and Ferraris, C., Simulation of expansion in cement based materials subjected to external sulfate attack, in *RILEM International Symposium: Advances in Concrete through Science and Engineering*, Evanston, IL, Comp. Weiss, Jason, and Shah, 2004.

Mobasher, B. and Li, C. Y., Mechanical properties of hybrid cement based composites, *ACI Materials Journal*, 93(3), 284–293, 1996.

Mobasher, B., Minor, G., Zenouzi, M., and Jalife, S. L., Thermal and mechanical characterization of contiguous wall systems for energy efficient low cost housing, in *Proceedings of the ASME-ES Fuel Cell 2011*, Washington, DC, August 7–10, 2011 (ESFuelCell2011-54952).

Mobasher, B., Peled, A., and Pahilajani, J., Distributed cracking and stiffness degradation in fabric-cement composites, *Materials and Structures*, 39(287), 317–331, 2006a.

Mobasher, B., Peled, A., and Pahilajani, J., Distributed cracking and stiffness degradation in fabric-cement composites, *Materials and Structures*, 39(3), 17–331, 2006b.

Mobasher, B. and Pivacek, A., A filament winding technique for manufacturing cement based cross-ply laminates, *Journal of Cement and Concrete Composites*, 20, 405–415, 1998.

Mobasher, B., Pivacek, A., and Haupt, G. J., Cement based cross-ply laminates, *Journal of Advanced Cement Based Materials*, 6, 144–152, 1997.

Mobasher, B., Stang, H., and Shah, S. P., Microcracking in fiber reinforced concrete, *Journal of Cement & Concrete Research*, 20, 665–676, 1990.

Neithalath, N. and Ramamurthy, K., Structure and properties of aerated concrete: A review, *Cement and Concrete Composites*, 20, 321–329, 2000.

Peled, A. and Mobasher, B., Cement based pultruded composites with fabrics, in *Proceedings of the Seventh International Symposium on Brittle Matrix Composites (BMC7)*, Warsaw, Poland, October 13–15, 2003.

Peled, A. and Mobasher, B., Pultruded fabric-cement composites, *ACI Materials Journal*, 102(1), 15–23, 2005.

Peled, A. and Mobasher, B., Properties of fabric-cement composites made by pultrusion, *Materials and Structures*, 39(8), 787–797, 2006.

Peled, A., Mobasher, B., and Sueki, S., Technology methods in textile cement-based composites, in K. Kovler, J. Marchand, S. Mindess, and J. Weiss (eds.), *RILEM Proceedings PRO 36: Concrete Science and Engineering, A Tribute to Arnon Bentur*, Evanston, IL, 2004, pp. 187–202.

Peled, A., Sueki, S., and Mobasher, B., Bonding in fabric-cement systems: Effects of fabrication methods, *Journal of Cement and Concrete Research*, 36(9), 1661–1671, 2006.

Perez-Pena, M., Mobasher, B., and Alfrejd, M. A., Influence of pozzolans on the tensile behavior of reinforced lightweight concrete, in *Materials Research Society, Symposium "O", Innovations in the Development and Characterization of Materials for Infrastructure*, Boston, MA, December 1991.

Pivacek A. and Mobasher, B., A filament winding technique for manufacturing cement based cross-ply laminates, *Journal of Materials in Civil Engineering*, 9(2), 55–57, 1997.

RILEM TC162-TDF, Test and design methods for steel fiber reinforced concrete: Recommendations. Bending test, *Materials and Structures*, 33(225), 3–5, 2000.

RILEM TC162-TDF, Test and design methods for steel fiber reinforced concrete; σ–ε design method, *Materials and Structures*, 36, 560–567, 2003.

Romano, G. Q., Silva, F. A., Toledo Filho, R. D., Fairbairn, E. M. R., and Battista, R., On the removal of steel fiber reinforced refractory concrete from thin walled steel structures using blast loading, in *Unified International Technical Conference on Refractories (Unitecr 07)*, Dresden, Germany, 2007a, pp. 509–512.

Roy, D. M., Alkali-activated cements: Opportunities and challenges, *Cement and Concrete Research*, 29, 249–254, 1999.

Silva, F. A., Toughness of non-conventional materials, MSc thesis, Department of Civil Engineering, PUC-Rio, Rio de Janeiro, Brazil, 2004.

Silva, F. A., Ghavami, K., and d'Almeida, J. R. M., Toughness of cementitious composites reinforced by randomly sisal pulps, in *Eleventh International Conference on Composites Engineering: ICCE-11*, Hilton Head Island, SC, 2004.

Silva, F. A., Mobasher, B., and Toledo Filho, R. D., Cracking mechanisms in durable sisal fiber reinforced cement composites, *Cement and Concrete Composites*, 31, 721–730, 2009.

Silva, F. A., Mobasher, B., and Toledo Filho, R. D., Fatigue behavior of sisal fiber reinforced cement composites, *Materials Science & Engineering A*, 527, 5507–5513, 2010a.

Silva, F. A., Zhu, D., Mobasher, B., Soranakom, C., and Toledo Filho, R. D., High speed tensile behavior of sisal fiber cement composites, *Materials Science & Engineering A*, 527, 544–552, 2010b.

Silva, F. A., Zhu, D., Mobasher, B., and Toledo Filho, R. D., Impact behavior of sisal fiber cement composites under flexural load, *ACI Materials Journal*, 108(2), 168–177, 2011.

Soranakom, C. and Mobasher, B., Closed-form moment-curvature expressions for homogenized fiber-reinforced concrete, *ACI Material Journal*, 104(4), 351–359, 2007a.

Soranakom, C. and Mobasher, B., Closed form solutions for flexural response of fiber reinforced concrete beams, *ASCE Journal of Engineering Mechanics*, 133(8), 933–941, 2007b.

Soranakom, C. and Mobasher, B., Correlation of tensile and flexural responses of strain softening and strain hardening cement composites, *Cement and Concrete Composites*, 30(6), 465–477, 2008.

Soranakom, C. and Mobasher, B., Flexural analysis and design of strain softening fiber-reinforced concrete, in G. J. Parra-Montesinos and P. Balaguru (eds.), *Antoine E. Naaman Symposium: Four Decades of Progress in Prestressed Concrete, FRC, and Thin Laminate Composites*, Los Angeles, CA, American Concrete Institute, Farmington Hills, MI, Special Publication SP-272, 2010, pp. 173–187.

Soranakom, C., Mobasher, B., and Bansal, S., Effect of material non-linearity on the flexural response of fiber reinforced concrete, in *Proceedings of the Eighth International Symposium on Brittle Matrix Composites BMC8*, Warsaw, Poland, 2006, pp. 85–98.

Toledo Filho, R. D., Silva, F. A., Fairbairn, E. M. R., and Melo Filho, J. A., Durability of compression molded sisal fiber reinforced mortar laminates, *Construction and Building Materials*, 23, 2409–2420, 2009.

Vandewalle, L., Test and design methods for steel fiber reinforced concrete proposed by RILEM TC 162-TDF, in *Proceeding of the International Workshop on Advances in Fiber Reinforced Concrete: From Theory to Practice*, Bergamo, Italy, September 24–25, 2004, pp. 3–12.

Zhu, D., Gencoglu, M., and Mobasher, B., Low velocity flexural impact behavior of AR glass fabric reinforced cement composites, *Cement and Concrete Composites*, 31, 379–387, 2009.

Index

T - #0097 - 111024 - C492 - 234/156/23 - PB - 9780367866914 - Gloss Lamination